Lecture Notes in Computer Science

Lecture Notes in Artificial Intelligence 15498
Founding Editor

Jörg Siekmann

Series Editors

Randy Goebel, *University of Alberta, Edmonton, Canada*
Wolfgang Wahlster, *DFKI, Berlin, Germany*
Zhi-Hua Zhou, *Nanjing University, Nanjing, China*

The series Lecture Notes in Artificial Intelligence (LNAI) was established in 1988 as a topical subseries of LNCS devoted to artificial intelligence.

The series publishes state-of-the-art research results at a high level. As with the LNCS mother series, the mission of the series is to serve the international R & D community by providing an invaluable service, mainly focused on the publication of conference and workshop proceedings and postproceedings.

Amir Hussain · Bo Jiang · Jinchang Ren ·
Mufti Mahmud · Erfu Yang · Aihua Zheng ·
Chenglong Li · Shuqiang Wang · Zhi Gao ·
Zhicheng Zhao
Editors

Advances in Brain Inspired Cognitive Systems

14th International Conference, BICS 2024
Hefei, China, December 6–8, 2024
Proceedings, Part II

Editors
Amir Hussain
Edinburgh Napier University
Edinburgh, UK

Jinchang Ren
Robert Gordon University
Aberdeen, UK

Erfu Yang
The University of Strathclyde
Glasgow, UK

Chenglong Li
Anhui University
Hefei, China

Zhi Gao
Wuhan University
Hubei, China

Bo Jiang
Anhui University
Hefei, China

Mufti Mahmud
King Fahd University of Petroleum
and Minerals
Dhahran, Saudi Arabia

Aihua Zheng
Anhui University
Hefei, China

Shuqiang Wang
Shenzhen Institutes of Advanced Technology,
Chinese Academy of Science
Shenzhen, China

Zhicheng Zhao
Anhui University
Hefei, China

ISSN 0302-9743 ISSN 1611-3349 (electronic)
Lecture Notes in Artificial Intelligence
ISBN 978-981-96-2884-1 ISBN 978-981-96-2885-8 (eBook)
https://doi.org/10.1007/978-981-96-2885-8

LNCS Sublibrary: SL7 – Artificial Intelligence

© The Editor(s) (if applicable) and The Author(s), under exclusive license
to Springer Nature Singapore Pte Ltd. 2025

This work is subject to copyright. All rights are solely and exclusively licensed by the Publisher, whether the whole or part of the material is concerned, specifically the rights of translation, reprinting, reuse of illustrations, recitation, broadcasting, reproduction on microfilms or in any other physical way, and transmission or information storage and retrieval, electronic adaptation, computer software, or by similar or dissimilar methodology now known or hereafter developed.
The use of general descriptive names, registered names, trademarks, service marks, etc. in this publication does not imply, even in the absence of a specific statement, that such names are exempt from the relevant protective laws and regulations and therefore free for general use.
The publisher, the authors and the editors are safe to assume that the advice and information in this book are believed to be true and accurate at the date of publication. Neither the publisher nor the authors or the editors give a warranty, expressed or implied, with respect to the material contained herein or for any errors or omissions that may have been made. The publisher remains neutral with regard to jurisdictional claims in published maps and institutional affiliations.

This Springer imprint is published by the registered company Springer Nature Singapore Pte Ltd.
The registered company address is: 152 Beach Road, #21-01/04 Gateway East, Singapore 189721, Singapore

If disposing of this product, please recycle the paper.

Preface

We are thrilled to welcome you to the 14th edition of the International Conference on Brain-Inspired Cognitive Systems (BICS 2024). Building on the success of previous years, this conference continues to serve as a key platform for exploring the latest advancements in brain-inspired computing, artificial intelligence, and cognitive systems. Since its inception in 2004, BICS has gathered researchers, practitioners, and thought leaders from around the world to share ideas and shape the future of intelligent systems.

Following the success of BICS 2023 in Kuala Lumpur, Malaysia, this year's event promised to continue that tradition of excellence. BICS 2024 covered a wide range of topics, from computational neuroscience and deep learning to brain-machine interfaces and cognitive computing. Our goal was to foster collaboration across disciplines and provide a space for meaningful discussions that will drive the next wave of innovation in these rapidly evolving fields.

After a single-blind review with submissions receiving on average 2.5 reviews each, 57 papers were accepted from 124 submissions. Of these, 56 papers can be found in these proceedings.

We would like to express our sincere gratitude to all contributors, speakers, and attendees who made this conference so impactful. With your continued support, we are confident that BICS 2024 was another milestone in advancing brain-inspired technologies and their applications.

Organization

General Chairs

Amir Hussain	Edinburgh Napier University, UK
Bo Jiang	Anhui University, China
Jinchang Ren	Robert Gordon University, UK

Advisory Board and Publicity Chairs

Cheng-Lin Liu	CAS, China
Bin Luo	Anhui University, China

Program Chairs

Mufti Mahmud	Nottingham Trent University, UK
Erfu Yang	University of Strathclyde, UK
Aihua Zheng	Anhui University, China
Chenglong Li	Anhui University, China
Shuqiang Wang	Chinese Academy of Sciences, China
Zhi Gao	Wuhan University, China

Publication Chairs

Junchi Yan	Shanghai Jiao Tong University, China
Qi Liu	University of Science and Technology of China, China
Zhicheng Zhao	Anhui University, China

Finance Chairs

Zhengzheng Tu	Anhui University, China
Ping Sun	Anhui University, China

Registrations Chairs

Wei Jia Hefei University of Technology, China
Yun Xiao Anhui University, China

Local Arrangement Chairs

Haifeng Zhao Anhui University, China
Futian Wang Anhui University, China
Yang Zhao Hefei University of Technology, China
Cunhang Fan Anhui University, China
Lingma Sun Hefei University, China

Contents – Part II

Multi-modal Dynamic Information Selection Pyramid Network for Alzheimer's Disease Classification 1
Yuanmin Ma, Yuan Chen, Yuqing Liu, Jie Chen, and Bo Jiang

Text-Guided Vision Mamba for Alzheimer's Disease Prediction Using ^{18}F-FDG PET 11
Die Zhou, Yuan Chen, Yuqing Liu, and Bo Jiang

EEG-Based Recognition of Knowledge Acquisition States in Second Language Learning 21
Shanlin Xi, Ziyu Li, and Xia Wu

A Study on the Neural Mechanism of the Spatial Position of Speech in Different Masking Types Affecting Auditory Attention Processing 31
Dawei Xiang, Yong Ma, and Yiming Yang

DSCF-DE: A Query-Based Object Detection Model via Dynamic Sampling and Cascade Fusion 41
Dengdi Sun, Wenhao Liu, and Zhuanlian Ding

MDFNet: Multi-dimensional Fusion Attention for Enhanced Image Captioning 52
Dengdi Sun, Xuetao Li, and Chaofan Mu

Dynamic Points Location of Professional Model Pose Based on Improved Network Stacking Model 62
Kaizhan Mai, Dazhi Li, Yuefang Gao, Pingping Mi, and Li Hao

A Redundancy Free Facial Acne Detection Framework Based on Multi-view Face Images Stitching 72
Ye Luo, Jianfei Wang, Linglin Zhang, Xinyu Liu, Ji Rao, Wantong Xu, Jianwei Lu, and Xiuli Wang

A New Device Placement Approach with Dual Graph Mamba Networks and Proximal Policy Optimization 82
Meng Han, Yan Zeng, Hao Shu, Xiaofei Lu, Jilin Zhang, Yongjian Ren, and Wangli Hao

Cross-Generational Contrastive Continual Learning for 3D Point Cloud
Semantic Segmentation 93
   Yuan He, Guyue Hu, and Shan Yu

TGAM-SR: A Sequential Recommendation Model for Long
and Short-Term Interests Based on TCN-GRU and Attention Mechanism 104
   Jiajing Zhang, Zhiya Shen, Jinlan Chen, and Qilang Li

Investigating ChatGPT Translation Hallucination
from an Embodied-Cognitive Translatology Perspective 117
   Hui Jiao, Xinwei Li, Jonathan Ding, and Xiaojun Zhang

A Study on Chinese Acronym Prediction Based on Contextual Thematic
Consistency 127
   Wan Tao, Xiaoran Wang, and Qiang Zhang

Learning Supportive Two-Stream Network for Audio-Visual Segmentation 138
   Hongfan Jiang, Tianyang Xu, Xuefeng Zhu, and Xiaojun Wu

Multi-exposure Driven Stable Diffusion for Shadow Removal 148
   Zheng Yan, Wenhao Tan, and Linbo Wang

Human Disease Prediction Based on Symptoms Using Novel Machine
Learning 159
   Ibukunoluwa Oluwabusayo Efunwoye, Mandar Gogate,
   Adeel Hussain, Bin Luo, Jinchang Ren, Fengling Jiang, Amir Hussain,
   and Kia Dashtipour

CAT-LCAN: A Multimodal Physiological Signal Fusion Framework
for Emotion Recognition 168
   Ao Li, Zhao Lv, and Xinhui Li

A Novel Thermal Imaging and Machine Learning Based Privacy
Preserving Framework for Efficient Space Allocation, Utilisation
and Management 178
   Maria Bruevich, Nilupulee A. Gunathilake, Mandar Gogate,
   Adeel Hussain, Bin Luo, Jinchang Ren, Amir Hussain, Fengling Jiang,
   and Kia Dashtipour

Training Feature-Aware GPU-Memory Allocation and Management
for Deep Neural Networks 188
   Qintao Zhang, Xin Li, Chengchuang Huang, Ying Zhu, Jilin Zhang,
   and Meng Han

TR-LDA: An Improved Potential Topic Recognition Model 201
   Anzhen Li, Shufan Qing, Weijie Qin, Liwen Qin, Jiawei Zhang,
   Meilin Shi, Jinchang Ren, and Mingchen Feng

Brain-Inspired Object Domain Adaptive Segmentation 211
   Mengyin Pang, Song Xu, Lina Wang, Zhenfei Liu, Meijun Sun,
   and Zheng Wang

Task Adaptive Feature Distribution Based Network for Few-Shot
Fine-Grained Target Classification 222
   Ping Li, Hongbo Wang, Jie Ren, Xin Mi, and Chao Shi

ST_TransNeXt: A Novel Pig Behavior Recognition Model 233
   Wangli Hao, Hao Shu, Xinyuan Hu, Meng Han, and Fuzhong Li

A Method for Predicting the RUL of HDDs Based on Bidirectional LSTM
and Transformer 243
   ZeHong Wu, Jinghui Qin, Zhijing Yang, and Yongyi Lu

Spatio-temporal Graph Learning on Adaptive Mined Key Frames
for High-Performance Multi-Object Tracking 252
   Futian Wang, Fengxiang Liu, and Xiao Wang

From Image to the Ground: Recover the Ground Location of Vehicles
from Traffic Cameras Using Neural Networks 262
   Xuzhen Wang, Wenzhong Wang, and Jin Tang

In-Depth Evaluation and Analysis of Hyperspectral Unmixing Algorithms
with Cognitive Models 273
   Shunan Deng, Jinchang Ren, Rongjun Chen, Huimin Zhao,
   and Amir Hussain

Effective Gas Classification Using Singular Spectrum Analysis
and Random Forest in Electronic Nose Applications 283
   Yuntao Wu, Jinchang Ren, Rongjun Chen, Huimin Zhao,
   and Amir Hussain

Author Index 295

Contents – Part I

A Lightweight Neural Network for SAR Ship Detection Based on YOLOv8 and Swin-Transformer 1
   Fei Gao, Chen Fan, Tianjin Liu, Jun Wang, and Amir Hussain

RA-BLS: A Sequential BLSs Integrated with Residual Attention Mechanism 10
   Yanqiang Wu, Jing Wang, and Wei Hu

EEG-Based Emotion Recognition Using Similarity Measures of Brain Rhythm Entropy Matrix 20
   Guanyuan Feng, Peixian Wang, Xinyu Wu, Ximing Ren, Chen Ling, Yuesheng Huang, Leijun Wang, Jujian Lv, Jiawen Li, and Rongjun Chen

Intensity Controllable Emotional Speech Synthesis Based on Valence-Arousal-Dominance 30
   Guopping Li and Yanxiang Chen

Unsupervised Person Re-identification with Random Occlusion and ContrastiveCrop 41
   Yang Jing, Gu Lingkang, Xia Zhouxiang, and Wu Mengqi

Dynamic Prompt Adjustment for Multi-label Class-Incremental Learning 52
   Haifeng Zhao, Yuguang Jin, and Leilei Ma

Using Decision Tree Classification to Identify Cost Drivers of Hospitalization Expenses for Elderly Patients 62
   Xiaojing Hu, Yudian Liu, and Shixi Liu

Adversarial Attacks on Facial Images Based on Attribute-Conditioned High-Camouflage Editing 72
   Jingjing Zhang, Huabin Wang, Dongxu Shang, Hongrui Yuan, and Liang Tao

A High Accuracy Text CAPTCHA Recognition Approach Through Opertimized Vision Transformer 82
   Wei Hao, Shoulai Shang, and Yepeng Zhang

LightMamba-UNet: Lightweight Mamba with U-Net for Efficient Skin Lesion Segmentation 93
   Wanzhen Hou, Shiwei Zhou, and Haifeng Zhao

Exploiting Memory-Aware Q-Distribution Prediction for Nuclear Fusion
via Modern Hopfield Network 104
   *Qingchuan Ma, Shiao Wang, Tong Zheng, Xiaodong Dai, Yifeng Wang,
   Qingquan Yang, and Xiao Wang*

Multi-modal Fusion Based Q-Distribution Prediction for Controlled
Nuclear Fusion 115
   *Shiao Wang, Yifeng Wang, Qingchuan Ma, Xiao Wang, Ning Yan,
   Qingquan Yang, Guosheng Xu, and Jin Tang*

Deformable Transformer for 3D Medical Image Segmentation 126
   Haifeng Zhao, Tianxia Yang, Minghui Xu, and Yanping Fu

On the Gap Between AI-Generated and Human-Written Patent Texts 136
   *Zhanhao Xiao, Wei Hu, Yanqiang Wu, Weiqi Chen, Huihui Li,
   and Xiaoyong Liu*

MRI-CT Brain Image Registration Based on SuperPCA
and Block-Matching Algorithm 147
   Wannan Zhang

Multi-teacher Knowledge Distillation with Triplet Loss for Cross-Modal
Object Tracking 155
   Yi Li, Lei Liu, Mengya Zhang, and Chenglong Li

Enhanced Comprehensive Competition Network for Domain Adaptive
Palmprint Recognition 166
   Congcong Jia, Xingbo Dong, Zhe Jin, and Lianqiang Yang

MBDR-V2: A Network for MRI Brain Tumor Image Segmentation
with Incomplete Modalities 177
   *Yanqi Hou, Longfeng Shen, Jiacong Chen, Liangjin Diao, Youle Xu,
   and Wei Zhao*

An Innovative Eco-Friendly Weighing System for Reusable Bags
Incorporating K210 and QR Code Technology 187
   Yubin Wei, Yufei Li, and Yiwen Zhang

Focal Consistency Network for Developmental Stage Classification
of Embryos with Time-Lapse Embryo Video Datasets 197
   Yiming Li, Hua Wang, Jingfei Hu, and Jicong Zhang

Chest X-ray Image Rib Segmentation via Disentanglement Enhancement
Network 208
   Lili Huang, Shiqi Li, Lingma Sun, and Chuanfu Li

Instance-Level 3D Model Reassembling from CLuttered Fragments 218
   Longteng Jiang, Yijian Liu, Feixiang Lu, Chenming Wu, and Xin Jin

Brain-Inspired Action Generation with Spiking Transformer Diffusion
Policy Model 229
   Qianhao Wang, Yinqian Sun, Enmeng Lu, Qian Zhang, and Yi Zeng

Single-Stage Dual-Task Joint Learning Framework for Hand Hygiene
Assessment 239
   Sizhe Qin, Zijian Tu, Deyu Su, and Zi Wang

Enhancing Few-Shot Learning in Spiking Neural Networks Through
Hebbian-Augmented Associative Memory 249
   Weiyi Li, Dongcheng Zhao, Yiting Dong, Guobin Shen, and Yi Zeng

Palmprint Texture Fusion Based on TinyViT for Recognition 259
   Fuchuan Huang, Cunyu Sheng, Jian He, and Wei Jia

Novel Device Placement Approach with Neighbor Effect Aware Graph
Mamba Networks 269
   Hao Shu, Wangli Hao, Meng Han, and Fuzhong Li

Research on Improved PointPillars Algorithm Based on Attention
Mechanism and Feature Fusion 280
   RunMei Zhang, AnLong Zhang, and Lei Yin

Author Index 291

Multi-modal Dynamic Information Selection Pyramid Network for Alzheimer's Disease Classification

Yuanmin Ma[1], Yuan Chen[2(✉)], Yuqing Liu[3], Jie Chen[4], and Bo Jiang[3,5]

[1] School of Artificial Intelligence, Anhui University, Hefei, China
[2] School of Internet, Anhui University, Hefei, China
ychen@ahu.edu.cn
[3] Institute of Artificial Intelligence, Hefei Comprehensive National Science Center, Hefei, China
[4] School of Electronic and Information Engineering, Anhui Jianzhu University, Hefei, China
[5] School of Computer Science and Technology, Anhui University, Hefei, China

Abstract. Alzheimer's disease (AD) is a common and dangerous disorder that primarily impacts older adults, early detection is crucial, and diagnostic tools like PET and MRI have an impact on offering detailed anatomical and metabolic insights, respectively. However, traditional methods usually directly concatenate the two modal data by channel, which fails to fully utilize the complementary information provided by MRI and PET data. Hence, this paper introduces a new multi-modal dynamic information selection framework to enhance the accuracy of AD classification. It includes three main modules: a dual ResNet50-based feature extraction module; a modal fusion module containing a feature pyramid for finer detail extraction and a cross-attention mechanism to integrate modal information; a dynamic information selection module that evaluates the data content across modalities to optimize decision-making. The outcome on the ADNI dataset confirm the efficiency of the proposed approach.

Keywords: Brain disease prediction · Alzheimer's disease · Multi-modal classification

1 Introduction

Alzheimer's disease (AD) is the most general form of dementia, resulting in permanent brain damage and diminished brain function [1]. Characterized by the loss of brain tissue and accumulation of amyloid and tau proteins [2], these changes will gradually impair cognitive functions and affect daily activities. Given the irreversible nature of AD and the lack of treatmens, early diagnosis is vital [3]. Timely detection along with medication and comprehensive interventions can significantly slow disease progression.

MRI [4] and PET are essential neuroimaging methods for diagnosing Alzheimer's disease (AD). Recently, some methods have achieved promising prediction accuracy by utilizing PET images. Nevertheless, these approaches [6–10]

are not entirely effective, frequently because of substantial noise in PET images, the unclear characteristics of important features, and the limited size of medical datasets that hinder feature extraction and the reliability of training results.

In recent years, the utilization of multi-modal data has been continuously growing compared to single-modal data, with modal fusion offering more comprehensive information. Fusion strategies can be categorized into early, intermediate, and late fusion stages. Early fusion [13,14] necessitates exact synchronization of data across various modalities, with one modality typically set as the standard for spatial correction. However, PET imaging characteristics can introduce noise early in the fusion process, potentially degrading image quality and affecting predictions. Intermediate fusion [16,17] involves integrating modal characteristics directly into the visual encoder, but may miss capturing complementary information between modalities like PET and MRI. Late fusion [11,15] processes data of each modality independently, which may prevent the integration of complex relationships and complementary data, leading to inconsistent final predictions between PET and MRI.

Hence, in this paper, a mixed modal fusion approach is designed, which specifically integrates mid-stage and late-stage fusion techniques. A simple ResNet [12] is utilized for feature extraction, with the extracted features refined further in a feature pyramid to capture critical details. These features are then processed through a cross-attention module to enhance the complementary information between the two modalities. Additionally, both the fusion and original features are integrated into a dynamic information selection module, which preserves modality-specific information while capturing commonalities. The resulting confidence weight is then applied to the raw features to derive the final classification result using a classification head.

The principal contributions of this paper are outlined as follows :

1) A multi-modal Alzheimer's disease classification model is proposed. which makes full use of MRI and PET data.
2) A strategic approach to dynamic information selection has been designed to identify key features from various modalities, thus enhancing the accuracy of classification results.
3) Results on a publicly available ADNI dataset containing 455 samples, confirm the efficiency of the proposed approach.

2 Proposed Method

The proposed multi-modal dynamic information selection pyramid network contains three primary modules: feature extraction module, modalities fusion module, and dynamic information selection module, as shown in Fig. 1. The details will be introduced.

2.1 Feature Extraction Module

Two encoders based on ResNet50 are used to obtain initial features from different modal data. Each convolutional block comprises a convolution layer Using a

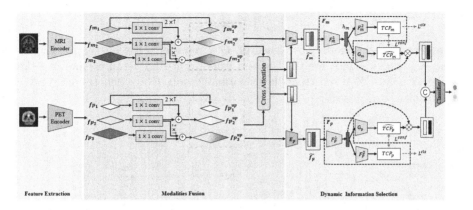

Fig. 1. The whole framework of the proposed method. The whole framework contains three modules. The feature extraction module employs ResNet as feature extraction network. In the modalities fusion module, cross-attention is used to enhance the interaction between PET and MRI features. In the dynamic information selection module, we adapt the dynamic information selection unit to dynamically adjust the feature weights, aiming to achieve better classification results.

stride of 2 and padding of 1, followed by a layer of batch normalization and a ReLU activation layer.

2.2 Modalities Fusion Module

The last three layers of feature maps extracted from the feature extraction module are used to construct the feature pyramid. In the process of bottom-up construction, for MRI branch, feature map $\boldsymbol{f}_{m_i}(i=1,2,3)$ is directly used without additional processing. The top-down fusion process is as follows:

$$\boldsymbol{f}^{up}_{m_i} = Upsample(Conv_{1\times1}\left(\boldsymbol{f}_{m_{i-1}}\right)) + Conv_{1\times1}\left(\boldsymbol{f}_{m_i}\right) \quad i=3,2, \qquad (1)$$

where + operation means adding element by element. Similarly, the feature maps from PET can also be calculated to get the pyramid network feature $\boldsymbol{f}^{up}_{p_i}$.

Drawing inspiration from [6], we add a cross attention module to fuse features from different modalities. In this way, the features of one modality can be exchanged with the features of another modality through the scale-point attention operation. The details are seen in Fig. 2. The features of feature pyramid for each modality are normalized through layers and projected into queries (Q), keys (K), and values (V) via various projection matrices. For a query from one modality, we use the key and the value from another modality to jointly calculate the scale-point attention. For example for MRI branch, the operation of the multi-modal cross attention can be shown as follows:

$$\widehat{feature}_M = Softmax\left(\frac{\boldsymbol{Q}_M \boldsymbol{K}_P^T}{\sqrt{C/h}}\right)\boldsymbol{V}_P, \qquad (2)$$

$$global_M = F_{MLP}\left(F_{LN}\left(\widehat{feature}_M + feature_M\right)\right), \qquad (3)$$

where $feature_M$ denotes MRI features, Q_M represents the query of MRI features, K_P and V_P indicate the key and value from PET features, and h is the number of heads. F_{LN} represents layer normalization, F_{MLP} denotes multi-layer perceptron.

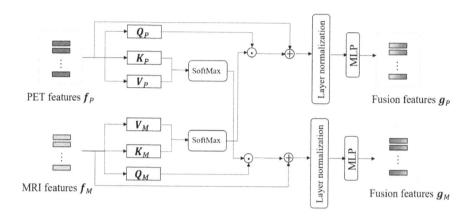

Fig. 2. Cross Attention Module: In Eqs. 2 and 3, $feature_M$ and $global_M$ referred to in the text correspond to f_m and g_m in the figure.

2.3 Dynamic Information Selection Module

For multi-modal data, the information content varies across samples [18,19]. Recognizing these variations enhances the ability to harness relevant information from multiple modalities for classification. We employ the True-Class-Probability [20] to measure the classification confidence associated with each modality. Low classification confidence in a modality indicates significant uncertainty and suggests limited informational value, whereas high confidence indicates greater certainty and more informative data. Maximum Class Probability (MCP) [22] is an effective indicator in assessing classification confidence. However, it usually leads to excessive confidence, especially for false confidence [18,19]. In this article, we call MCP as MCPL. For modality m, MCPL can be obtained as

$$MCPL_m = max[p_1^m, ..., p_n^m], \qquad (4)$$

where p_n^m represents the prediction probability for category n from the m modality.

The True Classification Probability (TCP) [20] index is another indicator, which can be used to obtain more reliable classification confidence. In contrast to MCP, which relies on the highest Softmax output for confidence measurement, TCP determines confidence using the Softmax output probability associated with the correct label. In this article, we call the TCP as TCPL. For modality m, given the prediction confidence $q^m(y \mid x^m) = [p_1^m, p_2^m, \ldots, p_N^m]$ and the true label y, $TCPL_m$ can be calculated as

$$TCPL_m = y \cdot q^m(y \mid x^m) = \sum_{n=1}^{N} y_n p_n^m, \tag{5}$$

where $q^m(y \mid x^m)$ denotes the prediction of the feature to the label in the m_{th} modality. N denotes the number of categories.

Features obtained from the cross-attention module and feature pyramid network are encoded into a uniform vector dimension and weighed using a sigmoid function based on their response values. The encoders E_m and E_p are tasked with assessing the informativeness of input features. These inputs include modal-specific features from the pyramid network and modal-common features from the cross-attention module. After processing these features through the encoders and applying a sigmoid activation, a weight score vector is generated. This vector is used to multiply the input features, creating a weighted feature representation, denoted as $\widetilde{f_m}$ and $\widetilde{f_p}$. Then, the weighted features are subsequently input into composite networks F_m and F_p, which are outlined within the dotted line box in Fig. 1. Take the MRI branch as example, The network F_m is composed of a feature extractor F_m^1 and a network of fully connected layers F_m^2. The extracted features h_m are processed by F_m^2 to output class probabilities $TCPL_m$. The total TCPL weight is as follows:

$$TCPL = [TCPL_1, TCPL_2, \ldots TCPL_W]. \tag{6}$$

In this paper, there are two modalities, hence, $W = 2$.

The cross-entropy loss is calculated with the actual label to optimize feature prediction.

$$L_{cls} = -\sum_{m=1}^{M} \sum_{n=1}^{K} y_n \log p_n^m, \tag{7}$$

where y_n is referred to the label, p_n^m shows the predicted value of the label for the n-th category of the m modality. Because the label information is unavailable during the test phase, a confidence network G_m is trained to fit the output of the F_m^2 during the training phase. MSE loss is used to allow the output of G_m to adapt to the outputs of F_m^2 as follows:

$$L_{conf} = \sum_{m=1}^{W} \left(\widehat{TCPL}_m - TCPL_m \right)^2, \tag{8}$$

where $\widehat{TCPL}$ is computed by the G_m network, which does not use the label, while TCPL is computed by F_m^2, which uses the label to determine the true class probability value. Then, during test phase, we can only use G_m. The final feature can be shown as follows:

$$\boldsymbol{F}_{Final} = \widehat{TCPL} * h_m = [\widehat{TCPL}_1 * h_{m1}, \ldots, \widehat{TCPL}_W * h_{mW}], \tag{9}$$

where h_m denotes the set of features obtained by the F_m^1 feature extraction layer. In this paper, there are two modalities, hence, $h_m = [h_{m1}, h_{m2}]$. Finally, features from the MRI and PET branches are concatenated and put into the final classifier to derive the disease classification outcomes. The cross entropy function is employed to maintain consistency between the predicted outcomes and the actual truth labels during the training stage.

3 Experiments

3.1 Dataset

We use the publicly accessible Alzheimer's Disease Neuroimaging Initiative dataset (ADNI, https://adni.loni.usc.edu/). In this group, there are 455 subjects in total, including 168 with Alzheimer's disease (AD), 162 with Mild Cognitive Impairment (MCI), and 125 who are Normal Controls (NC).

3.2 Implementation Details

Experiments were performed using the PyTorch library on Ubuntu 22.04 with an NVIDIA TITAN Xp GPU. The dataset was divided into ten segments, with nine used for training and one for validation, making sure there was no overlap with the test dataset. We used the SGD optimizer and the learning rate is 0.001, reduced by 90% every 20 epochs, across a total of 60 epochs and a batch size of 8. The evaluation encompassed four metrics, Accuracy (ACC) [25], Area Under the Curve (AUC), Sensitivity (SEN) [26], and Specificity (SPE) [27].

3.3 Performance Comparison

Our method is compared with several advanced methods in recent years, including MiSePyNet [5], PT-DCN [12], LLMF [24], MMGL [23] and WaveFusion [21]. Note that these methods are different in pre-processing dataset, hence, to be fair, we have reproduced these methods on our dataset with the same pre-processing steps. Three binary experiments were carried out: AD vs. NC, AD vs. MCI, and NC vs. MCI. The results of these experiments are displayed in Tables 1, 2 and 3.

Table 1 shows the outcomes of AD vs. NC. It can be confirmed that the proposed method outperforms the existing methods in three out of four metrics compared to previously proposed methods. Our model has different degrees of improvement in ACC, AUC, SPE and SEN. These performance improvements may be due to the large difference between AD and NC features, and the dynamic feature information selection module can better calculate the confidence weights of different features, thus forming feature vectors conducive to decision making. Table 2 and Table 3 list the outcomes of AD vs. MCI and NC vs. MCI. It can be found that the values of all metrics have decreased, because these two tasks are more demanding tasks than AD vs. NC. However, our method can also achieve

Table 1. Comparison with other advanced methods for AD VS. NC(%)

Method	ACC	AUC	SEN	SPE
PT-DCN(2022) [12]	87.50	87.30	88.89	85.71
MMGL(2022) [23]	81.25	82.53	72.22	92.86
LLMF(2023) [24]	87.50	88.10	83.33	92.86
WaveFusion [21]	84.31	85.31	77.77	92.86
MiSePyNet [5]	86.25	85.16	83.33	86.43
Ours	**90.63**	**90.87**	**88.89**	**92.86**

Table 2. Comparison with other advanced methods for AD VS. MCI(%)

Method	ACC	AUC	SEN	SPE
PT-DCN(2022) [12]	75.00	75.00	61.11	88.89
MMGL(2022) [23]	75.00	75.00	66.67	83.33
LLMF(2023) [24]	72.22	72.22	61.11	83.33
WaveFusion [21]	72.22	72.22	50.00	**94.44**
MiSePyNet [5]	72.22	72.22	55.56	88.89
Ours	**77.78**	**77.78**	**66.67**	88.89

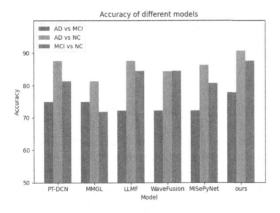

Fig. 3. Accuracy of different models across three tasks. In this section, we use feature cross module and pyramid of features to strengthen the representation of feature and we use dynamic information selection units to classify the result more accurately than other methods.

the best performance in all comparison methods, which further confirms the efficiency of our method in extracting AD disease features.

Figure 3 illustrates the accuracy of six models (PT-DCN, MMGL, LMLF, WaveFusion, MiSpeyNet, and "ours") more intuitively across three specific comparisons, i.e., AD vs. MCI, AD vs. NC, and MCI vs. NC. Performance of each

Table 3. Comparison with other advanced methods for MCI VS. NC(%)

Method	ACC	AUC	SEN	SPE
PT-DCN(2022) [12]	81.25	79.36	**94.44**	64.28
MMGL(2022) [23]	71.88	72.62	66.67	78.57
LLMF(2023) [24]	84.38	84.52	83.33	**85.71**
WaveFusion [21]	84.37	85.32	92.86	77.78
MiSePyNet [5]	80.63	80.48	81.67	79.29
Ours	**87.50**	**88.89**	90.63	78.57

model is color-coded: blue for AD vs. MCI, orange for AD vs. NC, and green for MCI vs. NC. Notably, our model exhibits superior accuracy, especially in the AD vs. NC comparison at 90.63%. These results verify our model's capability in effectively identifying the nuanced distinctions necessary for accurately classifying these pathological states.

The superior performance of our network over other models may stem from several theoretical advantages. Our fine-grained network is designed to capture delicate features and integrates a cross-attention mechanism that enhances the shared subtle features between the two modalities, providing additional information for decision-making. Furthermore, our dynamic information selection module learns the distribution of features and categories in feature space during training. In the testing phase, it dynamically adjusts weights for input features based on their informational value, enabling more precise prediction results through adaptive weight allocation.

3.4 Ablation Study

We conducted ablation experiments on ADNI dataset by constructing models without the feature pyramid (PF), without dynamic information selection (DIS), and without cross-attention (CA) to isolate the effects of each component.

From Table 4, it can be found that each module has certain performance benefits, especially dynamic information selection module, which verifies the effectiveness for AD disease prediction. Moreover, our model that integrates these modules achieves best results.

Table 4. Ablation study of the proposed model (%)

Component			AD vs NC				AD vs MCI				MCI vs NC			
CA	PF	DIS	ACC	AUC	SEN	SPE	ACC	AUC	SEN	SPE	ACC	AUC	SEN	SPE
×	×	×	86.75	85.65	90.55	80.75	72.22	72.22	55.56	88.89	81.25	81.75	85.71	77.78
×	×	✓	90.63	88.89	88.89	90.63	75.00	75.00	61.11	88.89	81.25	82.54	**92.86**	72.22
✓	×	×	81.25	80.96	78.57	83.33	72.22	72.22	61.11	83.33	84.38	83.73	88.89	78.57
×	✓	×	84.38	85.32	**92.86**	77.78	75.00	75.00	66.67	83.33	81.25	80.95	83.33	78.57
✓	✓	✓	**90.63**	**90.87**	88.89	**92.86**	**77.78**	**77.78**	66.67	88.89	**87.50**	**88.89**	90.63	78.57

4 Conclusion

A novel multi-modal dynamic information selection pyramid network is proposed for predicting AD. Dual branch encoders are used for feature extraction from MRI and PET data, a pyramid network is employed to further extract fine-grained features. Then, a dynamic information selection module is designed to better fuse features. A multi-layer perceptron evaluates feature confidence levels, which then influence their weighting and order in the concatenation process before classification. Ultimately, these weighted features are input into a classifier head. Testing on the ADNI dataset verifies our method's efficiency in predicting Alzheimer's disease.

Acknowledgement. This work is supported by the University Synergy Innovation Program of Anhui Province under Grant GXXT-2022-032 and GXXT-2022-014; the Anhui Province University Scientific Research Program Project 2024AH050052.

References

1. Alzheimer's Association: 2019 Alzheimer's disease facts and figures. Alzheimer's Dementia **15**(3), 321–387 (2019)
2. Fan, Y., Rao, H., Hurt, H., et al.: Multivariate examination of brain abnormality using both structural and functional MRI. Neuroimage **36**(4), 1189–1199 (2007)
3. Scheltens, P., De Strooper, B., Kivipelto, M., et al.: Alzheimer's disease. Lancet **397**(10284), 1577–1590 (2021)
4. Han, K., Luo, J., Xiao, Q., Ning, Z., et al: Light-weight cross-view hierarchical fusion network for joint localization and identification in alzheimer's disease with adaptive instance-declined pruning. Phys. Med. Biol. **66**(8), 085013 (2021)
5. Pan, X., Phan, T., Adel, M., et al.: Multi-view separable pyramid network for ad prediction at mci stage by 18f-fdg brain pet imaging. IEEE Trans. Med. Imaging **40**(1), 81–92 (2020)
6. Duan, J., Liu, Y., Wu, H., et al.: Broad learning for early diagnosis of alzheimer's disease using fdg-pet of the brain. Front. Neurosci. **17**, 1137567 (2023)
7. Chen, Y., Wang, H., Zhang, G., et al.: Contrastive learning for prediction of alzheimer's disease using brain 18f-fdg pet. IEEE J. Biomed. Health Inf. **27**(4), 1735–1746 (2023)
8. Duan, H., et al.: Eamnet: an alzheimer's disease prediction model based on representation learning. Phys. Med. Biol. **68**(21), 215005 (2023)
9. Liu, M., Cheng, D., Wang, K., et al.: Multi-modality cascaded convolutional neural networks for alzheimer's disease diagnosis. Neuroinformatics **16**, 295–308 (2018)
10. Yee, E., Popuri, K., Beg, M.F.: Quantifying brain metabolism from fdg-pet images into a probability of alzheimer's dementia score. Hum. Brain Mapp. **41**, 5–16 (2019)
11. Gao, X., Shi, F., Shen, D., et al.: Task-induced pyramid and attention GAN for multimodal brain image imputation and classification in alzheimer's disease. IEEE J. Biomed. Health Inform. **26**(1), 36–43 (2021)
12. He, K., Zhang, X., Ren, S., et al: Deep residual learning for image recognition. In: Proceedings of the IEEE Conference on Computer Vision and Pattern Recognition, pp. 770-778 (2016)

13. Venugopalan, J., Tong, L., Hassanzadeh, H.R., et al: Multimodal deep learning models for early detection of Alzheimer's disease stage. Sci. Rep. **11**(1), 3254 (2021)
14. Suk, H.-I., et al: Hierarchical feature representation and multimodal fusion with deep learning for AD/MCI diagnosis. Neuroimage **101**, 569–582 (2014)
15. Dyrba, M., et al.: Predicting prodromal Alzheimer's disease in subjects with mild cognitive impairment using machine learning classification of multimodal multi-center diffusion-tensor and magnetic resonance imaging data. Neuroimaging **25**(5), 738–747 (2015)
16. Lorenzi, M., et al.: Multimodal image analysis in Alzheimer's disease via statistical modelling of non-local intensity correlations. Sci. Rep. **6**(1), 22161 (2016)
17. Gray, K.R., et al: Random forest-based similarity measures for multi-modal classification of Alzheimer's disease. NeuroImage **65**, 167–175 (2013)
18. Hou, H., Zheng, Q., Zhao, Y., et al.: Neural correlates of optimal multisensory decision making under time-varying reliabilities with an invariant linear probabilistic population code. Neuron **104**(5), 1010–102 (2019)
19. Reuben, R., et al: How multisensory neurons solve causal inference. Proc. Natl. Acad. Sci. **118**(32), e2106235118 (2021)
20. Corbiére, C., et al.: Addressing failure prediction by learning model confidence. In: Advances in Neural Information Processing Systems, vol. 32 (2019)
21. Briden, M., WaveFusion squeeze-and-excitation: towards an accurate and explainable deep learning framework in neuroscience. In: 2021 43rd Annual International Conference of the IEEE Engineering in Medicine & Biology Society, pp. 1092–1095 (2021)
22. Han, Z., et al.: Multimodal dynamics: Dynamical fusion for trustworthy multimodal classification. In: Proceedings of the IEEE/CVF Conference on Computer Vision and Pattern Recognition, pp. 20707–20717 (2022)
23. Zheng, S., et al: Multi-modal graph learning for disease prediction. IEEE Trans. Med. Imaging **41**(9) 2207–2216 (2022)
24. Feng, Y., et al.: Large language models improve Alzheimer's disease diagnosis using multi-modality data. In: 2023 IEEE International Conference on Medical Artificial Intelligence (MedAI), pp. 61–66 (2023) ## acc
25. Baydargil, H.B., Park, J., Kang, D.Y.: Appl. Sci. **11**, 2178 (2021)
26. Chen, Y., et al: IEEE J. Biomed. Health Inf. **27**, 1735–1746 (2023)
27. Duan, J., et al.: Front. Neurosci. **17**, 1137567 (2023)

Text-Guided Vision Mamba for Alzheimer's Disease Prediction Using ^{18}F-FDG PET

Die Zhou[1], Yuan Chen[2(✉)], Yuqing Liu[3], and Bo Jiang[1,3]

[1] School of Computer Science and Technology, Anhui University, Hefei, China
[2] School of Internet, Anhui University, Hefei, China
ychen@ahu.edu.cn
[3] Institute of Artificial Intelligence, Hefei Comprehensive National Science Center, Hefei, China

Abstract. Recently, the number of Alzheimer's disease patients has increased, and the disease seriously affects their daily lives. Hence, more and more researchers have paid attention to this disease, and the diagnostic technology has been improved, particularly with the application of imaging technologies such as ^{18}F-FDG PET. Although many methods based on deep learning have made significant progress in this field, early effective diagnosis remains challenging. Existing research primarily relies on imaging data for disease prediction, but the information that imaging data can provide is limited. Hence, this paper proposes a text-guided method for Alzheimer's disease prediction. To be specific, vision Mamba is employed as image encoder and a CLIP encoder that can extract long text features is introduced as another branch. Then, an attention module is designed to make the text encoder provide additional semantic information for the image encoder, thus guiding attention distribution and enhancing cross-modal consistency. Finally, the prediction result can be obtained by the features after interaction. Experimental results demonstrate that the proposed method outperforms current state-of-the-art algorithms on the public ADNI dataset.

Keywords: Alzheimer's disease · ^{18}F-FDG PET · Disease prediction · Text guided

1 Introduction

Alzheimer's disease (AD) is a degenerative condition that leads to memory impairment and cognitive decline. Early diagnosis and effective diagnostic tools are crucial due to increasing life expectancy and the growing elderly population [1]. This technique is an imaging technique used to diagnose and evaluate neurological disorders, including AD and other cognitive impairments. ^{18}F-FDG PET is highly sensitive in detecting changes in tissue metabolism, it can identify abnormalities at an earlier stage [2–4]. Figure 1 shows several examples of PET images for different states of AD from ADNI dataset.

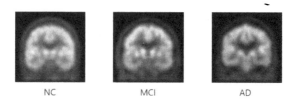

Fig. 1. PET images for different states of Alzheimer's disease from ADNI.

Convolutional neural networks (CNNs) have been extensively utilized for vison tasks including disease prediction due to their powerful capabilities. In brain disease prediction, methods using ^{18}F-FDG PET are typically divided into 2D CNN and 3D CNN approaches. 2D PET images derived from 3D PET scans help expand the dataset and reduce overfitting [5–7]. In contrast, 3D CNNs fully utilize spatial information by processing the entire 3D data, but they require more computational resources and training samples, which constraints their flexibility with limited data [8–11].

Recently, State Space Models (SSMs), like Mamba [12], are widely applied due to their low complexity, addressing some of the limitations found in Transformers. Liu et al. [13] proposed a vision backbone network, VMamba, specifically designed for processing visual data, which performed exceptionally well in natural image classification tasks. Subsequently, Yue et al. [14] proposed SS-Conv-SSM and successfully applied it to medical image classification tasks. Muthukumar et al. [15] applied Vision Mamba to 3D MRI image classification for detecting AD. Yang et al. [16] proposed an architecture named CMVim for Alzheimer's diagnosis using 3D PET scanns.

Moreover, in addition to using imaging data, text information, such as clinical reports, has been combined in medical image classification task, which can enhance the classification accuracy and model understanding. Some methods enhance medical image representation learning by incorporating textual insights during training, achieving fine-grained alignment between medical images and their corresponding reports, and thereby improving the capabilities of medical imaging tasks [17–19]. KAD [20] utilized existing medical knowledge to guide the pre-training of vision-language models, thus enhancing the ability of understanding and processing medical images and related texts. Vision-language pre-training frameworks for medical images have significantly improved model performance in medical image and text tasks [21, 22].

Inspired by these studies, this paper proposes a text-guided vision Mamba model for AD prediction. For brain PET images, we describe lesion characteristics in detail using text, which is then converted into vector representations by a text encoder inspired by CLIP [23]. An image encoder based on vision Mamba is employed to extract image features of PET data. Then a self-attention module is designed to integrate both text and image information, learning detailed text-image relationships and adjusting attention based on contextual content. Finally, the integrated features are used for prediction.

Main contributions of this paper are summarized as follows:

1) A novel text-guided vision Mamba model for Alzheimer's disease prediction is proposed, which contains image feature and text feature extraction branches, and a self-attention module for multi-modal information interaction and fusion.
2) Vision Mamba is employed to better extract PET imaging data features. Unlike using short category labels, more detailed long-form text descriptions are introduced in this paper, which is more conducive to assisting the imaging data to predict disease.
3) Experiments conducted on the ADNI dataset demonstrate that the proposed method achieves the best performance among all compared methods.

2 Proposed Method

Figure 2 shows the framework of the proposed method, which contains two branches. Details will be provided below.

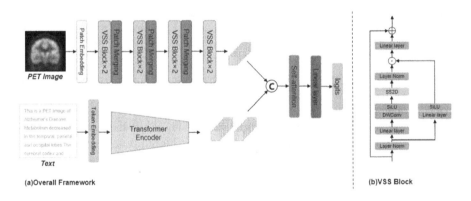

Fig. 2. The framework of the proposed method, which primarily consists of two branches, i.e., image feature extraction and text feature extraction.

2.1 Image Feature Extraction

Inspired by VM-UNet [24], vision Mamba is employed to extract image feature in our method. To be specific, the image encoder is primarily composed of three parts: patch embedding, patch merging, and Visual State Space (VSS) block [13]. First, the PET image $\boldsymbol{I}_{in} \in \mathbb{R}^{H \times W \times 3}$ is fed into the patch embedding layer, which divides the PET image into 4×4 non-overlapping patches while keeping the channel dimensions unchanged. Next, layer normalization is applied to the obtained features as:

$$\boldsymbol{I}_{patch} = \text{LayerNorm}(\text{PatchEmbed}(\boldsymbol{I}_{in})), \tag{1}$$

where PatchEmbed denotes the patch embedding layer, and LayerNorm denotes the layer normalization function.

Four stages are used in total, the first three stages each contain two VSS blocks and a patch merging layer, which doubles the channel dimensions, and the final stage consists of two VSS blocks. Finally, the image features $\boldsymbol{I} \in \mathbb{R}^{\frac{H}{32} \times \frac{W}{32} \times 8C}$ can be obtained. The process is detailed in Eqs. (2) and (3).

$$\boldsymbol{I}_{merge} = (\text{PatchMerging}(\text{VSS}(\text{VSS}(I_{patch}))))_{\times 3}, \qquad (2)$$

$$\boldsymbol{I} = (\text{VSS}(\text{VSS}(I_{merge}))), \qquad (3)$$

where VSS refers to the VSS block, PatchMerging refers to the patch merging layer. $\boldsymbol{I}_{merge}$ denotes the vector obtained after the first three stage.

As shown in Fig. 2, the VSS block starts with layer normalization of the input. The features are then processed through two branches. The right branch uses a linear layer followed by an activation function, while the left branch applies a linear layer, depthwise separable convolution (DW-Conv), and SiLU activation before passing through the 2D-Selective-Scan (SS2D) module [12]. The extracted features are then normalized. The final output is produced by element-wise multiplication, a linear layer mix, and a residual connection.

2.2 Text Feature Extraction

Many studies highlight that text offers rich semantic information. Therefore, we use text to describe disease features in brain PET images. For example, for PET images of AD patients, we use statements like "Metabolism decreased in the temporal, parietal, and occipital lobes," "The hippocampus showed the accumulation of amyloid plaques," and "Parts of the brain showed atrophy" to describe such images. To improve diagnostic accuracy with detailed descriptions, we use Long-CLIP [25] as the text encoder, which supports longer inputs than CLIP. The text is first converted into tokens as:

$$\boldsymbol{T}_{token} = \text{Tokenize}(\text{Text}), \qquad (4)$$

where Text represents the long text inputs. Tokenize is an operation that embeds the text into word tokens. $\boldsymbol{T}_{token}$ denotes the obtained word tokens. These tokens are subsequently fed into the text encoder,

$$\boldsymbol{T} = \text{LongCLIP}(\boldsymbol{T}_{token}), \qquad (5)$$

where LongCLIP denotes the text encoder of Long-CLIP. $\boldsymbol{T}$ denotes the obtained text features.

Note that the improved text encoder supports text inputs of up to 248 tokens. Detailed pathological descriptions can provide additional context, helping the model more accurately understand and analyze abnormalities and lesions in the images.

2.3 Fusion and Prediction

The extracted image features and text features are concatenated and passed to a self-attention module. This module effectively merges the image features and text features, enabling the model to more effectively understand the connections and interactions between them.

$$\boldsymbol{F} = \text{Concat}(\boldsymbol{I}, \boldsymbol{T}), \tag{6}$$

$$\boldsymbol{F}_{att} = \text{softmax}\left(\frac{(\boldsymbol{F}\boldsymbol{W}^Q)(\boldsymbol{F}\boldsymbol{W}^K)^T}{\sqrt{d_k}}\right)(\boldsymbol{F}\boldsymbol{W}^V), \tag{7}$$

where $\boldsymbol{W}^Q$, $\boldsymbol{W}^K$, and $\boldsymbol{W}^V$ are the matrices for transforming the input feature $\boldsymbol{F}$ into the query vector $\boldsymbol{Q}$, key vector $\boldsymbol{K}$, and value vector $\boldsymbol{V}$, respectively. $\sqrt{d_k}$ is a scaling factor. $\boldsymbol{F}_{att}$ represents the final attention output.

Next, iterate over each image attribute and text attribute, extract the feature of each attribute from the input tensor, process these features using the corresponding weight layers, and then combine all the processed attributes into a complete output.

$$\boldsymbol{Logits} = \text{Concat}([\text{Linear}(\boldsymbol{F}_{att}^1), \ldots, \text{Linear}(\boldsymbol{F}_{att}^k)]), \tag{8}$$

where k denotes the number of attribute.

2.4 Loss Function

Let $\boldsymbol{Y} = (y_1, y_2, \cdots, y_B)$ denotes the ground-truth labels for PET slices of a batch size. Cross entropy is used as the loss function, defined as follows,

$$\mathcal{L} = -\sum_{b=1}^{B}[y_b log(\hat{y}_b) - (1-y_b)log(1-\hat{y}_b)], \tag{9}$$

where B represents the size of batchsize and $\hat{y}_b$ represents the predicted result.

3 Experiments

3.1 Dataset

Data Acquisition. This paper uses the publicly available Alzheimer's Disease Neuroimaging Initiative (ADNI, https://adni.loni.usc.edu/) dataset, which includes extensive neuroimaging data such as magnetic resonance imaging (MRI) and positron emission tomography (PET) scans.

We selected PET imaging data from ADNI for 455 individuals, comprising 168 with Alzheimer's disease (AD), 162 with Mild Cognitive Impairment (MCI), and 125 Normal Control (NC) subjects. Detailed information about subjects are listed in the Table 1.

Table 1. Details of ADNI dataset

ADNI	AD	NC	MCI
Number of subjects	168	125	162
Female/male	81/87	66/59	67/95
Age(Mean ± Standard Deviation)	74.68 ± 8.35	77.14 ± 6.38	73.72 ± 8.14

Data Pre-processing. PET imaging data can be affected by noise, motion artifacts, scanner discrepancies, and other factors that impact image quality. Therefore, preprocessing to address these issues is crucial, as low-quality images can significantly affect disease prediction accuracy.

First, AC-PC correction is applied, setting the origin at the midpoint of the AC-PC line. Second, head movement correction is performed. Third, the PET image is registered to a clearer MRI image. Fourth, the PET image is normalized to the ICBM152 brain template (193 × 229 × 193). Fifth, the PET image is smoothed. Finally, to increase data volume, 3D PET data is sliced into 2D images using the nibabel library in Python.

3.2 Implementation Details

Experiments were performed on a Linux machine with an NVIDIA GeForce RTX 3090 graphics card and 24 GB of RAM. The proposed architecture was implemented using the PyTorch library. 10-fold cross-validation was utilized for network training. The dataset was split into ten folds, with nine used for training and one for validation. There is no overlap between our test set and training set. The model is trained using the Adam optimizer with a learning rate of $1e^{-4}$, over 80 epochs, and a batch size of 8. Four commonly evaluation metrics are used, i.e., Accuracy (ACC), Area Under the Curve (AUC), Sensitivity (SEN), and Specificity (SPE).

3.3 Performance Comparison

Our method is compared with several advanced techniques, including 2D CNN and 3D CNN based methods. The 2D CNN methods include AlexNet [26], ResNet50 [27], BLADNet [7] and EAMNet [28]. The 3D CNN-based approaches include MiSePyNet [10] and PT-DCN [29]. Differences in data selection, preprocessing, and dataset partitioning among various methods can lead to discrepancies, even when all the data come from ADNI, potentially resulting in unfairness. Hence, to ensure a fair comparison, we reproduced these methods on our dataset using the same partitioning scheme. Two binary experiments are conducted, i.e., AD vs. NC and AD vs. MCI. Results are presented in Tables 2 and 3.

For AD vs. NC classification, the results are shown in Table 2. According to the data in the table, although the SEN metric is not the best, our method shows the best performance in the other three metrics, and the SEN metric also achieves the second-best performance. Our method exceeds the second-best

Table 2. Comparison with other deep learning methods for AD VS. NC(%)

Method	Modality	ACC	AUC	SEN	SPE
MiSePyNet(2021)	3D	76.25	75.16	83.89	66.43
PT-DCN(2022)	3D	85.00	84.76	**86.67**	82.86
AlexNet	2D	80.94	81.63	76.11	87.14
ResNet50	2D	83.13	83.41	81.11	85.71
BLADNet(2023)	2D	84.36	85.32	77.78	92.86
EAMNet(2023)	2D	83.13	83.65	79.44	87.86
ours	2D	**89.06**	**89.56**	85.56	**93.57**

Table 3. Comparison with other deep learning methods for AD VS. MCI(%)

Method	Modality	ACC	AUC	SEN	SPE
MiSePyNet(2021)	3D	62.50	62.50	**74.44**	50.56
PT-DCN(2022)	3D	69.17	69.17	62.22	76.11
AlexNet	2D	64.17	64.17	52.78	75.56
ResNet50	2D	60.83	60.83	52.78	68.89
BLADNet(2023)	2D	63.89	63.89	61.11	66.67
EAMNet(2023)	2D	65.00	65.00	52.22	77.78
ours	2D	**73.61**	**73.61**	68.89	**78.33**

values by 4.06% in ACC, 4.24% in AUC, and 0.71% in SPE. Compared to the second-best 3D CNN-based method PT-DCN, our method improves the ACC, AUC, and SPE metrics by 4.06%, 4.80%, and 10.71%, respectively. The most significant improvement is in SPE, indicating that the proposed method excels in recognizing negative samples.

Table 3 shows that our method outperforms other deep learning approaches in the AD vs. MCI classification task, achieving the highest ACC, AUC, and SPE metrics. It surpasses the second-best values by 4.44%, 4.44%, and 0.55%, respectively, and achieves the second best performance in SEN metric. Compared to the second-best 3D CNN-based method PT-DCN, our method improves the ACC, AUC, SEN, and SPE metrics by 4.44%, 4.44%, 6.67%, and 2.22%, respectively. The most significant improvement is in SEN, indicating that our method is notably better at recognizing positive samples compared to PT-DCN.

To better compare performance across methods, we present a bar chart illustrating the accuracy of the proposed approach alongside four advanced methods for AD prediction tasks, as depicted in Fig. 3. Our method consistently achieves the highest accuracy across both tasks.

3.4 Ablation Study

Table 4 presents the experimental results using different data modalities (image only and image plus text) for classification. For the first task, the model com-

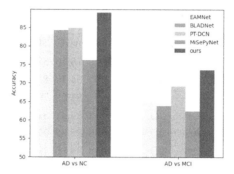

Fig. 3. Accuracy comparison for different methods.

Table 4. Performance of using different modalities

Task	Modality	ACC	AUC	SEN	SPE
AD vs. NC	image	86.56	86.55	**86.67**	86.43
	image + text	**89.06**	**89.56**	85.56	**93.57**
AD vs. MCI	image	70.56	70.56	**70.00**	71.11
	image + text	**73.61**	**73.61**	68.89	**78.33**

bining image and text data outperforms the image-only model in three out of four performance metrics, showing a significant improvement. For the AD vs. MCI task, after incorporating image and text data, the accuracy and specificity improve to 73.61% and 78.33%, respectively. Overall, incorporating textual information strengthens the model's classification ability, particularly enhancing the accuracy of NC recognition.

4 Conclusion

This paper presents a text-guided vision Mamba model for Alzheimer's disease prediction. The model is structured with two branches: one is an image encoder based on Mamba that extracts image features from PET imaging data, and the other is a text encoder based on Long-CLIP that extracts text features related to disease descriptions. Finally, image and text features are fused through a self-attention module, enhancing cross-modal consistency. Experimental results demonstrates that the proposed method can effectively predict AD disease. It also shows that integrating text data into brain disease prediction models is crucial as it complements the semantic information provided by imaging data.

Acknowledgement. This work is supported by the University Synergy Innovation Program of Anhui Province under Grant GXXT-2022-032 and GXXT-2022-014; the Anhui Province University Scientific Research Program Project 2024AH050052.

References

1. Theofilas, P., et al.: Probing the correlation of neuronal loss, neurofibrillary tangles, and cell death markers across the alzheimer's disease braak stages: a quantitative study in humans. Neurobiol. Aging **61**, 1–12 (2018)
2. Jagust, W., Gitcho, A., Sun, F., Kuczynski, B., Mungas, D., Haan, M.: Brain imaging evidence of preclinical alzheimer's disease in normal aging. Ann. Neurol. Official J. Ame. Neurol. Assoc. Child Neurol. Soc. **59**(4), 673–681 (2006)
3. Mosconi, L., Berti, V., Glodzik, L., Pupi, A., De Santi, S., de Leon, M.J.: Preclinical detection of alzheimer's disease using fdg-pet, with or without amyloid imaging. J. Alzheimers Dis. **20**(3), 843–854 (2010)
4. Nobili, F., et al.: for the Prescription of FDG-PET for Dementing Neurodegenerative Disorders, E.E.T.F., Festari, C., et al.: European association of nuclear medicine and european academy of neurology recommendations for the use of brain 18f-fluorodeoxyglucose positron emission tomography in neurodegenerative cognitive impairment and dementia: Delphi consensus. European journal of neurology **25**(10), 1201–1217 (2018)
5. Baydargil, H.B., Park, J., Kang, D.Y.: Anomaly analysis of alzheimer's disease in pet images using an unsupervised adversarial deep learning model. Appl. Sci. **11**(5), 2178 (2021)
6. Chen, Y., et al.: Contrastive learning for prediction of alzheimer's disease using brain 18f-fdg pet. IEEE J. Biomed. Health Inf. **27**, 1735–1746 (2022)
7. Duan, J., Liu, Y., Wu, H., Wang, J., Chen, L., Chen, C.L.P.: Broad learning for early diagnosis of alzheimer's disease using fdg-pet of the brain. Front. Neurosci. **17**, 1137567 (2023)
8. Liu, M., Cheng, D., Wang, K., Wang, Y.: Multi-modality cascaded convolutional neural networks for alzheimer's disease diagnosis. Neuroinformatics **16**, 295–308 (2018)
9. Yee, E., Popuri, K., Beg, M.F.: Quantifying brain metabolism from fdg-pet images into a probability of alzheimer's dementia score. Hum. Brain Mapp. **41**, 5–16 (2019)
10. Pan, X., et al.: Multi-view separable pyramid network for ad prediction at mci stage by 18f-fdg brain pet imaging. IEEE Trans. Med. Imaging **40**, 81–92 (2020)
11. Santi, L.A.D., Pasini, E., Santarelli, M.F., Genovesi, D., Positano, V.: An explainable convolutional neural network for the early diagnosis of alzheimer's disease from 18f-fdg pet. J. Digit. Imaging **36**, 189–203 (2022)
12. Gu, A., Dao, T.: Mamba: Linear-time sequence modeling with selective state spaces. ArXiv **abs/2312.00752** (2023)
13. Liu, Y., et al.: Vmamba: visual state space model. ArXiv **abs/2401.10166** (2024)
14. Yue, Y., Li, Z.: Medmamba: vision mamba for medical image classification. ArXiv **abs/2403.03849** (2024)
15. Gurung, A., Ranjan, P., et al.: Vision mamba: cutting-edge classification of alzheimer's disease with 3d MRI scans. arXiv preprint arXiv:2406.05757 (2024)
16. Yang, G., Du, K., Yang, Z., Du, Y., Zheng, Y., Wang, S.: Cmvim: contrastive masked vim autoencoder for 3d multi-modal representation learning for ad classification. ArXiv **abs/2403.16520** (2024)
17. Zhang, Y., et al.: Text-guided foundation model adaptation for pathological image classification. In: MICCAI, pp. 272–282 (2023)
18. Huang, W., Zhou, H., Li, C., Yang, H., Liu, J., Wang, S.: Enhancing representation in radiography-reports foundation model: a granular alignment algorithm using masked contrastive learning. arXiv preprint arXiv:2309.05904 (2023)

19. Lei, Y., Li, Z., Shen, Y., Zhang, J., Shan, H.: Clip-lung: textual knowledge-guided lung nodule malignancy prediction. In: International Conference on Medical Image Computing and Computer-Assisted Intervention, pp. 403–412. Springer (2023)
20. Zhang, X., Wu, C., Zhang, Y., Xie, W., Wang, Y.: Knowledge-enhanced visual-language pre-training on chest radiology images. Nat. Commun. **14**(1), 4542 (2023)
21. Lin, W., et al.: Pmc-clip: Contrastive language-image pre-training using biomedical documents. In: International Conference on Medical Image Computing and Computer-Assisted Intervention, pp. 525–536. Springer (2023)
22. Dai, T., Zhang, R., Hong, F., Yao, J., Zhang, Y., Wang, Y.: Unichest: conquer-and-divide pre-training for multi-source chest x-ray classification. IEEE Trans. Med. Imaging **43**, 2901–2912 (2023)
23. Radford, A., et al.: Learning transferable visual models from natural language supervision. In: International Conference on Machine Learning, vol. 139, pp. 189–2038748-8763 (2021)
24. Ruan, J., Xiang, S.: Vm-unet: vision mamba unet for medical image segmentation. arXiv preprint arXiv:2402.02491 (2024)
25. Zhang, B., Zhang, P., Dong, X., Zang, Y., Wang, J.: Long-clip: unlocking the long-text capability of clip. arXiv preprint arXiv:2403.15378 (2024)
26. Krizhevsky, A., Sutskever, I., Hinton, G.E.: ImageNet classification with deep convolutional neural networks. Commun. ACM **60**, 84–90 (2012)
27. He, K., Zhang, X., Ren, S., Sun, J.: Deep residual learning for image recognition. In: 2016 IEEE Conference on Computer Vision and Pattern Recognition (CVPR), pp. 770–778 (2015)
28. Duan, H., bin Wang, H., ping Chen, Y., Liu, F., Tao, L.: Eamnet: an alzheimer's disease prediction model based on representation learning. Phys. Med. Biol. **68**(21), 215005 (2023)
29. Gao, X., Shi, F., Shen, D., Liu, M.: Task-induced pyramid and attention gan for multimodal brain image imputation and classification in alzheimer's disease. IEEE J. Biomed. Health Inform. **26**, 36–43 (2021)

EEG-Based Recognition of Knowledge Acquisition States in Second Language Learning

Shanlin Xi[1], Ziyu Li[2], and Xia Wu[2(✉)]

[1] School of Artificial Intelligence, Beijing Normal University, Beijing 100875, China
[2] School of Computer Science and Technology, Beijing Institute of Technology, Beijing 100081, China
wuxia@bit.edu.cn

Abstract. Decoding information related to language cognition from Electroencephalogram (EEG) signals can reveal learners' genuine responses to knowledge. Existing language learning research using EEG experimental paradigms often focuses on abstract stimuli, which undermines the accuracy of reflecting real experiences in the actual learning process. This study presents a more natural EEG experimental paradigm for second language(L2) learners, recording EEG signals in video-based language learning scenarios. Data analysis from 12 participants confirms the paradigm's validity, allowing us to establish a mapping model between EEG signals and learners' states of language acquisition, achieving an accuracy of 0.95. These findings demonstrate the feasibility of decoding cognitive states from real-world scenarios, highlighting the proposed paradigm's significant research potential in second language learning and offering insights into the brain's linguistic cognitive processes.

Keywords: EEG · language cognition · second language learning · language acquisition state recognition

1 Introduction

In the process of second language learning, adapting learners' language learning ability and recognizing whether they have acquired a certain knowledge point are particularly important for designing and adjusting language learning programs. Due to the scientific understanding of human language cognition, current curricula and testing programs based on teaching experience and linguistic have not been satisfactory, and teacher-driven feedback on students' learning status has even hindered the learning of languages, especially second languages, to a certain extent [11]. Therefore, people are more and more interested in objective methods to determine learners' language acquisition status and language acquisition ability level based on learners' EEG signals. EEG signals are rich in information but difficult to decode. Fortunately, the rapid development of artificial intelligence technology in recent years has opened up a possibility, which shows great potential in the field of language cognition [2].

The sources of stimuli in natural scenes are complex, and previous studies often adopt the experimental paradigm design of abstracted scenes to control the stimulus properties and reduce the difficulty of signal analysis, which also makes it impossible to collect data during the learning process [7]. Moreover, most of the previous studies have used Chinese and English or artificial machine language as the learning corpus with little attention paid to the learning of other languages. The responses of the subjects in the abstract paradigm experiments are also greatly deviated from the real scenarios, and it is still necessary to collect real responses in natural situations.

To address these limitations, this study simulates a state-of-the-art video learning scenario and designs a new EEG experimental paradigm for language learning that collects physiological data on learners' short-term language learning in familiar scenarios. The key points of the paradigm are: 1) recording the closest to real learning process responses without limiting the stimulus form; 2) recording the EEG signals in the whole experimental process, including the course learning process; and 3) choosing a minority language as the learning object to enrich the language learning EEG data.

To verify the rationale of the paradigm design and enable the discrimination of learners' cognitive states during language learning from EEG signals, we recruit 12 participants to conduct German learning experiments based on the designed paradigm. We then analyze the collected behavioral and EEG data. The results show that the distributions of learners' reaction times and accuracy measures align with the learning ability results obtained from traditional behavioral assessments. Furthermore, there is a strong correlation between power-related EEG features and the learners' acquisition states, confirming the feasibility of the paradigm. Additionally, leveraging the high correlation between EEG features and acquisition states, we successfully develop a model to recognize the learner's knowledge acquisition state, achieving an optimal accuracy of 0.95.

2 Method

2.1 Paradigm Design

To simulate the real-life scenario of online short-term video-based language learning as closely as possible, we randomly recruit 12 subjects (7 males and 5 females, all college students, average age 20.83, all with prior study of only English and Chinese), set up the experimental environment in a familiar classroom to exclude discomfort, and isolate the monitoring equipment from the subjects' view to prevent influence from the data collection tools. The experiment takes place when the subjects are energetic and in good condition, with the beginning and end controlled by the subjects themselves using the keyboard. The experiment records the subjects' brain neural activities and behavioral responses during the learning process, including the video course, the preparation before the course, and the test after the study.

The EEG signals are recorded using an EGI 128-conductor high-density EEG cap, while the learning materials are presented and timed with high precision using E-prime 3.0 software. Subjects control the experiment's progress and respond to test questions using the space bar (left-handed) and the arrow keys (right-handed) on the keyboard.

To fully track the subjects' learning status, each subject participates in two learning experiments spaced one week apart. In each experiment, the subjects are asked to learn an introductory German lesson lasting about 10 min.

First Learning Experiment. The flow and experimental paradigm of the first learning experiment is shown in Fig. 1. The experiment is divided into three main phases: the first phase is pre-course preparation. In this phase, participants first read through the test questions, which are designed based on the knowledge included in the subsequent video lesson. By previewing these test questions, participants familiarize themselves with the content to be studied, thus preparing for the formal learning session. The second phase is the teaching session, where participants watch a 10-minute introductory video on German. This ensures that the total duration of the experiment remains reasonable, minimizing the risk of mental fatigue due to an overly long experiment. The third phase is the test session, in which participants respond to the pre-test questions based on their learning progress. Responses are given via keystrokes. To eliminate the potential bias caused by guessing, it is explicitly agreed with the participants prior to the experiment that if they cannot provide a definite answer to a question, that question will be skipped. The experiment consists of 25 test questions, and the order of the questions is randomized both before and after the learning phase.

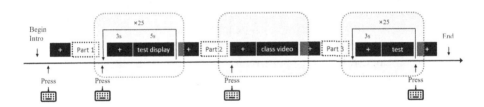

Fig. 1. The flow and experimental paradigm of the first learning experiment.

Second Learning Experiment. The second experiment is conducted one week after the first experiment. As shown in Fig. 2, the experimental procedure is largely the same as that of the first experiment, with the main difference being the addition of a test during the preparatory phase before the study: specifically, at the beginning of the second study, the test from the first course is repeated. This session serves to review (memorize and recall) the content from the first study. The video for the second session is approximately 6 min long, which reduces the knowledge volume compared to the first study. Consequently, the number of test questions is reduced to 15. This adjustment ensures that the overall length of the experiment is similar to that of the first experiment, thereby minimizing potential fatigue for the subjects. The second test covers content from both courses, integrating knowledge points from the first and second study.

During the reading preparation and teaching phases of both courses, only the subjects' EEG signals are recorded. In the testing phase, additional data is collected, including the subjects' key press feedback, reaction times, and statistics on whether they answer the test questions correctly. Furthermore, considering that the subjects are all college students with a background in English learning, information regarding their English proficiency is also recorded, with their consent, as a supplementary indicator of their ability to learn a foreign language.

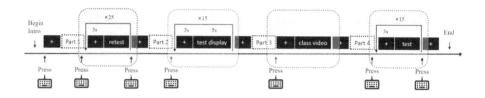

Fig. 2. The flow and experimental paradigm of the second learning experiment.

Only EEG signals and behavioral information of the subjects during the two pre-study reading preparation sessions and the lesson sessions were recorded. The information collected in the test session also included the subjects' key press feedback and reaction time, and statistics on whether the subjects answered the test questions correctly or not. In addition, given that the subjects were all college students with English learning backgrounds, information on their English proficiency was collected after obtaining their own consent, as an external evaluation index of foreign language learning.

2.2 Data Preprocessing

The acquired signals are downsampled and filtered to reduce the sampling rate to 250 Hz while retaining data in the 0.05–100 Hz band. Industrial frequency interference is removed using a 50 Hz notch filter [9]. The filtered, downsampled signal is then decomposed into independent components using independent component analysis (ICA). Noisy components, such as those from ECG, eye movements, and EMG, are manually selected and removed based on the subject's spontaneous physiological activities in the original EEG signal [6].

Additionally, test session data from two learning experiments are extracted to assess the feasibility of identifying knowledge acquisition states from EEG. The first 7 s of each trial in each session are selected as epochs, with the correct and incorrect responses to test questions labeled accordingly. Finally we obtain a total of 780 test data segments.

2.3 Feature Engineering

EEG signals provide information related to cognitive activities in the brain, reflected in the signal features of each EEG channel. To investigate the relationship between these features and language knowledge acquisition, we extract the power of each channel in different frequency bands (δ: 0.05–4 Hz, θ: 4–8 Hz, α: 8–12 Hz, β: 12–30 Hz, and γ: 30–100 Hz), along with other time and frequency domain features, such as the entropy of the signal distribution in the frequency domain. The EEG data is recorded using a 128-channel high-density EEG cap, with each electrode positioned in different brain regions, providing diverse insights into brain cognitive activities.

Feature extraction targets the signals from each channel, and to reduce the dimensionality of the feature matrix for classifier input, features are grouped into sets. This approach limits the number of features per training session and helps distinguish their relevance to the cognitive task of language learning. It also aids in identifying the features most indicative of language learning-related cognitive states. The extracted EEG features and their groupings are presented in Table 1.

Table 1. Specific information on feature sets.

Feature set number	Feature included
①	peak-to-peak value, skewness, kurtosis, variance, instability index
②	Zero crossings, C0 complexity, Hjorth parameters (activity, movement, complexity), Shannon entropy
③	δ(0.05–4 Hz), θ(4–8 Hz), α(8–12 Hz), β(12–30 Hz), γ(30–100 Hz) band power, total power, spectral entropy
④	②+③
⑤	①+②
⑥	①+②+②

2.4 Model

The amount of data collected for validating the experimental paradigm is relatively small. Given this limitation and the need for efficient classification, traditional machine learning algorithms and ensemble learning methods are the primary choice for analysis. To compare the classification performance across different algorithms and feature sets, the following ten classification algorithms (listed Table 2.) are selected to classify the data collected in this experiment:

Table 2. Table of selected machine learning algorithms.

Classifier category	Name of the classification algorithm
Traditional Classification Algorithms	SVM(Support Vector Machine)
	Gaussian NB
	KNN(K-Nearest Neighbors)
	Logistic Regression
	Decision Tree
Classification algorithm with dimensionality reduction techniques	LDA (Linear Discriminant Analysis)
	QDA (Quadratic Discriminant Analysis)
Ensemble Learning Algorithms	Random Forest
	Gradient Boosting
	Adaboost(Adaptive Boosting)

3 Experiments and Results

3.1 Experiment Settings

Among the 780 test data segments, 552 are correct and 218 are incorrect, yielding a class distribution ratio of 5:2. This imbalance may lead to overfitting, with the classification model likely favoring the majority class. To mitigate the effects of this imbalance, the SMOTE (Synthetic Minority Over-sampling Technique) method was applied to oversample the minority class and balance the two classes [1]. After applying SMOTE, the adjusted dataset contains 1104 samples, with an equal ratio of 1:1 between the two classes.

The new dataset is split into a training set and a test set in a 4:1 ratio. Ten classifiers are trained to predict whether the learner answered the questions correctly or not. The optimal parameters for each machine learning algorithm are determined using 5-fold cross-validation and grid search (lattice analysis) to fine-tune the models.

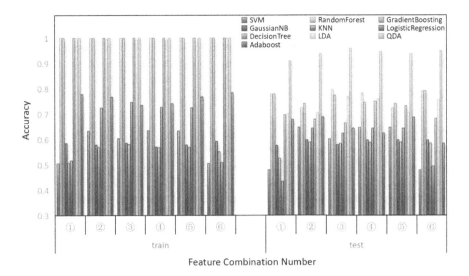

Fig. 3. The flow and experimental paradigm of the second learning experiment.

3.2 Results

Model Training. After cross-validation and grid analysis, the best performance of the ten algorithms is shown in Fig. 3. The results reveal that several algorithms with poor classification performance are primarily constrained by high feature dimensions and limited data volume, as indicated by the variations in feature combinations ①-⑥. The decision tree, three ensemble learning algorithms, and two classification algorithms with dimensionality reduction techniques demonstrate nearly 100% accuracy, which is considered optimal. Among these, Gradient Boosting and Quadratic Discriminant Analysis (QDA) stand out for their classification accuracy and generalization performance. Both reach 100% accuracy on the training set, with Gradient Boosting achieving around 0.75 accuracy on the test set, while QDA reaches approximately 0.95. The superior performance of QDA is likely due to its inherent dimensionality reduction properties. With an accuracy of 95%, close to 100%, QDA demonstrates the feasibility of recognizing a language learner's knowledge acquisition state from EEG data.

Features Screening. The performance differences driven by features are evident across several classifiers, particularly in SVM, Random Forest, Logistic Regression, KNN, QDA, and LDA, as shown in Fig. 4. Most classifiers achieve notably higher classification accuracy for feature sets ② and ③ (see Table 1 for details), with set-③ standing out in particular. All feature sets containing spectral features (such as spectral entropy, α-band power, total power, etc.) are classified relatively well by the classifiers. This suggests that, among the 18 features extracted in this study, power spectrum-related features are particularly relevant for cognitive recognition tasks in language learning. This finding is consistent with current research on brain cognition, which shows that information from β, γ, and other frequency bands in brain waves is strongly correlated with human cognitive functions [3,4,10]. It also supports the scientific validity and reasonableness of the experimental paradigm design used in this study's small language learning task.

Behavioral Data Statistical Analysis. Statistics on the scores and average response time of the twelve subjects across the three test parts of the two experiments, along with the English language proficiency assessment as reflected in the $CET-4$ and $CET-6$ scores, are shown in Fig. 5. The scores on the three tests differed slightly, but the overall ranking of the subjects' scores remained consistent. The second retest demonstrated substantially lower accuracy for all subjects compared to the first test, aligning with the experimental design's expectation that memory retention would decline over the one-week interval between tests [15]. Subjects generally exhibited longer reaction times on both mixed tests of course content compared to the first test. This observation supports the exper-

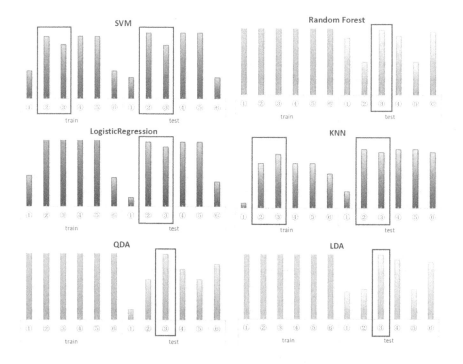

Fig. 4. The performance difference brought by features reflected in SVM, Random Forest, Logistic Regression, KNN, QDA and LDA (The horizontal coordinates are the feature group ordinal numbers, delineated by training and test sets, and the heights of the bar graphs all indicate the classifier accuracy.).

imental hypothesis that the brain responds more quickly to familiar content, with reaction times increasing as the complexity and volume of the content rise [8]. A comparison of the distribution of subjects' English proficiency with the distributions of reaction times and accuracy revealed some similarities. This suggests that individual behavioral differences, such as reaction time and accuracy, may reflect variations in language proficiency [5]. These findings further validate the experimental paradigm employed in the study.

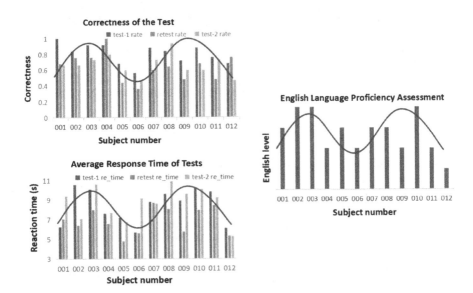

Fig. 5. Subjects' accuracy and reaction time for the three parts of the test in the two experiments, and the distribution of subjects' English proficiency levels (all three reflect a high degree of similarity in their distributions.).

4 Conclusion

This paper introduces an EEG-based experimental paradigm for language learning and validates its feasibility through a series of experiments. The experiments simulate an online video learning environment, recording both behavioral and physiological signals to capture language learning data that closely mirrors real-world scenarios. Preliminary results show that the paradigm is effective, with strong potential for extracting language-related cognitive information from EEG data, even amid the complex stimuli of real-world learning.

Future work will expand the dataset by collecting more learner data to enhance its diversity and representativeness, improving model reliability. This expanded dataset will also allow for deeper investigation into the language learning process and better assessment of second language learning abilities, advancing cognitive modeling and educational technologies.

Acknowledgements. This work was supported by the Sub-project of National Key Research and Development Program of China (Grant No. 2023YFC3305603). The authors declare that there is no conflict of interest.

References

1. Chawla, N.V., Bowyer, K.W., Hall, L.O., Kegelmeyer, W.P.: Smote: synthetic minority over-sampling technique. J. Artif. Int. Res. **16**(1), 321–357 (2002)
2. Défossez, A., Caucheteux, C., Rapin, J., Kabeli, O., King, J.R.: Decoding speech perception from non-invasive brain recordings. Nat. Mach. Intell. **5**(10), 1097–1107 (2023)
3. Kanta, V., Pare, D., Headley, D.B.: Closed-loop control of gamma oscillations in the amygdala demonstrates their role in spatial memory consolidation. Nat. Commun. **10**(1), 3970 (2019)
4. Klimesch, W.: Eeg alpha and theta oscillations reflect cognitive and memory performance: a review and analysis. Brain Res. Rev. **29**(2), 169–195 (1999)
5. Luque, A., Morgan-Short, K.: The relationship between cognitive control and second language proficiency. J. Neurolinguistics **57**, 100956 (2021). https://doi.org/10.1016/j.jneuroling.2020.100956, https://www.sciencedirect.com/science/article/pii/S0911604420301160
6. Mannan, M.M.N., Kamran, M.A., Jeong, M.Y.: Identification and removal of physiological artifacts from electroencephalogram signals: a review. IEEE Access **6**, 30630–30652 (2018). https://doi.org/10.1109/ACCESS.2018.2842082
7. Massam, W.E.: Investigating effects of contextualized science curricular experiences on students' learning and their teachers' teaching in Tanzania. Ph.D. thesis, University of British Columbia (2019). https://doi.org/10.14288/1.0377644, https://open.library.ubc.ca/collections/ubctheses/24/items/1.0377644
8. Nucci, L., et al.: Reaction time and cognitive strategies: the role of education in task performance. Learn. Motiv. **82**, 101884 (2023)
9. Pise, A.W., Rege, P.P.: Comparative analysis of various filtering techniques for denoising EEG signals. In: 2021 6th International Conference for Convergence in Technology (I2CT), pp. 1–4 (2021). https://doi.org/10.1109/I2CT51068.2021.9417984
10. Schmidt, R., Herrojo Ruiz, M., Kilavik, B.E., Lundqvist, M., Starr, P.A., Aron, A.R.: Beta oscillations in working memory, executive control of movement and thought, and sensorimotor function. J. Neurosci. **39**(42), 8231–8238 (2019)
11. Yu, S., Xu, H.: Feedback and assessment in second language education during the COVID-19 and beyond. Asia Pac. Educ. Res. **30**(6), 483–486 (2021)

A Study on the Neural Mechanism of the Spatial Position of Speech in Different Masking Types Affecting Auditory Attention Processing

Dawei Xiang, Yong Ma(✉), and Yiming Yang(✉)

Jiangsu Normal University, Xuzhou, China
{may,yangym}@jsnu.edu.cn

Abstract. In complex communicative scene, listeners can selectively focus their attention on the position of the target speaker based on their communicative intentions, suppress sound input from the same or different spatial positions and effectively process the information of the target sound. However, it is currently unclear how different types of interfering sounds affect the allocation of auditory attention resources when listener selectively processes the target sound, and how attention regulates cortical responses from top to bottom to locate and perceive the target sound. This article mainly focuses on how the spatial position of speech and the number of speakers in the masking speech affect auditory attention. The experimental results indicate that the interaction between the spatial position of speech and the number of masking speech affects the efficiency of processing target speech. The separation of masking and the spatial position of the target speech can promote the perception of the target speech, manifested as an enhancement of cortical response. However, the advantage of spatial position separation in promoting target perception is also affected by the number of masking speech. In addition, this paper also found that the brain has the advantage of spatial location selectivity and attention resource allocation in processing target speech input from different directions.

Keywords: masking type · spatial position of speech · auditory attention · neural mechanism

1 Introduction

The "cocktail party effect" refers to the ability of the human brain to focus attention on sound information in specific spatial location and perform related semantic processing in complex sound environments, while suppressing interfering sounds from other directions (Cherry 1953), thereby ensuring efficient perception and processing of sound information at target locations. Therefore, on the one hand, research on the cocktail party effect focuses on the factors that affect target speech perception (Erol et al. 2023; Jonghwa et al. 2023), on the other hand, focusing on the neural basis of the listener focusing on the target sound and suppressing interfering sounds in complex communicative environments (Heather et al. 2021; Subong et al. 2021; Christina et al. 2021).

The process by which the acoustic or informational content of a sound is disturbed by other sounds is called sound masking. If the frequencies of sound signals are the same or similar, masking is most likely to occur (Mayer 1894; Wegel and Lane 1924). The number of masking sounds, the similarity in acoustics and content between masking sounds and target sounds can all affect the quality of target sound received by the listener (Brouwer et al. 2012; Vermiglio et al. 2019; Mathew et al. 2021).When the number of masks is within a certain range, the semantic information in the masking sound will directly affect the target semantic information, which is called information masking. Due to the fact that the semantic content in the two masking quantities can be clearly recognized by the listener, the semantic comprehensibility is high, and it can directly affect the information content in the target sound. Therefore, the two masking sound quantities have a significant impact on the perception of the target sound (Brouwer et al. 2012). As the number of masking sound increases, the degree of semantic information recognition in the masking sound gradually decreases, resulting in poorer masking effects, similar to energy masking (Song et al. 2011; Rosen et al. 2013; Humes et al. 2017; Alageel et al. 2017; Mathew et al. 2021).

The spatial relationship between sounds can also affect the effectiveness of masking release, thereby affecting the efficiency of target sound perception (Jakien et al. 2017; Ellinger et al. 2017; Andreeva et al. 2019). Previous studies have shown that spatial separation between masking and target sound promotes the listener's perception of the target sound (Das et al. 2018), and the larger the angle of spatial separation between masking and target sounds, the better the masking release effect, the clearer the perception of the target sound and the faster the processing speed of semantic contents (Su et al. 2022). In summary, both the number of masking sounds and the spatial position relationship of the sounds can affect the perception effect of the target sound.

The EEG components N1, N2 and P2 have been confirmed to be related to auditory semantic processing, and attention focusing enhances cortical representation in the brain (Diane et al. 2020; Christina et al. 2021). The N1, N2 and P2 refer to the negative and positive phase peaks induced at around 100 ms, 200 ms and 800 ms relative to baseline in specific brain regions when auditory stimuli are presented. In the masking sound conditions, N1 and P2 are regulated by nerves, and the separation of sound positions enhances the neural response of N1 and P2. However, when the masking effect is strong, the amplitude difference of N1 between different conditions is small (Erol et al. 2023). The amplitude of N2 increases with the allocation of attention resources (Christina et al. 2021).

Selective attention is an important form of attention that focuses more cognitive resources on the target sound while suppressing other interfering sounds under the condition of the interaction between target search and interference suppression (Wöstmann et al. 2019). Auditory attention is divided into two types: top-down and bottom-up, corresponding to the dorsal attention system and ventral attention system. The brain regions related to higher-order cognitive processing are mainly the dorsal attention system, so in the "cocktail party" scene, the neural mechanism of the dorsal attention system is the focus of our study.

2 Experimental Design and Results

2.1 Experimental Subjects

A total of 37 college students aged 18–26 were recruited for the experiment, including 18 males and 19 females.

3 Experimental Procedure

The details are shown in Fig. 1.

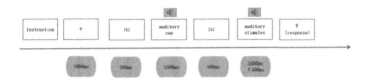

Fig. 1. Experimental procedure.

3.1 Preliminary Results

Behavioural Outcomes. The results for correct rate and reaction time for each condition in the behavioural data are shown in Table 1 and Fig. 2.

Table 1. Correct rate

Type of condition	Correct rate	
	average value	Standard deviation
Condition1	0.87	0.10
Condition2	0.89	0.10
Condition3	0.95	0.09
Condition4	0.94	0.14
Condition5	0.91	0.07
Condition6	0.88	0.09
Condition7	0.97	0.09
Condition8	0.96	0.09

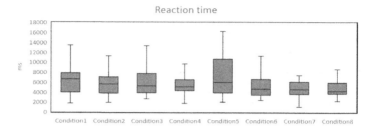

Fig. 2. Reaction time. Condition 1 is that the target speech is on the right side, the target and masking speech positions are consistent, and there are two speakers in masking. Condition 2 is that the target speech is on the left side, the target and masking speech positions are consistent, and there are two speakers in masking. Condition 3 is that the target speech is on the right side, the target and masking speech positions are separated, and there are two speakers in masking. Condition 4 is that the target speech is on the left side, the target and masking speech positions are separated, and there are two speakers in masking. Condition 5 is that the target speech is on the right side, the target and masking speech positions are consistent, and there are four speakers in masking. Condition 6 is that the target speech is on the left side, the target and masking speech positions are consistent, and there are four speakers in masking. Condition 7 is that the target speech is on the right side, the target and masking speech positions are separated, and there are four speakers in masking. Condition 8 is that the target speech is on the left side, the target and masking speech positions are separated, and there are four speakers in masking.

A three-way statistical analysis of correct rate revealed a significant main effect of masking and speech position relationship ($F(1,36) = 24.640$, $p < 0.001$). A three-way statistical analysis of reaction time revealed a significant main effect of target speech position ($F(1,36) = 4.579$, $p = 0.039$) and a significant main effect of masking and target speech position relationship ($F(1,36) = 6.137$, $p = 0.018$).

Analysis of EEG Results. As shown in Fig. 3., preliminary statistical analysis was conducted on the N1 component. It was found that within a specific region of interest, the main effect of the relationship between masking and target speech position was

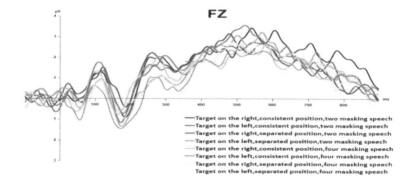

Fig. 3. N1 component at FZ electrode point.

significant (F (1,36) = 4.260, p = 0.046). When the masking and target speech positions were consistent, the amplitude of N1 was significantly enhanced. The main effect of the number of masking speech is significant (F (1,36) = 4.353, p = 0.044). When the number of masking speech is two, the amplitude of N1 is significantly enhanced.

As shown in Fig. 4., preliminary statistical analysis was conducted on the N2 component and it was found that the main effect in the brain region was significant (F (1,51) = 16.383, p < 0.001). The interaction effect among the target speech position, the relationship between masking and target speech position and the brain region was significant (F (1,59) = 3.908, p = 0.033). Simple effects showed that in a specific brain region, when the masking and target speech positions were separated, the N2 amplitude induced by the target sound position on the left increased (p = 0.010). When the target speech was on the right and the masking and target speech positions were consistent, the N2 difference between different brain regions was more significant. When the target speech is on the left side, regardless of whether the masking and target speech positions are consistent or separated, there is a significant difference in N2 amplitude between different brain regions. The interaction effect among the position relationship between masking and target speech, the number of masking speech and brain regions are significant (F (1,56) = 5.679, p = 0.010). Simple effects indicate that regardless of whether the masking and target speech positions are separated or the number of speakers in masked speech, there are significant differences between different brain regions.

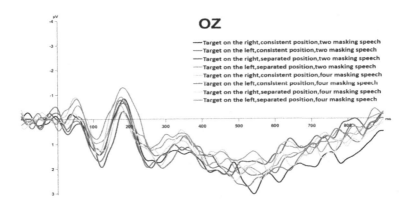

Fig. 4. N2 component at OZ electrode point.

As shown in Fig. 5., preliminary statistical analysis was conducted on the P2 component, and it was found that the main effect of the target speech position was significant (F (1,36) = 4.946, p = 0.033). When the target speech position was on the left side, a larger P2 was induced (p = 0.012). The interaction effect between the position relationship between masking and target speech and the number of masking speech is significant (F (1,36) = 6.386, p = 0.016). Simple effects indicate that when masking and target speech positions are separated, a larger P2 amplitude is induced when the number of speakers in masking speech is four (p = 0.002). When the number of masking speech is four, separating the masking and target speech positions will induce a larger P2 amplitude (p = 0.008).

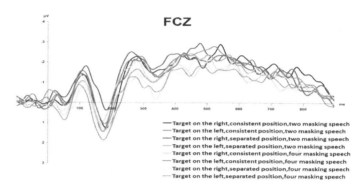

Fig. 5. P2 component at FCZ electrode point.

4 Discussion

Our research found that when the masking and target speech positions are consistent, the N1 amplitude is significantly enhanced. This is because when the masking and target speech positions are consistent, the masking effect is more significant. In order to more accurately locate the target speech, suppress the masking speech and process the target speech information, the brain needs to invest more attention resources (Christina et al. 2021). When the semantic content in masking speech can be clearly perceived, masking speech has a strong interference effect on auditory attention. Therefore, in order to perceive the target speech more clearly, the brain needs to invest more attention resources to locate the target speech and suppress the interfering speech. Therefore, the cortical response of the brain is stronger (Brouwer et al. 2012; Christina et al. 2021).

When attention is focused on the left position, the P2 amplitude is significantly enhanced. This indicates that the brain has spatial position selectivity and processing priority when processing target speech from different directions. The degree of cognitive control and cognitive resources invested in processing speech from different positions vary (Abrams et al. 2008). When masking and target speech positions are separated, the neural response of the brain is stronger, manifested by a significant increase in EEG amplitude (Jakien et al. 2017; Das et al. 2018). When the number of masking speech is four, similar to energy masking, the masking effect is poor. When the number of masking speech is four, the comprehensibility of semantic information in the masking speech is low and the semantic content in the masking speech would not distract the auditory attention used to locate the target speech. When the number of masking speech is two, semantic extraction in the masking speech is easier and semantic perception in the target speech is clearer. Therefore, there is greater interference with the attention used to locate the target speech and the attention resources required for locating the target speech will be reduced, resulting in a weaker response of the cortex to the target speech (Song et al. 2011; Rosen et al. 2013; Humes et al. 2017; Alageel et al. 2017; Andrew et al. 2022; Mathew et al. 2021).

When the positions of masking and target speech are separated, the brain is more inclined to process the target speech on the left position, manifested as N2 component amplitude enhancement. The right hemisphere of the brain has processing advantages. That is that the right hemisphere of the brain can process information input from both the left and right positions. The left hemisphere of the brain can only process information from the right position (Abrams et al. 2008). When the target speech is input from the left side, the separation of masking and target speech location results in the brain not only having to allocate attention resources to the target speech, but also being affected by masking speech from the same location, which weakens attention resources. Compared to speech input from the right side, it is more difficult to locate the target speech (Subong et al. 2021; Erol et al. 2023).

Unlike previous studies, we not only found that the number of masking speech and the position relationship between masking and target speech affect the listener's target perception processing, but also further discovered that the position of target speech input also affects the brain's response mechanism, mainly manifested in N1, P1, and P2 components. Through experimental results, we found that when the target sound is input at a certain position on one side, the number of masking speech and the positional relationship between the masking and the target speech have a more significant impact on the perception of spatial speech than when input at the other position. In complex communication scene, when faced with simultaneous input of speech information from multiple directions, the brain's processing reactions and cognitive resources invested in it are different, mainly manifested in two aspects. Firstly, when faced with target speech from two directions, the neural response is significantly higher when the target speech is input from one position than the other, indicating that the brain has agency and positional differences in processing speech from different directions. Secondly, the brain also has a hemisphere main effect when processing target speech. One side of the brain has significant differences in the amplitude of various EEG components under different conditions, while the other side has less significant differences. This also indicates the lateralization feature of brain processing (Gamble and Waldorf 2015; Erol et al. 2023).

Language cognition research mainly focuses on the neural mechanisms of the brain in processing language, explaining the brain's language ability, and exploring the neural encoding and decoding processes in language processing. Computational linguistics mainly focuses on how to transform the brain's processing mechanism of language into a model that can be recognized and represented by machines through training, and construct language models through a series of algorithm applications. Therefore, the conclusions of language cognition research can promote existing computational linguistics models and help improve the construction of language deep learning networks (Wang et al., 2022). This study can promote the research on auditory attention decoding (AAD). Auditory attention decoding can decode the spatiotemporal information and target objects of auditory attention from neural activated features (Vandecappelle et al. 2021). Through canonical correlation analysis (CCA) (de Cheveigné et al. 2018), Bayesian spatial state modeling (Miran et al. 2018) and convolutional neural networks

(CNNs; Vandecappelle et al. 2021), it can explain large language models and deep learning networks related to auditory spatial attention and be applied to neural guided artificial ear worms. The theoretical and practical prospects are enormous. By characterizing the neural activation characteristics of the brain in different auditory environments, this study found that the decoding of auditory spatial attention varies in different masking scenes. The neural activation characteristics under different conditions can obtain the spatial allocation characteristics of cognitive resources, as well as how the brain regulates cortical responses from top to bottom when facing masking speech in different directions to help listeners suppress interfering speech and locate target speech signals. These cognitive processing signals of auditory attention can help adjust and improve attention decoding models such as CNNs and EEG-Graph nets and train language models similar to BERT (Devlin et al. 2019; Ding et al. 2022), promoting machines to perceive and understand natural language more accurately and autonomously.

5 Conclusions

The position of target speech input, the number of masking speech and the positional relationship between masking and target speech all affect the processing of target speech and the suppression of interfering speech. The cognitive resources invested by listeners under different conditions also have significant differences. The advantage of spatial separation in promoting masking release is influenced by the number of masking speech, mainly because the recognition and comprehensibility of semantic content in different numbers of masking speech have differences, and the masking effect is also different. In summary, when the relationship between masking and target speech location affects auditory attention, it is also influenced by the comprehensibility of semantic content in masked speech, as well as the priority and positional differences in the attention resources allocated by the brain to target speech input from different positions. Our experiment not only confirmed that the number of masking speech and spatial position relationship affect the perceptual processing of target speech, but also further obtained the results of the interaction between the two affecting the brain mechanism, and creatively discovered that the position of target speech also affects auditory attention. Future research can focus on the distribution of alpha topographic maps under different conditions and the differences in alpha amplitude to explore the processing of attention in depth, and further decode auditory attention based on these EEG data.

Acknowledgements. This research was supported by the National Basic Research Program of China (973) (2014CB340502).The author would like to thank Professor Jia Maoshen of Beijing University of Technology for his technical support.

References

Abrams, D.A., Nicol,T., Zecker, S., Kraus, N.: Right-hemisphere auditory cortex is dominant for coding syllable patterns in speech. J. Neurosci. **28**(15), 3958–3965 (2008)

Alageel, S., Sheft, S., Shafiro, V.: Linguistic masking release in young and older adults with age-appropriate hearing status. J. Acoust. Soc. Ame. **142**(1), EL155–EL161 (2017)

Andreeva, I.G., Dymnikowa, M., Gvozdeva, A.P., Ogorodnikova, E.A., Pak, S.P.: Spatial separation benefit for speech detection in multi-talker babble-noise with different egocentric distances. Acta Acust. Acust. Acust. Acust. **105**, 484–491 (2019)

Andrew F., McDermott, J.H.: Deep neural network models of sound localization reveal how perception is adapted to real-world environments. Nat. Hum. Behav. **6**, 111–133(2022)

Brouwer, S., Van Engen, K.J., Calandruccio, L., Bradlow, A.R.: Linguistic contributions to speech-on-speech masking for native and non-native listeners: Language f-amiliarity and semantic content. J. Acoust. Soc. Am. Acoust. Soc. Am. **131**(2), 1449–1446 (2012)

Cherry, E.C.: Some experiments on the recognition of speech, with one and with two ears. J. Acoust. Soc. Am.Acoust. Soc. Am. **25**(5), 975–979 (1953)

Christina H., Michael-Christian S., Stephan G., Jo¨rg L.: Short-term audiovisual spatial training enhances electrophysiological correlates of auditory selective spatial attention. Front. Neurosci. **15**, 645702 (2021)

Das, N., Bertrand, A., Francart, T.: EEG-based auditory attention detection: boundary conditions for background noise and speaker positions. J. Neural Eng. **15**(6), 066017(2018)

Ding, X., Chen, B., Du, L., Qin, B., Liu, T.: CogBERT: cognition-guided pre-trained language models. In: Proceedings of the 29th International Conference on Computational Linguistics, Gyeongju, Republic of Korea. International Committee on Computational Linguistics, pp. 3210–3225 (2022)

Devlin, J., Chang, M-W., Lee, K., Toutanova, K.: Bert: pre-training of deep bidirectional transformers for language understanding. arxiv preprint arxiv.1810.04805 (2019)

de Cheveigné, A., Wong, D.D., Di Liberto, G.M., Hjortkjær, J., Slaney, M., Lalor, E.: Decoding the auditory brain with canonical component analysis. Neuroimage **172**, 206–216 (2018)

Diane Baier and Ulrich Ansorge,Can subliminal spatial words trigger an attention shift? Evidence from event-related-potentials in visual cueing. Vis. Cognition. 10–32 (2020)

Ellinger, R. L., Jakien, K.M., Gallun, F.J.: The role of interaural differences on speech intelligibility in complex multi-talker environments. J. Acoust. Soc. Am. **141**(2), EL170–EL176 (2017)

OzmeralID, E.J., Menon, K.N.: Selective auditory attention modulates cortical responses to sound location change for speech in quiet and in babble. PLOS ONE **18**(1) (2023)

Gamble, M.L., Woldorff, M.G.: The temporal cascade of neural processes underlying target detection and attentional processing during auditory search.Cerebral Cortex. **25**(9), 2456–2465 (2015)

Daly, H.R., Pitt, M.A.: Distractor probability influences suppression in auditory selective attention. Cognition. **216**, 104849 (2021)

Humes, L.E., Kidd, G.R., Fogerty, D.: Exploring use of the coordinate response measure in a Multitalker babble paradigm. J. Speech Lang. Hearing Res. **60**(3), 741 (2017)

Jakien, K.M., Kampel, S.D., Stansell, M.M., Gallun, F.J.: Validating a rapid, automated test of spatial release from masking. Am. J. Audiol. **26**(4), 507 (2017)

Jonghwa, J.P., Seung-Cheol, B., Myung-Whan, S., Jongsuk, C., Sung, J.K., Yoonseob, L.: The effeect of topic familiarity and volatility of auditory scene on selective auditory attention. Hear. Res. **433**, 108770 (2023)

Mathew, T., John, J.G., Qian-Jie, F.: Interactions among talker sex, masker number, and masker intelligibility in speech-on-speech recognition. JASA Exp. Lett. **1**(1), 015203 (2021)

Mayer, A.M.: Research in acoustics. Loud.Edinb. Dubl. Phil. Mag. Ser. **5**, 259–288 (1984)

Miran, S., Akram, S., Sheikhattar, A., Simon, J.Z.: Babadi, B.: Real-time tracking of selective auditory attention from M/EEG: a bayesian filtering approach. Front. Neurosci. **12**, 262 (2018)

Rosen, S., Souza, P., Ekelund, C., Majeed, A.A.: Listening to speech in a background of other talkers: effects of talker number and noise vocoding. J. Acoust. Soc. Am. **133**(4), 2431–2443 (2013)

Song, J.H., Skoe, E., Banai, K., Kraus, N.: Perception of speech in noise: neural correlates. J. Cogn. Neurosci.Cogn. Neurosci. **23**(9), 2268–2279 (2011)

Su, E., Cai, S., Xie, L., Li, H., Schultz, T.S.: TAnet: a spatiotemporal attention network for decoding auditory spatial attention from EEG. IEEE Trans. Biomed. Eng. **69**(7), 2233–2242 (2022)

Subong, K., Caroline, E., Inyong, C.: Neurofeedback training of auditory selective attention enhances speech-in-noise perception. Front. Hum. Neurosci.Neurosci. **15**, 676992 (2021)

Vandecappelle, S., Deckers, L., Das, N., Ansari, A.H., Bertrand, A., Francart, T.: EEG-based detection of the locus of auditory attention with convolutional neural networks. Elife **10**, e56481 (2021)

Vermiglio, A.J., Herring, C.C., Heeke, P., Post. C.E., Fang, X.: Sentence recognition in steady-state speech-shaped noise versus four-talker babble. J. Am. Acad. Acad. Audiol. **30**(01), 054–065 (2019)

Wang, S., Ding, N., Lin, N., Zhang, J., Zong, C.: Language cognition and language computing - language understanding between humans and machines. Chin. Sci. Inf. Sci. **52**(10), 1748–1774 (2022)

Wegel, R.L., Lane, C.E.: The auditory masking of one sound by another and its probable relation to the dynamics of the inner ear. Phys. Rev. **23**, 266–285 (1924)

Wöstmann, M., Alavash, M., Obleser, J.: Alpha oscillations in the human brain implement distractor suppression independent of target selection. J. Neurosci. **39**(49), 9797–9805 (2019)

DSCF-DE: A Query-Based Object Detection Model via Dynamic Sampling and Cascade Fusion

Dengdi Sun[1,2], Wenhao Liu[1], and Zhuanlian Ding[3(✉)]

[1] School of Artificial Intelligence, Anhui University, Hefei 230601, China
[2] Jianghuai Advance Technology Center, Hefei 230031, China
[3] School of Internet, Anhui University, Hefei 230039, China
dingzhuanlian@163.com

Abstract. In recent years, query-based object detection methods have made significant progress in several fields, but still face the challenges of high computational complexity and slow convergence, especially when dealing with complex scenes and high resolution images. In addition, the layer-by-layer dependency structure of the model may lead to cascading errors, affecting stability and detection performance. To address these issues, we design a new detection model, Query-based Dynamic Sampling and Cascade Fusion Object Detector (DSCF-DE). The model is improved in two main aspects. First, DSCF-DE employs a dynamic sampling strategy that reduces redundant attention computations in the decoder, reduces the reliance on complex attention mechanisms, significantly reduces the computational burden, and enhances the query representation through spatial and channel blending. Secondly, DSCF-DE introduces a cascade fusion module, which utilizes the history query cache to complement and correct the current query, effectively mitigating the cascade error during training and enhancing the convergence speed and stability of the model. The experimental results verify the effectiveness of DSCF-DE.

Keywords: Object Detection · Dynamic Sampling · Cascading Error · Query Fusion

1 Introduction

Object detection is a key task in computer vision that aims to accurately locate and classify objects from images or videos. Traditional convolution-based object detection methods, such as Faster R-CNN and YOLO, use a one or two stage detection framework. These methods typically first generate candidate regions and then perform classification and bounding box regression on these regions. Despite significant progress in accuracy, these methods still rely on manually designed components.

In recent years, the proposal of the DETR [1] (Detection Transformer) model has marked an important change in the field of object detection. Using an end-to-end self-attention mechanism, DETR simplifies the detection framework, removes reliance on hand designed components of a classical convolutional detector, and drives consistency and efficiency in object detection. This approach not only improves the model's ability to adapt to complex scenarios but also provides new ideas for subsequent research. However, the DETR model also faces the problems of high computational complexity and long training time, especially when dealing with large scale datasets. This challenge stems mainly from the computational overhead of the global attention mechanism, which makes the efficiency of the model limited in practical applications.

Subsequent DETR variants (e.g., Deformable DETR, Efficient DETR, and SMCA DETR) attempt to reduce the computational overhead and improve the performance by tuning the global attention mechanism. For example, Deformable DETR introduces the concept of deformable convolution to enhance the ability to capture local features, while Efficient DETR reduces the computational burden through sparse attention mechanisms. However, these improvements are usually accompanied by additional complexity that further increases the difficulty of model training and inference, especially in application scenarios that require real time processing.

In addition, DETR and its variants may suffer from cascading errors during training. This phenomenon is manifested specifically in the fact that the prelude layer, although capable of generating correct output, may lead to the generation of erroneous results after the training of subsequent layers. The overdependence of the subsequent layers on the output of the prelude layer makes the error of the prelude layer gradually amplified in the transmission process, forming a negative feedback loop. This cascade effect not only reduces the convergence speed of the model, but also significantly affects the detection accuracy, limiting the effectiveness of the model in complex scenarios. Therefore, solving the problem of cascade error becomes an important challenge in improving the stability and performance of the model [4,15].

In order to solve the above problems, we propose a lightweight query-based object detection model named Dynamic Fusion of Sampling and Cascading Detector (DFCS-DE), which is mainly composed of two modules: dynamic sampling and cascade fusion. In the dynamic sampling module, an initial position is first generated through the interaction of the feature graph and the query, and then the training process is based on this initial position for dynamic sampling. This dynamic strategy significantly reduces the amount of computation while effectively retaining the focus on key features, thus improving computational efficiency and detection performance. By optimizing the selection and utilization of features, our model is able to significantly reduce the consumption of computational resources while maintaining high accuracy.

The cascade fusion module, on the other hand, utilizes cached historical queries to supplement and correct the current query during multiple rounds of training, effectively mitigating the impact of cascading errors. In this way, our model can gradually optimize the accuracy of the current query, thus reducing the negative impact of cascading effects. Experimental results show that our approach not only reduces the computation and training time but also significantly improves the performance and stability of the model. As shown in Fig. 1, in the COCO 2017 dataset, DFCS-DE achieves commendable performance in a shorter training cycle, and the training curve is smoother compared to other models, indicating improved stability [14]. These results show that DFCS-DE provides new ideas and directions for research in the field of object detection.

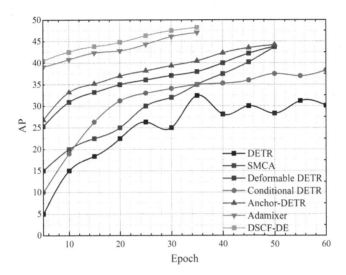

Fig. 1. DFCS-DE model training and performance comparison on the COCO 2017 dataset. DFCS-DE demonstrates stable performance improvement throughout training and consistently outperforms other detection models.

2 Related Works

In recent decades, single stage detectors [2], multistage detectors, and anchor-based detectors have adopted the dense assumption [5,15]. Such detectors apply dense a priori information, such as anchor points or anchor frames, to the feature map to fully detect or classify foreground objects [7]. These methods have been shown to be effective in object detection by refinement of each region of the detection process. The DETR model was subsequently proposed in 2020, and this method has been a great advancement in the field of object detection. Following the introduction of DETR, several variants [3,9] have emerged to further explore the application of Transformers in object detection:

- Deformable DETR [16]: introduces a deformable attention mechanism to enhance detection performance for small objects by dynamically selecting key positions.
- Condition DETR [10]: provides explicit conditional information for each query vector, accelerating convergence and improving accuracy.

- Anchor DETR [13]: combines traditional anchor methods with an improved object query process for a more accurate identification of small objects.

These variants collectively aim to improve detection performance and efficiency in various scenarios [2,4,6].

On this basis, we investigate and propose a novel query-based object detection model, DSCF-DE, which aims to address the challenges faced by the existing DETR family of models [11,16]. We optimize the classical architecture by no longer strongly relying on the attention mechanism to aggregate relevant features but adopting a dynamic sampling strategy to efficiently aggregate global features, thus simplifying the model structure and reducing the computational burden. A cascade fusion module is also introduced to alleviate the cascade error problem by fusing query information from different stages, thus improving the overall stability of the model. With these improvements, our model DSCF-DE becomes more efficient and reliable.

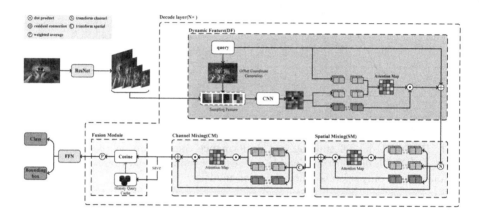

Fig. 2. Overview of our DSCF-DE for query-based object detector.

3 Methodology

In this paper, we discuss the dynamic sampling strategy and cascade fusion module in DSCF-DE. First, we review the DETR series of coding methods based on the attention mechanism [12] as shown in Fig. 2 and analyze their limitations in terms of feature extraction efficiency and computational resource consumption. Then, we elaborate the dynamic sampling strategy to illustrate how to efficiently sample relevant features in multiscale feature maps to avoid repeated computation in multiple decoding layers, thus achieving lightweight feature aggregation. To further enhance the feature representation capability, we introduce the spatial and channel mixing module to enhance the fusion of sample features. Finally, we introduce the cascade fusion module to emphasize its importance in multiple rounds of training, explaining how it enhances model stability by suppressing cascade error propagation, speeds up convergence, and ultimately improves detection performance.

3.1 Attentional Encoding Review

DETR utilizes a self-attention mechanism to capture global dependencies, but the computational overhead is huge when dealing with high resolution images and multiscale features. Although subsequent variants of DETR have improved by introducing other attentional mechanisms, they still do not fully address the problems of resource consumption and long training time. This motivates us to explore other feature aggregation strategies in depth, and we finally propose a dynamic sampling strategy, which effectively optimizes the computational efficiency and training time.

3.2 Dynamic Sampling

DSCF-DE uses resnet as the backbone and utilizes the C2-C5 stage features as the multiscale features. DSCF-DE introduces a dynamic sampling method that embeds location coding in the query. Each query generates a set of offset vectors that are combined with the location information of the query to determine the specific sampling location. Then, features are extracted from feature maps at different scales and projected to the same dimension to update the query. Next, the information carried by these sampled features is fused using a spatial and channel mixing module to enhance the image feature representation and passed to the decoder to perform the object detection task.

Offset Vector Generation. For each query, we generate a set of offset vectors associated with that query. These offset vectors depend on the content of the query and will vary across feature maps at different scales. The generating formula is as follows:

$$P_{initial} = Attention(q, f_{original}), \quad (1)$$

$$\Delta P = MLP(q \cdot f_{local}), \quad (2)$$

where $P_{initial}$ denotes initial position, $f_{original}$ is the initial feature map. ΔP denotes offset vector, q is the query, f_{local} is features around the initial location.

Sampling Coordinate Calculation. Each query vector consists of a position encoding that represents the initial position information on the feature map. The exact sampling coordinates for each query are computed by adding the position encoding to the corresponding offset vector, element by element. These coordinates indicate the exact location on the feature map where the feature is to be extracted. The calculation is as follows:

$$P_{sampled}^{s} = (P_{initial} + \Delta P)R^{s}, \quad (3)$$

where $P_{sampled}^{s}$ denotes the sampling position of scale s. R^s is the downsample ratio of scale s with respect to the original image.

Feature Sampling. We extract features from the feature maps of the corresponding scales based on the offset coordinates that have been computed. Since feature maps of different scales have different dimensions, we convert these sampled features to a uniform dimension by projection mapping. As follows:

$$Sample^{out} = linear(Sample_s^{input}), \qquad (4)$$

where *input* denotes the channel dimension on scale s, *out* is the transformed dimension.

Sampled Feature Mixing. We perform spatial and channel attention mixing of features from different scales. This enables a better combination of the unique information of each scale feature, which further enhances the semantic representation of the features.

Spatial Mixing and Channel Mixing. The query and sampled features are further enhanced with spatial and channel fusion operations for feature representation. Through this fusion, the model is able to capture richer contextual information and details, which significantly enhances the overall feature representation and ultimately improves the detection performance. The mixing is as follows:

$$SM(q) = Softmax(\frac{QK_s^T}{\sqrt{d}})V_s, \qquad (5)$$

$$CM(q) = Softmax(\frac{Q_{SM}K_c^T}{\sqrt{d}})V_c, \qquad (6)$$

where Q is the query, K_s and V_s is the sampled feature converted to the spatial dimension. K_c and V_c is the sampled feature converted to the channel dimension, d is the scaling factor.

3.3 Query Fusion Module

The sampled acquired features are passed to the query fusion module after spatial and channel fusion. At each stage, the features processed by the spatial and channel fusion modules (SM and CM) are saved into the memory cache for subsequent query fusion. Specifically, the current query qc is weighted and fused with the historical queries in the memory cache by performing cosine similarity computation. The weight of the current query qc is set to β, and the weight of the historical query sums to 1-β. In this way, the historical query is fused with the current query to update the query information. The fusion is as follows:

$$\alpha_c = \beta, \qquad (7)$$

$$\alpha_i = softmax(cosine(q_c, q_h^i)) \cdot (1 - \beta), \qquad (8)$$

$$q_{fused} = \alpha_c \cdot q_c + \sum_{i=1}^{n} \alpha_i \cdot q_h^i, \qquad (9)$$

where a_c is the weight of the current query q_c, q_c is the current query, q_h^i is the historical query, a_i denotes the weight of the fused query, q_{fused} denotes fused query.

4 Experiment

4.1 Implementation Details

Dataset. In order to evaluate the effectiveness of the DSCF-DE, we conduct extensive experiments on MS COCO 2017 dataset [8]. Following the common practice, we use trainval35k subset consisting up of 118K images to train our models and use minival subset of 5K images as the validation set. Average Precision (AP) is an important metric for evaluating the performance of a model. AP calculates the average recall precision. AP_{50} and AP_{75} denote the average precision when the intersection and concurrency ratio (IoU) thresholds are 0.5 and 0.75. AP_S, AP_M, and AP_L denote the average precision for small, medium, and large objects, respectively, and all metrics have IoU thresholds of 0.5.

Configurations. We use feature maps C2–C5 from the resnet. The number of decoder stages is set to 6 also following the common practice of query-based detectors. We use AdamW as our optimizer with weight decay 0.0001. The initial learning rate is 2.5×10^{-5}.

4.2 Comparison with Other Query-Based Detectors

To evaluate the detection performance of DSCF-DE, we conducted experiments on the COCO dataset using ResNet50 and ResNet101 as backbones. As shown in Table 1, our model outperforms others across key performance metrics, indicating a significant improvement in detection accuracy. Additionally, our model exhibits

Table 1. Performance comparison of different query-based object detection models.

Model	backbone	query	epochs	GFLOPs	COCO 2017					
					AP	AP_{50}	AP_{75}	AP_S	AP_M	AP_L
DETR	ResNet-50	100	500	187	43.3	63.1	45.9	22.5	47.3	61.1
SMCA	ResNet-50	300	50	152	43.7	63.6	47.2	24.2	47.0	60.4
Deformable DETR	ResNet-50	300	50	173	43.8	62.6	47.7	26.4	47.1	58.0
Conditional DETR	ResNet-50	300	108	195	45.1	65.4	48.5	25.3	49.0	62.2
Anchor-DETR	ResNet-50	300	50	151	44.2	64.7	47.5	24.7	48.2	60.6
Adamixer	ResNet-50	100	36	142	47.0	66.0	51.1	30.1	50.2	61.8
DSCF-DE (ours)	ResNet-50	100	36	**131**	**47.7**	**67.1**	**52.0**	**30.6**	**51.1**	**63.0**
DETR	ResNet-101	100	500	253	44.9	64.7	47.7	23.7	49.5	62.3
SMCA	ResNet-101	300	50	218	44.4	65.2	48.0	24.3	48.5	61.0
Deformable DETR	ResNet-101	300	50	244	42.2	63.1	44.7	21.5	45.7	60.3
Conditional DETR	ResNet-101	300	108	262	45.9	66.8	49.5	27.2	50.3	63.3
Anchor-DETR	ResNet-101	300	50	214	42.5	61.5	45.6	24.6	45.1	59.2
Adamixer	ResNet-101	100	36	208	48.0	67.0	52.4	30.0	51.2	63.7
DSCF-DE (ours)	ResNet-101	100	36	**201**	**48.9**	**68.0**	**53.6**	**30.8**	**52.5**	**64.8**

lower computational complexity, completing convergence in just 36 epochs compared to classical attention-based models. These results validate our model's effectiveness in optimizing computational efficiency and accelerating the training process.

4.3 Ablation Experiment

To validate the effectiveness of the dynamic sampling and cascade fusion modules in DSCF-DE, we conducted experiments by systematically removing or replacing these modules to quantify their contributions to model performance.

Table 2. The effect of sampling features at different scales of feature maps.

sampling	AP	AP_{50}	AP_{75}	AP_S	AP_M	AP_L
only C2 feature	22.6	33.4	23.9	15.7	24.1	36.8
only C3 feature	33.5	45.3	34.2	20.3	34.7	43.8
only C4 feature	42.2	59.1	47.6	25.1	45.8	58.1
only C5 feature	41.6	60.3	48.2	24.2	46.7	58.9
dynamic feature	**47.7**	**67.1**	**52.0**	**30.6**	**51.1**	**63.0**

Table 2 demonstrates the impact of dynamic sampling with different scale feature maps on model detection performance. We conducted experiments on backbone's C2-C5 layers and multi-scale dynamic sampling, and evaluated the contribution of different scale feature maps to the model training effect by comparing the performance on key performance indicators. The experimental results show that the dynamic sampling approach combined with multi-scale feature maps can significantly improve the detection accuracy of the model.

Table 3. Design in our spatial and channel mixing procedure.

SM	CM	AP	AP_{50}	AP_{75}	AP_S	AP_M	AP_L
		46.6	66.0	50.9	29.1	50.2	61.5
✓		47.1	66.7	51.5	29.4	50.6	62.6
	✓	47.0	66.5	51.3	29.6	50.5	62.4
✓	✓	**47.7**	**67.1**	**52.0**	**30.6**	**51.1**	**63.0**

Table 3 shows the impact of the spatial mixing (SM) and channel mixing (CM) modules on AP metrics, using ResNet50. The results indicate that while

SM and CM each improve performance individually, the best results are achieved when both are applied together, confirming their synergistic effect in enhancing feature representation and boosting detection performance.

Table 4. Effects of using the fusion module at different stages.

Dataset	stage	AP	AP_{50}	AP_{75}	AP_S	AP_M	AP_L
COCO	2	47.4	66.2	52.0	29.0	51.1	63.0
	2,3	48.0	66.9	52.4	29.4	51.6	63.1
	2,3,4	48.3	67.2	53.0	30.2	52.0	63.7
	4,5,6	48.5	67.4	53.2	30.3	52.1	64.1
	2,3,4,5	48.6	67.5	53.2	30.4	52.2	64.5
	3,4,5,6	**48.9**	**68.0**	**53.6**	**30.8**	**52.5**	**64.8**
	2,3,4,5,6	48.7	67.8	53.3	30.5	52.3	64.5

Table 4 presents the impact of applying the fusion module at different training stages on DSCF-DE's performance, using ResNet101. The results show that early stage fusion yields limited improvements, while medium and late stage fusion significantly enhances performance. Notably, applying the fusion module only in the late stages is less effective than its gradual introduction from the middle stages onward, indicating that phased fusion mitigates cascade errors more effectively and improves training stability.

4.4 Visualization

As shown in Fig. 3, we demonstrate the detection results of the DFCS-DE model in three groups: a, b, and c, for small, medium, and large objects, respectively. These results visualize the model's detection capabilities at different object scales. The small object detection results demonstrate the model's robustness in complex backgrounds, while the medium and large object detection results highlight DFCS-DE's adaptability and accuracy for objects of varying sizes.

(a) small objects   (b) medium objects

(c) large objects

Fig. 3. Detection results of the DFCS-DE model for different object sizes. (a) Small objects (b) Medium objects (c) Large objects.

5 Conclusion

In this paper, we make two main contributions by innovating based on the shortcomings of existing models. We design the DSCF-DE architecture to effectively reduce the computational effort through the dynamic sampling module and further enhance the semantic information through spatial and channel blending. In addition, the cascade fusion module introduced by DSCF-DE effectively mitigates the cascade error in multiple rounds of training and improves the stability and efficiency of the model. The superiority of our model is verified through experiments.

Acknowledgements. This work was supported by NSFC (62076005, 61906002), the Dreams Foundation of Jianghuai Advance Technology Center (NO. 2023-ZM01Z015), NSF of Anhui Province (2008085MF191, 2008085QF306), and the University Synergy Innovation Program of Anhui Province (GXXT-2021-002).

References

1. Carion, N., Massa, F., Synnaeve, G., et al.: End-to-end object detection with transformers. In: ECCV, pp. 213–229. Springer (2020)

2. Chen, H., He, T., Tian, Z., et al.: Fcos: fully convolutional one-stage object detection. In: ICCV (2020)
3. Hou, X., Liu, M., Zhang, S., et al.: Salience detr: enhancing detection transformer with hierarchical salience filtering refinement. In: CVPR, pp. 17574–17583 (2024)
4. Jia, D., Yuan, Y., He, H., et al.: Detrs with hybrid matching. In: CVPR, pp. 19702–19712 (2023)
5. Kong, T., Sun, F., Liu, H., et al.: Foveabox: beyond anchor-based object detection. TIP **29**, 7389–7398 (2020)
6. Li, F., Zhang, H., Liu, S., et al.: Dn-detr: accelerate detr training by introducing query denoising. In: CVPR, pp. 13619–13627 (2022)
7. Lin, T.: Focal loss for dense object detection. arXiv preprint arXiv:1708.02002 (2017)
8. Lin, T.Y., Maire, M., Belongie, S., et al.: Microsoft coco: common objects in context. In: ECCV, pp. 740–755. Springer (2014)
9. Liu, S., Li, F., Zhang, H., et al.: Dab-detr: dynamic anchor boxes are better queries for detr. arXiv preprint arXiv:2201.12329 (2022)
10. Meng, D., Chen, X., Fan, Z., et al.: Conditional detr for fast training convergence. In: ICCV, pp. 3651–3660 (2021)
11. Sun, P., Zhang, R., Jiang, Y., et al.: Sparse r-cnn: end-to-end object detection with learnable proposals. In: CVPR, pp. 14454–14463 (2021)
12. Vaswani, A.: Attention is all you need. In: NIPS (2017)
13. Wang, Y., Zhang, X., Yang, T., et al.: Anchor detr: query design for transformer-based detector. In: AAAI, vol. 36, pp. 2567–2575 (2022)
14. Ye, M., Ke, L., Li, S., et al.: Cascade-detr: delving into high-quality universal object detection. In: ICCV, pp. 6704–6714 (2023)
15. Zhu, C., Chen, F., Shen, Z., et al.: Soft anchor-point object detection. In: ECCV, pp. 91–107. Springer (2020)
16. Zhu, X., Su, W., Lu, L., et al.: Deformable detr: deformable transformers for end-to-end object detection. arXiv preprint arXiv:2010.04159 (2020)

MDFNet: Multi-dimensional Fusion Attention for Enhanced Image Captioning

Dengdi Sun[1,2], Xuetao Li[1(✉)], and Chaofan Mu[1]

[1] School of Artificial Intelligence, Anhui University, Hefei 230601, China
lixuetao4549@163.com
[2] Institute of Artificial Intelligence, Hefei Comprehensive National Science Center, Hefei 230026, China

Abstract. Attention mechanisms have become critical elements in deep learning models, especially for tasks such as Image Captioning (IC) and Visual Question Answering (VQA). However, current attention modules predominantly focus on spatial dependencies, often overlooking the critical aspect of multi-dimensional perception fusion, which is vital for achieving a deep and comprehensive semantic understanding. To address this limitation, we propose a novel design, namely Multi-Dimensional Fusion Transformer Network (MDFNet). MDFNet integrates Swin Transformers with shifted window partitioning for feature map division and spatial/channel-wise attention, effectively aggregating global and local features for enhanced multi-modal understanding and accurate scene interpretation. This method enables the model to more effectively grasp complex relationships within the data, leading to enhanced multi-modal reasoning and a more accurate interpretation of complex scenes. Extensive experiments demonstrate the effectiveness of MDFNet, achieving a new state-of-the-art performance with a CIDEr score of 135.2 on the COCO Karpathy test split. In addition to its superior performance in image captioning, MDFNet also enhances the model's overall robustness, representing a significant advancement in the evolution of attention mechanisms for multimedia tasks.

Keywords: Image Caption · Multi-dimension · MDFNet

1 Introduction

Image Captioning (IC) is a vital and complex task in the field of computer vision, focused on generating accurate, coherent, and contextually appropriate sentences that describe a given image. The task goes beyond simply identifying objects, aiming to capture relationships between various elements in the scene and convey them through natural language. IC leverages the capabilities of both computer vision and natural language processing, which is crucial for applications like aiding visually impaired individuals, optimizing image search engines, and enriching human-computer interaction. Despite recent advancements, encompassing deep learning methods such as convolutional neural networks (CNNs) and various attention mechanisms, IC remains a complex problem. Challenges persist in describing novel objects and understanding intricate relationships within diverse and unfamiliar visual contexts.

In recent years, attention mechanisms have played a central role in multimodal semantic reasoning, playing a critical role in capturing correlations between different modalities. For instance, visual attention and self-attention have become essential components of image captioning (IC) models. These modules not only aid in extracting visual information relevant to language but also enhance cross-modal interactions. However, most of the existing attention mechanisms in IC predominantly focus on spatial dependencies, which limits their ability to effectively learn new object categories and attributes [13]. This limitation is significant, as the ability to comprehend new objects and their attributes is essential for comprehensive image understanding and cross-modal reasoning.

To address these limitations, we propose a novel network architecture named Multi-Dimensional Fusion Transformer Network (MDFNet), which integrates the hierarchical processing capabilities of the Swin Transformer [11] with the detailed focus provided by Channel-wise Attention. The Swin Transformer is well-regarded for its ability to manage complex spatial hierarchies in images, while Channel-wise Attention enhances the model's sensitivity to specific feature channels, improving its ability to discern and utilize fine-grained details within the image.

The MDFNet is further refined by incorporating the Swin-Channel Attention Fusion(SCAF) module, which serves as a fusion mechanism to combine the outputs of Spatial attention and Channel-wise attention. The SCAF module ensures that the model can effectively integrate the spatial and channel-specific information, resulting in a more resilient and precise caption generation process. In summary, our contributions are threefold:

- We identify and address the limitations of conventional attention mechanisms in capturing fine-grained dependencies, and propose MDFNet as a more effective approach for enhancing semantic modeling in image captioning.
- We propose the Channel-wise Attention Block (CAB) to capture channel dependencies, incorporating two new designs: a reduction-reconstruction structure and gating-based attention, aimed at reducing training overhead. Additionally, we investigate a joint approach to model both spatial self-attention and channel attention mechanisms.
- Our provided method is validated through comprehensive experiments on the MS COCO dataset, showcasing substantial advancements compared to state-of-the-art models in image captioning and various multimodal tasks.

2 Related Work

Recent progress in Image Captioning has been primarily fueled by the development of sophisticated attention mechanisms and the introduction of various deep learning architectures [12,21]. Conventional IC methods generally adopt an encoder-decoder framework, using Convolutional Neural Networks (CNNs) to convert images into feature vectors, which are then decoded by Recurrent Neural Networks (RNNs) to generate descriptive sentences. However, this approach often struggles to capture the intricate dependencies present in intricate visual environments, especially when it comes to handling subtle details.

Attention mechanisms have been incorporated to overcome this constraint. by aenabling models to concentrate on particular areas of an image during the caption generation process. Early approaches, such as spatial attention, assign different weights to different parts of the image, helping the model to concentrate on the most relevant regions. For instance, works like [10] utilize visual attention to enhance the performance of IC models by directing focus towards the most salient parts of an image. However, these models often fall short in capturing channel-wise dependencies, which are crucial for distinguishing subtle visual features.

In recent years, transformer-based models have emerged as a strong alternative to RNNs in sequence tasks, with self-attention excelling at capturing long-range dependencies. Models like [2,17] have improved image captioning (IC) by dynamically focusing on different image regions, though they often neglect local features, affecting fine-grained tasks. Hybrid models, such as the dual-level attention in [4] and the grid-region feature integration in [5], address some of these limitations by enhancing diverse visual information processing. However, a gap remains in modeling both spatial and channel-wise dependencies, as many methods overlook crucial local and channel-specific details essential for high-quality IC.

In response to these limitations, we introduce an innovative network architecture that integrates the hierarchical processing capabilities of the Swin Transformer with the channel-wise attention mechanism. Our approach, MDFNet is designed to utilize the advantages of both spatial and channel-wise attention, ensuring that the model can effectively capture and integrate visual features.

3 Method

In this part, we introduce our proposed MDFNet, designed to comprehensively capture dependencies both spatially and across channels. As illustrated in Fig. 1, the MDFNet consists of three fundamental components: the Shifted Window Partition(SWP), the Spatial Attention Block (SAB), the Channel-wise Attention Block (CAB), and the Swin-Channel Attention Fusion (SCAF). By incorporating SCAM into the standard Transformer architecture, our goal is to enrich the exploration of visual information and boost the effectiveness of multimodal reasoning in image captioning tasks.

3.1 Shifted Window Partition (SWP)

As shown in Fig. 1, we adopt the window partitioning strategy inspired by the Swin Transformer to segment the overall feature map and extract channel and spatial features in different modules. Specifically, we implement two distinct partitioning methods across various encoders: non-overlapped window partitioning and shifted window partitioning.

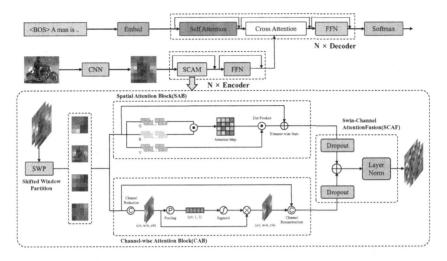

Fig. 1. Overview of our SCAM for image captioning

In the non-overlapped window partitioning method, the feature map is divided into non-overlapping windows, where each window processes local features independently. This approach significantly reduces computational complexity by limiting interactions within small, fixed regions. On the other hand, the shifted window partitioning method introduces a shift between consecutive layers, allowing windows to overlap partially across the boundaries. This overlapping facilitates information exchange between adjacent windows, thereby enhancing the model's ability to understand global context and long-range dependencies. By employing these two partitioning methods in combination across different encoder layers, we ensure that the model can efficiently capture features at both local and global levels across multiple scales.

3.2 Spatial Attention Block(SAB)

To capture spatial dependencies in images and sentences, we employ Self-Attention as shown in Fig. 1. The decision to use Self-Attention is motivated by the relatively small size of the spatial domain compared to the channel domain(e.g., grids of 8×8 for images and approximately 15 words for sentences). Therefore, the computation with quadratic complexity involved in Self-Attention is computationally feasible and offers robust modeling capabilities to accurately capture spatial dependencies.

Consider the features of the input image or text as $S \in \mathbb{R}^{N \times C}$, where N indicates the spatial domain length, and C refers to the channel domain length.

The inputted feature is first projected into three representations: Q, K, and V. Each head of Q, K, and V can be computed as follows:

$$Q_i = SW_i^Q + b^Q, \quad K_i = SW_i^K + b^K, \quad V_i = SW_i^V + b^V, \tag{1}$$

where $W_i^Q \in \mathbb{R}^{C \times d_k}$, $W_i^K \in \mathbb{R}^{C \times d_k}$, and $W_i^V \in \mathbb{R}^{C \times d_v}$ represent the parameter metrices. Additionally, $b^Q \in \mathbb{R}^{d_k}$, $b^K \in \mathbb{R}^{d_k}$, and $b^V \in \mathbb{R}^{d_v}$ denote the bias parameters. Here, d_k and d_v correspond to the dimensions of Q, K, and V for each attention head. The computation of the map of spatial attention for each head is as follows:

$$\text{head}_i = \text{Att}(Q_i, K_i, V_i) = \text{Softmax}\left(\frac{Q_i K_i^\top}{\sqrt{d_k}}\right) V_i, \tag{2}$$

where $\text{head}_i \in \mathbb{R}^{N \times d_v}$. The resulting feature of spatial attention can be represented as follows:

$$\tilde{S} = \text{Concat}(h_1, h_2, \ldots, h_h) W^O + b^O, \tag{3}$$

where h refers to the count of heads, $W^O \in \mathbb{R}^{hd_k \times C}$ is the mapping matrix, $b^O \in \mathbb{R}^C$ is the corresponding bias parameter, and $\tilde{S} \in \mathbb{R}^{N \times C}$ represents the resulting map of spatial attention features.

3.3 Channel-wise Attention Block (CAB)

To apply CAB, the one-dimensional visual feature $V \in \mathbb{R}^{N \times C}$ is first reshaped into a two-dimensional feature $P \in \mathbb{R}^{H \times W \times C}$, where C denotes the count of channels in the representation. In this context, N (with $N = H \times W$) indicates the size of the spatial feature, and H and W correspond to the height and width dimensions, respectively. To minimize the number of parameters, we apply a channel reduction step prior to modeling the channel-wise dependencies. We apply a 1×1 convolutional kernel to reduce the count of channels to $\frac{1}{r}$ (with $r = 16$ being a common configuration in experiment):

$$U = F_{\text{red}}(P) = \delta(f^{1 \times 1}_{C \to C/r}(P)), \tag{4}$$

where $\delta(\cdot)$ signifies the ReLU activation function, and $f^{1 \times 1}_{C \to C/r}(\cdot)$ indicates a convolution operation with a kernel size of 1×1, where the input has a dimension of C and the output has a dimension of C/r.

In CAB, the objective is to assign attention weights to each channel individually. To prevent introducing redundant parameters and computations in later stages, we first condense information from the spatial domain. While more sophisticated approaches such as [6,20] utilize advanced methods to aggregate spatial information, we opt for a simpler solution by applying global average pooling (GAP) to integrate spatial information.

$$Y_c = F_{\text{gap}}(U_c) = \frac{1}{H \times W} \sum_{i=1}^{H} \sum_{j=1}^{W} U_c(i, j), \tag{5}$$

where $U_c \in \mathbb{R}^{H \times W}$ and $Y_c \in \mathbb{R}^1$ correspond to the information of the c-th channel from the feature $U \in \mathbb{R}^{H \times W \times C}$ and $Y \in \mathbb{R}^{1 \times C/r}$, respectively.

In order to lessen the substantial computational burden of the scaled dot-product, we utilize a basic gating mechanism to determine the attention weights for each channel, as described below:

$$Y' = F_{\text{att}}(Y, W_1, W_2) = \sigma(\delta(YW_1)W_2), \tag{6}$$

where $W_1 \in \mathbb{R}^{C/r \times C/r}$ and $W_2 \in \mathbb{R}^{C/r \times C/r}$ represent the projection matrices. The function $\delta(\cdot)$ denotes the ReLU function, and $\sigma(\cdot)$ stands for the sigmoid function. The channel feature is generated by applying the attention across channels using a map $Y' \in \mathbb{R}^{1 \times C/r}$ to the reduced visual feature $U \in \mathbb{R}^{H \times W \times C/r}$:

$$\hat{U} = F_{\text{com}}(U, Y') = U * Y', \tag{7}$$

where $*$ denotes channel-wise multiplication, and $\hat{U} \in \mathbb{R}^{H \times W \times C/r}$ refers to the channel feature. In Equation 4, we reduce the channel domain's dimensionality. The output of CAB in the encoder is obtained by restoring the channel domain through a similar process, as described below:

$$\hat{T} = F_{\text{rec}}(\hat{U}) = f^{1 \times 1}_{C/r \to C}(\hat{U}), \tag{8}$$

where $\hat{T} \in \mathbb{R}^{H \times W \times C}$ denotes the involved feature, and $f^{1 \times 1}_{C/r \to C}(\cdot)$ refers to a convolution operation with a 1×1 kernel.

3.4 Swin-Channel Attention Fusion (SCAF)

To maintain simplicity in the model, we combine the two attention-based feature maps using an element-wise summation:

$$S_{\text{dual}} = LN\left(\rho(\hat{T}) + \rho(\hat{P})\right), \tag{9}$$

where $\hat{T}$ and $\hat{P}$ denote the feature maps obtained from spatial and channel-wise attention, correspondingly. Here, $\rho(\cdot)$ denotes the dropout function, applied with a probability of 0.1 to zero out elements, while $LN(\cdot)$ stands for Layer Normalization.

4 Experiments

In this section, we carried out numerous experiments on the popular MS COCO dataset to assess the effectiveness of our proposed algorithm. In Sect. 3.1, we present the relevant details of the MS COCO dataset. In Sect. 3.2, we provide the specific implementation aspects of the experiment. In Sect. 3.3, we present and analyze the results of our experiments. and In Sect. 3.4 we conducted an analysis of the effectiveness of each module.

4.1 Dataset

The primary experiments are carried out on the extensively employed MS COCO dataset, which consists of 123,287 images, each paired with five captions. This dataset is divided into 82,783 images for training and 40,504 images for validation. We use the standard Karpathy split for evaluation, allocating 5,000 images each for validation and testing, with the rest used for training.

4.2 Implementation Details

In our study, the model is configured with a dimension d_{model} of 512, utilizing 8 attention heads and consisting of 3 layers for the transfomer modules. To accurately gain the spatial relationships within the grid features, we incorporate relative positional encodings into the self-attention mechanism (SA), with the distances between grids being measured via the Chebyshev Distance. The training process is run for a maximum of 60 epochs, with the initial 25 epochs to pre-train using cross-entropy and the remaining 35 epochs devoted to self-critical training. During the initial phase, a gradual learning rate warm-up is employed over the first 4 epochs, progressively increasing the learning rate to 1×10^{-4}, followed by adjustments to 2×10^{-5} and 4×10^{-6} for the subsequent stages. For feature map segmentation, a non-overlapping partition strategy is applied in the first and third encoders. In contrast, the second encoder utilizes a shifted window partitioning method to better capture global features. We use a pre-trained Faster R-CNN model for visual features extracting. In accordance with standard evaluation criteria, five metrics-BLEU-N [16], METEOR [3], ROUGE-L [9], CIDEr [19] and SPICE [1]-are simultaneously utilized to evaluate our model's performance.

Table 1. Comparison with State-of-the-Art on the Karpathy Test Split in a Single-Model Setting. All values are reported as percentage (%).

Model	BLEU-1	BLEU-2	METEOR	ROUGE-L	CIDEr	SPICE
Transformer	80.07	38.6	29.1	58.5	130.1	22.7
AOANet	80.2	38.9	29.2	58.8	129.8	22.4
M^2Transformer	80.8	39.1	29.2	58.6	131.2	22.6
XTransformer	80.9	39.7	29.5	59.1	132.8	23.4
LSTNet	82.2	40.6	29.6	59.6	132.0	23.5
RSTNet	81.1	39.3	29.4	58.8	133.3	23.3
SDATR	81.3	39.7	29.5	59.1	134.5	23.4
DSNT	81.5	39.7	29.5	59.2	134.8	23.6
Ours	**81.8**	**39.8**	**29.74**	**59.24**	**135.2**	**23.8**

4.3 Performance Comparison

In this section, we assess our SCAM model using the MS COCO dataset for evaluation and compare its performance against state-of-the-art models in offline testing. The models included are: Transformer [18], AOA [8], M2Transformer [5], XTransformer [15], LSTNet [14], RSTNet [22], SDATR [13], and DSNT [7]. Table 1 shows a comprehensive performance comparison of these models and ours on the COCO Karpathy test split. For a fair evaluation, all results are reported for single models without using ensemble methods.

GT: A black Honda motorcycle parked in front of a garage.

Trans: A motorcycle is parked there.

Ours: A black motorcycle parked on the grass.

Fig. 2. Examples of image captioning results.

The results demonstrate that our proposed model outperforms the other SOTA models across most metrics. Specifically, our approach achieves a CIDEr score of 135.2%, surpassing the best-performing model, DSNT, by 0.4%. Additionally, our model achieves BLEU-1 and BLEU-2 scores of 81.8% and 39.8%, respectively, which signifies a notable improvement in linguistic precision and consistency. Moreover, it leads in METEOR and ROUGE-L metrics, reflecting its effectiveness in capturing sentence-level semantics and structure. Our model also achieves the highest SPICE score, highlighting its strength in semantic content matching.In summary, our findings indicate that our model outperforms current leading approaches.

Table 2. Ablation study on the use of channel and split in model.

Combination	BLEU-1	BLEU-2	METEOR	ROUGE-L	CIDEr	SPICE
Trans + Channel	79.6	37.4	28.3	58.1	133.5	22.4
Trans + Split	80.4	38.6	29.2	58.5	132.1	22.7
Ours	**81.8**	**39.8**	**29.74**	**59.24**	**135.2**	**23.8**

4.4 Ablation Study

Tables 2 and 3 present the detailed data on module combinations and module ablation, respectively. Table 2 presents the experimental results of different combinations of the Transformer. We can observe that the split method outperforms

Table 3. Various combinations of split, spatial and channel-wise attention. 'r' refers to channel scaling parameter. '&' refers to exploiting both attentions in parallel.

Combination	BLEU-1	BLEU-2	METEOR	ROUGE-L	CIDEr	SPICE
Channel(r = 2)	79.6	37.4	28.3	58.1	131.2	22.4
Channel(r = 4)	79.6	37.6	28.3	58.2	131.3	22.4
Channel(r = 8)	79.5	37.7	28.5	58.3	132.4	22.5
Channel(r = 16)	80.2	38.2	28.7	58.5	133.5	22.9
Channel(r = 32)	80.0	38.1	28.5	58.3	132.1	22.6
Split(Non-ovlp)	80.4	38.6	29.2	58.5	131.5	22.7
Split(Ovlp)	80.3	38.3	29.1	58.3	131.8	22.4
Split(Ours)	80.8	39.2	29.3	58.6	132.1	22.8

the channel-wise attention in terms of performance. The ablation study in Table 3 highlights the impact of various design choices on model performance, specifically focusing on the channel scaling parameter r and the split partitioning strategy. The channel scaling parameter r influences the model's capacity and computational cost. The study shows that performance improves with increasing r up to 16, where the model achieves optimal results across most metrics. However, further increasing r to 32 offers no significant gains, indicating that larger values may introduce unnecessary complexity without corresponding performance benefits.

In 'S (Non-ovlp),' all three encoders use non-overlapping window partitions, ensuring clear segmentation. 'S (Ovlp)' applies overlapping partitions in all encoders, enhancing feature continuity. 'S (Ours)' combines both: non-overlapping partitions in the first and third encoders, and overlapping in the second, balancing local feature clarity and global continuity for better performance.

5 Conclusion

In this study, we introduce the Multi-Dimensional Fusion Transformer Network (MDFNet) for image captioning, which utilizes window partitioning on feature maps. Comprehensive experiments carried out on the MS COCO dataset highlight the efficiency and advantages of the proposed SCAM. Significantly, our method sets a new benchmark for performance on the MS COCO dataset.

Acknowledgment. This work was supported by NSFC (62076005, 61906002, U20A20398), NSF of Anhui Province (2008085MF191, 2008085QF306), and the University Synergy Innovation Program of Anhui Province (GXXT-2021-002).

References

1. Anderson, P., Fernando, B., Johnson, M., etc.: Spice: semantic propositional image caption evaluation. In: ECCV, pp. 382–398. Springer (2016)
2. Anderson, P., He, X., Buehler, C., et al.: Bottom-up and top-down attention for image captioning and visual question answering (2017)
3. Banerjee, S., Lavie, A.: Meteor: an automatic metric for mt evaluation with improved correlation with human judgments. In: Proceedings of the acl Workshop on Intrinsic and Extrinsic Evaluation Measures for Machine Translation and/or Summarization, pp. 65–72 (2005)
4. Cornia, M., Baraldi, L., Cucchiara, R.: Show, control and tell: a framework for generating controllable and grounded captions. In: IEEE/CVF (2020)
5. Cornia, M., Stefanini, M., Baraldi, L., et al.: Meshed-memory transformer for image captioning. In: Proceedings of the IEEE/CVF Conference on Computer Vision and Pattern Recognition, pp. 10578–10587 (2020)
6. He, K., Zhang, X., Ren, S., et al.: Spatial pyramid pooling in deep convolutional networks for visual recognition. IEEE Trans. **37**(9), 1904–1916 (2015)
7. Hu, J., Yang, Y., An, Y., et al.: Dual-spatial normalized transformer for image captioning. Eng. Appl. Artif. Intell. **123**, 106384 (2023)
8. Huang, L., Wang, W., Chen, J., et al.: Attention on attention for image captioning. In: IEEE/CVF, pp. 4634–4643 (2019)
9. Lin, C.Y.: Rouge: a package for automatic evaluation of summaries. In: Text Summarization Branches Out, pp. 74–81 (2004)
10. Liu, W., Chen, S., Guo, L., et al.: Cptr: full transformer network for image captioning (2021)
11. Liu, Z., Lin, Y., Cao, Y., et al.: Swin transformer: hierarchical vision transformer using shifted windows (2021)
12. Lu, J., Xiong, C., Parikh, D., et al.: Knowing when to look: adaptive attention via a visual sentinel for image captioning. In: IEEE (2017)
13. Ma, Y., Ji, J., Sun, X., et al.: Knowing what it is: semantic-enhanced dual attention transformer. IEEE Trans. Multimedia **25**, 3723–3736 (2022)
14. Ma, Y., Ji, J., Sun, X., et al.: Towards local visual modeling for image captioning. Pattern Recogn. **138**, 109420 (2023)
15. Pan, Y., Yao, T., Li, Y., et al.: X-linear attention networks for image captioning. In: IEEE (2020)
16. Papineni, K., Roukos, S., Ward, T., et al.: Bleu: a method for automatic evaluation of machine translation. In: Proceedings of the 40th annual meeting of the Association, pp. 311–318 (2002)
17. Rennie, S.J., Marcheret, E., Mroueh, Y., et al.: Self-critical sequence training for image captioning. In: IEEE (2016)
18. Vaswani, A., Shazeer, N., Parmar, N., et al.: Attention is all you need. arXiv (2017)
19. Vedantam, R., Lawrence Zitnick, C., Parikh, D.: Cider: consensus-based image description evaluation. In: CVPR, pp. 4566–4575 (2015)
20. Woo, S., Park, J., Lee, J.Y., et al.: Cbam: convolutional block attention module. In: ECCV, pp. 3–19 (2018)
21. Xu, K., Ba, J., Kiros, R., et al.: Show, attend and tell: neural image caption generation with visual attention. Comput. Sci. 2048–2057 (2015)
22. Zhang, X., Sun, X., Luo, Y., et al.: Rstnet: captioning with adaptive attention on visual and non-visual words. In: IEEE/CVF, pp. 15465–15474 (2021)

Dynamic Points Location of Professional Model Pose Based on Improved Network Stacking Model

Kaizhan Mai, Dazhi Li, Yuefang Gao, Pingping Mi, and Li Hao[✉]

South China Agricultural University, Guangzhou 501642, China
{gaoyuefang,mipingping}@scau.edu.cn, haoliscau@163.com

Abstract. In the fields of advertising and fashion, the pose of a professional model plays a pivotal role in highlighting the distinct features and aesthetics of apparel. Models rely on observing images, videos, or classroom teaching to train their body poses. This process often faces challenges in grasping the nuances of movements and lacking digital and standardized feedback, thus hindering the development of correct pose habits. Additionally, the clothing, hairstyles, styling, and actions exhibited by models are diverse and complex, making it difficult for existing human pose estimation methods to accurately capture and locate the display posture of models. Addressing these limitations, this paper constructs a dataset of 12,284 images, encompassing professional models in various poses, and defines 20 key points across 11 critical pose areas. Based on this, an improved stacked network model is proposed to accurately locate keypoints of models. The model consists of a backbone network composed of multiple stacked V modules and integrates attention mechanisms and depth-adaptive intermediate supervision to enhance the model's expressive capability, training speed, and localization precision. Comprehensive experiments conducted on the constructed model pose dataset demonstrate the superiority of the proposed algorithm, and further ablation study analysis verifies the effectiveness of each module within the proposed method.

Keywords: Model pose estimation · Stacked network · Attention mechanism

1 Introduction

In the realm of advertising and fashion performance, professional models use exquisite body poses to display the characteristics and aesthetics of clothing. Body poses can enhance the visual appeal of clothing and assist audiences in better understanding the style and designing concept of the apparel [1]. With the rapid advancement of information technology, digital technology offers new perspectives and methods that make model training more scientific. However, existing methods like motion capture can record body movements and poses with high precision [2, 3], yet their widespread application is limited by high costs, complex technical demands, and specific environmental constraints.

With the development of deep learning, automatic human pose estimation has achieved significant progress, leading to the proposal of various deep pose estimation models, such as DeepPose [4], Stacked Hourglass Network [5], HRNet [6], SCAI [7], and so on. These models extract information layer by layer through multi-layer neural networks and learn the feature representation of human poses from a large volume of annotated images or video data. To further enhance the accuracy and robustness of human pose estimation, some studies [8, 9] have attempted to introduce techniques like feature fusion and attention mechanisms to bolster the model's ability to learn critical features. Although these methods have achieved precise skeletal point location information, they largely rely on specific datasets for training. For model practitioners, the pose learning process is often non-standardized, and the need for varied model performance poses is highly complex. In practice, models typically rely on observing and imitating images and videos of professional models, which lack a systematic feedback mechanism, making it difficult to form accurate and consistent pose habits. Moreover, the diversity of clothing, hairstyles, styling, and actions involved in model displays makes it challenging for existing human pose estimation methods to accurately capture these complex poses. Furthermore, due to insufficient training data, deep blurriness, and occlusion issues, considerable challenges exist concerning accuracy and error control when handling models' complex and variable poses.

To address these challenges, this paper explores the digital expression methods of poses in model images, aiming to quantify different pose strengths. To this end, a professional model pose dataset encompassing 12,284 images has been constructed, comprehensively illustrating a variety of model poses, including standing, sitting, supporting, and leaning. Based on this dataset, an improved stacked network model is proposed to precisely capture key points of model poses. Specifically, the backbone network of this model is composed of a series of stacked V modules and incorporates attention mechanisms and depth-adaptive intermediate supervision strategies to further enhance the model's performance.

The main contributions of this paper are as follows: (1) A novel improved stacked network model is proposed, combining attention mechanism and adaptive intermediate supervision to improve the localization precision of model pose key points; (2) A professional model pose dataset with 12,284 images is constructed, defining 20 key points, supporting various pose analyses, and providing a foundation for the digital expression of model poses.

2 Related Work

Human pose estimation aims to detect and locate the pose and parts of human body from images [10–12]. It can be broadly classified into regression-based [4, 13] and detection-based methods [14–16] depending on the CNN output representation. Regression-based methods predict keypoint coordinates directly from input images through end-to-end training. For example, Sun et al. [17] proposed a structure-aware regression method, which improved the ResNet50 architecture by using a skeletal representation. The approach incorporates body structure information for more stable joint position determination. Some works [18, 19] have attempted to merge heatmap and regression methods to enhance keypoint localization accuracy. Recent studies [20, 21]utilized Transformer for pose estimation, employing self-attention mechanism to efficiently capture keypoint information. While regression approaches are straightforward, they are sensitive to noise and occlusion, making it challenging for models to effectively utilize spatial relationships between keypoints.

Detection-based pose estimation methods identify and locate human keypoints from images or videos, thereby inferring the overall body pose. These methods effectively utilize heatmaps to represent the ground truth positions of joints. By leveraging intermediate supervision, pose estimation models can mitigate the issue of diminishing input information. The Stacked Hourglass Network has demonstrated considerable success in this regard. Recently, High-Resolution Network (HRNet) and its variants [1, 22] have achieved accuracy improvements through multi-scale feature fusion. To enhance heatmap prediction accuracy, Qu et al.[23] minimized the distance between characteristic functions derived from predicted and ground truth heatmaps instead of relying on the overall L2 loss. SSPCM [24] addressed the issues of pseudo-label noise and complexity by incorporating a position inconsistency correction module and engaging in interactive training with two teacher models.

In contrast to the aforementioned methods, our proposed approach focuses on enhancing the model's expressive capability by stacking network models and employing a deep adaptive intermediate supervision strategy. Furthermore, we integrate both channel and spatial self-attention mechanisms to effectively capture the spatial and appearance features of keypoints.

3 Model Pose Dataset

3.1 Model Pose Dataset Construction

Fig. 1. Single-model Static Planar Pose Images and Dynamic Point Position Setting. (A) Model Key Points; (B) Common Human Key Points

To ensure the professionalism and accuracy of the dataset, we collaborate with the College of Arts at South China Agricultural University. We collected over 30,000 single-model static planar pose images from fashion magazines and other professional model domains. These images underwent a meticulous process of classification, selection, and filtering processes, resulting in a highly valuable dataset of 12,284 images. As shown in Fig. 1, the dataset encompasses various planar poses. Among these, 872 images feature male models, while 11,412 feature female models. Specifically, the female model images include 10,475 standing poses, 342 leaning and resting poses, 11 walking poses, 369 sitting poses, and 215 complex special poses.

3.2 Dataset Characteristics

The Relative positions of key points alone do not adequately capture the diverse pose variations and dynamic movements in model performances. Therefore, we incorporate dynamic values on top of key points to enhance the description of specific body parts. The relative position of key points signifies the pose's shape, while dynamic values indicate the relative force exerted and pressure endured compared to a standard pose for a given body part.

Unlike existing human pose datasets (Fig. 1(B)), our model dataset includes not only key skeletal points but also core areas frequently involved in modeling performances, such as the waist and abdomen. However, these areas are often obscured by clothing, making their features less apparent. Points 10–14 (Fig. 1(A)), representing extensive muscle areas, require professional knowledge for precise annotation, necessitating subjective evaluation by experts, posing significant challenges to dataset construction.

3.3 Keypoints Quantity and Position Setting

Since model poses involve the coordination of different body parts, we employed the point-line-plane theory from human aesthetics to deconstruct and analyze various body parts in planar modeling poses, ultimately determining 11 core areas: chin, neck, shoulder, elbow, hand, waist, abdomen, hip, knee, ankle, and toe (Fig. 1(A)). Apart from the chin and abdomen, each area is symmetrical, totaling 20 key points. These key points are not direct skeletal positions but relate more to the dynamic changes and mechanical effects throughout model performances, thus termed "dynamic points." To quantitatively evaluate poses, we assign a relative weight to each of the 20 model pose key points, accurately describing their specific status during performances. The range [1, 10] reflects the relative force exerted and pressure withstood compared to a standard pose. Corresponding to the concept of dynamic points, the weight value is named the "dynamic value," serving as a key indicator for quantifying dynamic expressions in model poses.

4 Proposed Method

4.1 Overview

The structure of our keypoint localization model is illustrated in Fig. 2. The proposed model primarily employs multiple stacked V-modules as the backbone network, incorporating attention mechanism and deep adaptive intermediate supervision to enhance expressiveness, training speed, and improve localization accuracy. Specifically, the input images are processed through a Conv1 module and then enter the stacked V-module network. Each V-module is followed by a 1×1 convolution, reducing the feature map from 256 channels to the number of keypoints, i.e., 20 channels. A depth-variable loss function is calculated here for intermediate supervision of network parameters. After the final V-module is reduced to 20 channels through a 1×1 convolution, an attention mechanism is added, with the ultimate model output determined by the heatmap from the final V-module.

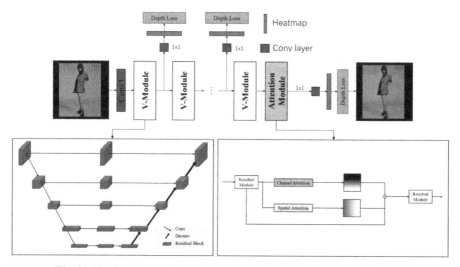

Fig. 2. The framework for automatic localization of model dynamic points

4.2 V-Module

The V-module structure is shown in Fig. 2. Each V-module maintains a consistent design, utilizing residual modules to ensure effective gradient backpropagation. A symmetric structure aligns the resolution scaling of encoders and decoders, simplifying the neural network module's code implementation.

The design of the residual module is akin to ResNet's, with sequential 1×1, 3×3, and 1×1 convolutions, followed by the addition of the shortcut and original feature map's residuals. For downsampling, max pooling is first applied, then a residual module reduces the feature map resolution by half. After four downsampling operations, a 4×4

feature map is obtained. In the encoder part, a 4 × 4 deconvolution kernel with stride 2 and padding 1 enhances the heatmap resolution through ReLU activation, doubling the resolution. Throughout this process, feature maps of the same resolution are processed through the residual module and added to deconvolved feature maps to generate a new feature layer, continuing until the feature map resolution matches the heatmap.

4.3 Attention Module

To boost the model's performance in capturing and processing human keypoint information, we designed a stepwise attention mechanism module. This adjusts the feature map adaptively to allow the model to focus more accurately on the variations of different dynamic point areas. Its structure is depicted in Fig. 2. The initial feature map is processed through a residual module, then duplicated. The first copy undergoes global average pooling, with the channel weights applied to the original feature map, resulting in channel attention features. The second copy uses a 1 × 1 convolution to create an attention map, which is applied to the original feature map, yielding spatial attention features. These are summed with the original feature map and processed through a residual module for final heatmap prediction.

4.4 Depth-Adaptive Intermediate Supervision

Following each stacked module, a 1 × 1 convolution outputs keypoint predictions, guided by a supervisory loss function that facilitates network learning. Intermediate supervision aids network convergence and robustness. To optimize the role of intermediate supervision in model training, we designed a depth-adaptive intermediate supervision module to ensure optimal supervision effects. The loss function is as follows:

$$L_p = \sum_{k=1}^{n} \alpha k L_k \quad (1)$$

where n is the stack count, α is the weight factor, k represents each V-module's depth as an integer. L_k denotes the heatmap loss for each intermediate supervision. The heatmap loss used here is MSELoss, with the calculation formula:

$$L_k = \frac{1}{N} \sum_{i=1}^{N} (y_{k,i} - t_{k,i})^2 \quad (2)$$

where N is the number of joints, $y_{k,i}$ and $t_{k,i}$ separately represents the network output and ground truth heatmap for the i keypoint from the k stacked module output.

5 Experiments

5.1 Experimental Setup

Experiments are conducted on Ubuntu 22.04, PyTorch 2.1.0, with a GeForce RTX 3060. The batch size is set to 48 for networks with 2 stacked V-modules and 32 for those with 4. The optimizer used is Adam with an initial learning rate of 0.001 and a decay factor of 0.1.

5.2 Evaluation Criterion

We evaluate the accuracy of model keypoint localization using Average Precision (AP) and Average Recall (AR). Following the COCO dataset's setup, Object Keypoints Similarity (OKS) is used to measure prediction accuracy, with sigma values determined for each keypoint based on our dataset's distribution. For larger muscle areas like the waist and hips, the sigma values are set at 1.37, 1.37, and 1.07, reflecting a wider acceptable error range, while smaller areas like the chin and neck require greater precision, with values of 0.25 and 0.35, respectively.

AP is the mean accuracy of keypoint detection across various OKS thresholds, calculated as:

$$AP = 1/n \sum_{i}^{n} p(i) \qquad (3)$$

where n is the number of OKS thresholds, and $p(i)$ is the precision at the i threshold.

AR is the mean coverage rate of keypoint detection across different OKS thresholds and maximum detection numbers, expressed as:

$$AR = 1/(n*m) \sum_{i}^{n} \sum_{j}^{m} r(i,j) \qquad (4)$$

where n is the number of OKS thresholds, m is the number of maximum detection, and $r(i, j)$ is the recall at the i threshold and j maximum detection.

We compute the average of 10 AP and AR values across thresholds ranging from 0.5 to 0.95, in increments of 0.05, and use the values at 0.5 and 0.75 as additional evaluation standards.

5.3 Comparisons with the Existing Methods

To verify the effectiveness of our proposed algorithm, we compare it with current representative human keypoint localization algorithms, such as Stacked Hourglass [21], SimpleBaseline [12], HRNet [1] and their variants on the model dataset. The results are shown in Table 1. (Our(x4)) achieves the best performance across all metrics, with AP, AP.5, AP.75, AR, AR.5, and AR.75 at 0.820, 0.991, 0.801, 0.855, 0.992, and 0.859, respectively. Compared to the second best HRNet model, ours shows improvements of 5.11% in AP and 4.63% in AR at the 0.75 keypoint similarity threshold. The result can be attributed to reducing input information loss through stacked network models and deep adaptive intermediate supervision.

The SimpleBaseline model with ResNet101 backbone performs well in AP and AP.75, reaching 0.798 and 0.802, but is weaker in AP.5 and AR.5, likely due to the lack of intermediate supervision leading to some keypoint prediction errors. Our model surpasses SimpleBaseline with a 2.76% increase in AP and 2.40% in AR. The reason can be attributed to the fact that we employ channel and spatial attention mechanisms to more accurately capture keypoint region features.

Additionally, the Stacked Hourglass(x4) and SimpleBaseline50 model with ResNet50 backbone both achieve an AP of 0.784. Our model (Ours (x4)) improves average precision by 4.5% and AR from 0.838 to 0.855, a 2.0% increase. For the stricter

Table 1. Keypoint Localization Results of Different Models

Methods	AP	AP.5	AP.75	AR	AR.5	AR.75
Stacked Hourglass(x2)	0.761	0.990	0.686	0.815	0.992	0.774
Stacked Hourglass(x4)	0.784	0.990	0.712	0.838	0.992	0.797
SimpleBaseline50	0.784	0.990	0.709	0.838	0.992	0.796
SimpleBaseline101	0.798	0.980	0.802	0.835	0.988	0.859
HRNet-W32	0.748	0.973	0.679	0.813	0.986	0.779
HRNet-W48	0.807	0.990	0.762	0.847	0.990	0.821
Ours(x2)	0.789	0.990	0.724	0.838	0.994	0.799
Ours(x4)	**0.820**	**0.991**	**0.801**	**0.855**	**0.992**	**0.859**

AP.75 standard, improvements of 12.5% in AP and 10.98% in AR.75 are observed over Stacked Hourglass.

For lightweight models, the double-stacked hourglass model (Stacked Hourglass (x2)) performs poorly. Compared to it, our model (Ours (x2)) improves AP by 3.6% and AR by 2.8%. More stringent improvements are seen with AP.75 increasing by 5.54% and AR.75 by 10.98%. These results indicate that our model provides more accurate localization on the model pose keypoint dataset.

Fig. 3. Visualization of Keypoint Predictions by Different Models. From left to right: Our method, SimpleBaseline, HRNet, Stacked Hourglass, and Ground truth.

Figure 3 illustrates the visualized keypoint localization. It shows that models excluding ours, such as SimpleBaseline, HRNet, and SHN, fail to correctly identify 9 points from the legs to the abdomen, along with incorrect classifications of keypoints 12 (abdomen), 13 (right hip), 14 (left hip), 15 (right knee), 16 (left knee), 17 (right ankle), 18 (left ankle), 19 (right toes), and 20 (left toes). Our model accurately aligns with the Ground truth, demonstrating superior localization under complex backgrounds and clothing.

5.4 Ablation Study

To analyze the contributions of each model improvement, we incrementally enhance parts of the network from Stacked Hourglass (x4) as a baseline. The models and results are in Table 2. VNet serves as our backbone, VNet(x4) + ISAD includes depth-adaptive intermediate supervision, and Ours (VNet(x4) + ISAD + Attn) further adds the attention module.

Table 2 shows that VNet(x4) achieves an AP of 0.802 and AP.75 of 0.789, improving Stacked Hourglass under the same conditions by 2.30% and 10.81%. AR.75 improves by 4.27%, indicating the backbone network's enhancements provide better point localization. The VNet(x4) + ISAD model further increases AP by 1.50% and AR.75 by 2.53%, demonstrating that the improved intermediate supervision loss function guides learning more effectively. Moreover, Ours(VNet(x4) + ISAD + Attn) enhances AP by 0.74% and AR by 1.06%, demonstrating the attention strategy to capture keypoint region features accurately.

Table 2. Ablation Experiment Results of our proposed model

Models	AP	AP.5	AP.75	AR	AR.5	AR.75
Stacked Hourglass(x4)	0.784	0.990	0.712	0.838	0.992	0.797
VNet(x4)	0.802	0.990	0.789	0.841	0.990	0.831
VNet(x4) + ISAD	0.814	0.990	0.793	0.846	0.992	0.852
Ours(VNet(x4) + ISAD + Attn)	**0.820**	**0.991**	**0.801**	**0.855**	0.992	**0.859**

6 Conclusion

This paper proposes a novel keypoint localization model for precise model pose keypoint position. The backbone network, composed of stacked V-modules, integrates attention mechanism and depth-adaptive intermediate supervision to enhance expressiveness and improve keypoint localization accuracy. Extensive experiments on our model pose dataset validate the proposed algorithm's effectiveness. This work offers a novel, low-cost digital presentation method for the modeling industry and presents an intriguing and challenging application scenario for human pose estimation.

References

1. Al-Halah, Z., et al.: Modeling fashion influence from photos. IEEE Trans. Multimedia **23**, 4143–4157 (2020)
2. Mahmood, N., et al.: AMASS: Archive of motion capture as surface shapes. In: IEEE Conference on Computer Vision, pp. 5442–5451 (2019)
3. Ortega, B.P., Olmedo, J.M.J.: Application of motion capture technology for sport performance analysis. Retos: Nuevas Tendencias en Educación Física, Deporte y Recreación **32**, 241–247 (2017)

4. Toshev, A., Szegedy, C.: Deeppose: human pose estimation via deep neural networks. In: IEEE Conference on Computer Vision and Pattern Recognition, pp. 1653–1660 (2014)
5. Newell, A., Yang, K., Deng, J.: Stacked hourglass networks for human pose estimation. In: European Conference on Computer Vision, pp. 483–499 (2016)
6. Sun, K., Xiao, B., Liu, D., et al.: Deep high-resolution representation learning for human pose estimation. In: IEEE Conference on Computer Vision and Pattern Recognition, pp. 5693–5703 (2019)
7. Kan, Z., Chen, S., Zhang, C., et al.: Self-correctable and adaptable inference for generalizable human pose estimation. In: IEEE Conference on Computer Vision and Pattern Recognition, pp. 5537–5546 (2023)
8. Cai, Y., Wang, Z., Luo, Z., et al.: Learning delicate local representations for multi-person pose estimation. In: European Conference on Computer Vision, pp. 455–472 (2020)
9. Luo, Y., Gao, X.: Lightweight human pose estimation based on self-attention mechanism. In: The World Conference on Intelligent and 3D Technologies, pp. 237–246 (2023)
10. Zheng, C., et al.: Deep learning-based human pose estimation: a survey. ACM Comput. Surv. **56**(1), 1–37 (2023)
11. Purkrabek, M., Matas, J.: Improving 2D human pose estimation in rare camera views with synthetic data. In: IEEE Conference on Automatic Face and Gesture Recognition, pp. 1–9 (2024)
12. Papaioannidis, C., Mademlis, I., Pitas, I.: Fast single-person 2D human pose estimation using multi-task convolutional neural networks. In: IEEE Conference on Acoustics, Speech and Signal Processing, pp. 1–5 (2023)
13. Pfister, T., Simonyan, K., Charles, J., et al.: Deep convolutional neural networks for efficient pose estimation in gesture videos. In: Asian Conference on Computer Vision, pp. 538–552 (2015)
14. Rafi, U., Leibe, B., Gall, J., et al.: An efficient convolutional network for human pose estimation. In: British Machine Vision Conference, pp. 1–11 (2016)
15. Jin, S., Xu, L., Xu, J., et al.: Whole-body human pose estimation in the wild. In: European Conference on Computer Vision, pp. 196–214 (2020)
16. Xiao, B., Wu, H., Wei, Y.: Simple baselines for human pose estimation and tracking. In: European Conference on Computer Vision, pp. 466–481 (2018)
17. Sun, X., Shang, J., Liang, S., et al.: Compositional human pose regression. In: IEEE Conference on Computer Vision, pp. 2602–2611 (2017)
18. Luvizon, D.C., Tabia, H., Picard, D.: Human pose regression by combining indirect part detection and contextual information. Comput. Graph. **85**, 15–22 (2019)
19. Nibali, A., He, Z., Morgan, S., et al.: Numerical coordinate regression with convolutional neural networks. arXiv preprint arXiv:1801.07372 (2018)
20. Li, K., Wang, S., Zhang, X., et al.: Pose recognition with cascade transformers. In: IEEE Conference on Computer Vision and Pattern Recognition, pp. 1944–1953 (2021)
21. Yoshitake, Y., Nishimura, M., et al.: Transposer: transformer as an optimizer for joint object shape and pose estimation. arXiv preprint arXiv:2303.13477 (2023)
22. Cheng, B., Xiao, B., Wang, J., et al.: HigherHRNet: Scale-aware representation learning for bottom-up human pose estimation. In: IEEE Conference on Computer Vision and Pattern Recognition, pp. 5386–5395 (2020)
23. Qu, H., Cai, Y., Foo, L. G., et al.: A characteristic function-based method for bottom-up human pose estimation. In: IEEE Conference on Computer Vision and Pattern Recognition, pp. 13009–13018 (2023)
24. Huang, L., Li, Y., Tian, H., et al.: Semi-supervised 2D human pose estimation driven by position inconsistency pseudo label correction module. In: IEEE Conference on Computer Vision and Pattern Recognition, pp. 693–703 (2023)

A Redundancy Free Facial Acne Detection Framework Based on Multi-view Face Images Stitching

Ye Luo[1], Jianfei Wang[2], Linglin Zhang[3], Xinyu Liu[1], Ji Rao[1], Wantong Xu[1], Jianwei Lu[4,5(✉)], and Xiuli Wang[3(✉)]

[1] School of Computer Science and Technology, Tongji University, Shanghai 201804, China
{yeluo,xinyuliu,jirao,wantongxu}@tongji.edu.cn

[2] Suzhou Chien-Shiung Institute of Technology, Taicang 215411, China
jianfeiwang@scsit.edu.cn

[3] Institute of Photomedicine, Shanghai Skin Disease Hospital, Tongji University School of Medicine, Shanghai 200043, China
{linglinzhang,wangxiuli_1400023}@tongji.edu.cn

[4] College of Rehabilitation Science, Shanghai University of Traditional Chinese Medicine, Shanghai, China
jwlu33@shutcm.edu.cn

[5] Engineering Research Center of Traditional Chinese Medicine Intelligent Rehabilitation, Ministry of Education, Shanghai, China

Abstract. Facial acne is a prevalent skin disease worldwide. Misjudging the severity of its condition can lead to permanent facial damage or negatively impact patients' self-esteem. With the rapid advancements in computer vision techniques, automated methods for detecting or grading facial acne in digital images are attracting growing attention. However, in clinical practice, multi-view (left, front, and right) face images used for accurate diagnosis pose a significant challenge due to redundant acne detection on the overlapped image regions that can mislead diagnostic results. In this paper, we propose a novel facial acne detection framework to detect various kinds of acne for accurate acne diagnosis or prognosis prediction. Specifically, multi-view facial images are first stitched by aligning key facial feature points, and then the proposed acne detection network YoloV5-Acne is then applied to the stitched images for patient-level acne detection. Experimental results on a collected multi-view face image dataset (ACNE-Shanghai) and a public acne dataset (ACNE04) validate the effectiveness of the proposed method on various evaluation metrics. The code and the dataset will be released for the research purpose once the paper gets published.

Keywords: Face Image · Facial Acne Detection · Multi-view Image Stitching · Acne Treatment

---

Y. Luo, J. Wang and L. Zhang—Equal Contribution.
This paper is supported by the General Program of the National Natural Science Foundation of China (NSFC) under Grant 62276189.

1 Introduction

Acne, the most common skin disease characterized by comedos, papules, pustules, and nodules, has a prevalence of 90% during adolescence [1]. Scarring from these acne lesions often appears on the face and has a negative effect on the general functional and social well-being of patients [2]. Therefore, it is crucial to accurately grade the severity of facial acne, enabling the development of a precise and early treatment plan [10]. In clinical practice, the traditional method of grading facial acne severity involves manually counting the number of lesions on the face [15], which is labor-intensive, tedious, and subjective.

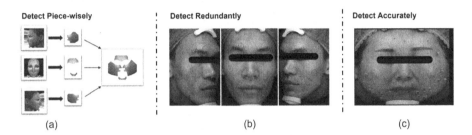

Fig. 1. Two existing acne detection methods ((a) and (b)) and our proposed method (c) for multi-view face images. In (a) and (b), acne are either missing or repetitively detected across different pieces or images of faces. While in (c), thanks to the image stitching, we get the accurate acne detection result for each individual patient.

With the rapid development of computer vision techniques, several automatic acne assessment methods have been proposed [6,11]. In [11], a robust acne detection model using label distribution learning was introduced, but it overlooks the fact that acne assessment is performed at the patient level, not per individual image. This effect is illustrated by Fig. 1 (b). On the other hand, [6] manually decomposes multi-view face images into facial regions (forehead, lower jaw, left and right sides) and then combines them for acne assessment (as shown in Fig. 1 (a)). While this approach reduces duplicate acne detection, it risks underestimating acne due to missed detection in overlapped facial regions. Combining image stitching with deep learning-based object detection presents unique challenges and opportunities in the context of facial acne detection. In 2020, Zhang et al. [7] introduced an unsupervised learning-based image registration method that learns a mask within the network framework, utilizing only reliable regions of the image for homography matrix estimation. In 2019, Wu et al. [11] approached acne image analysis through Label Distribution Learning (LDL), which assigns each instance a label distribution that describes the degree to which each label applies to it. Later, in 2022, Quan et al. [8] developed AcneDet, an AI system based on deep learning that automatically analyzes facial images from smartphones, performing two tasks: detecting acne lesions and grading acne severity by the number of acne detected. In 2023, Wang et al. [9,24] proposed a fully automated approach using deep learning models to provide valuable diagnostic evidence for dermatologists in acne detection and severity quantification. However, none of the existing methods focus on accurately detecting various categories of facial acne from each patient's whole face image.

To address the limitations of existing methods, we propose a redundancy-free acne detection framework based on multi-view face image stitching. In order to avoid the redundant acne detection on every face image, we propose to first stitch these images which are with large view angle translations (e.g. every ninety degree) into a whole face image. An improvement of YoloV5 based acne detection network (YoloV5-Acne) is then applied to the stitched image for further acne prognosis assessment. Experimental results on a realistic dataset collected from Shanghai Skin Disease Hospital and a public available dataset ACNE04 validate the effectiveness of the proposed method for acne detection and its potential for patient-centered acne assessment.

2 The Proposed Method

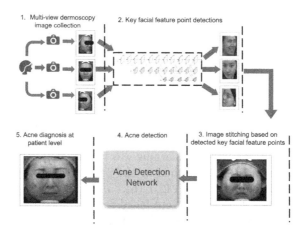

Fig. 2. The proposed framework of our redundancy-free acne detection method for each individual patient.

The overall framework of our proposed method is shown in Fig. 2. In the following Subsections, we will introduce the main steps of our method, including multi-view face image stitching and the acne detection network.

2.1 Multi-view Face Image Stitching

Considering that the images in our dataset are high-definition, traditional methods, we utilize HRNet [4] for feature point detection, which can maintain high-resolution representations throughout the process. HRNet starts from a high-resolution sub-network and progressively adds high-to-low resolution sub-networks in parallel. In total, 98 key facial points are detected via HRNet for each image. Figure 3(a) shows the key facial features of the left and front face images learned by HRNet, with feature points indexed in green dots and matched in red lines.

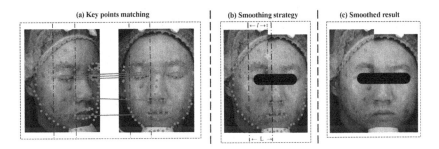

Fig. 3. Illustration of feature matching between two views, the smoothing strategy, and the smoothed result. (Color figure online)

After obtaining matched keypoint pairs, we implement image registration and stitching by estimating the homography matrix [16], which describes the transformation between two images based on pixels. However, since the multi-view face images are not in the same image plane and the view change between consecutive images is close to 90°C, we use an approximated homography matrix for registration. For simplification, we select a region between two eyebrow feature points, whose size is $L \times H$, as the candidate region for homography matrix estimation. Here, $L = |x_{eb}^l - x_{eb}^r|$, where $|.|$ denotes the absolute value, and x_{eb}^l and x_{eb}^r represent the x coordinates of the leftmost and rightmost eyebrow feature points, respectively, while H is the height of the original image. Visualization of the matching process between feature points is shown in Fig. 3(a), illustrating the large overlap between views. Finally, consecutive views (e.g., left-front, front-right) are warped using the estimated homography matrices. We implemented the keypoint extraction on multi-view facial images using a pre-trained HRNet network, followed by an image registration-based stitching method to combine the three views into a single image. The model was trained using the Adam optimizer for 100 epochs with a batch size of 4, an initial learning rate of 0.001, and a weight decay coefficient of 0.0005.

After wrapping two images P_1 and P_2, artifacts may appear in the overlapping region. To address this, we perform smoothing in the overlapped region using linearly weighted summation. Given a position (x, y), the pixel value $P(x, y)$ is calculated as follows:

$$P(x,y) = \left(1 - \frac{l}{L}\right) P_1(x,y) + \frac{l}{L} P_2(x,y), \quad (1)$$

where L is the distance between the left and right borders of the overlapped region, and $l = |y - y_{eb}^t|$. Here, y_{eb}^t is the y coordinate of the leftmost eyebrow feature point when stitching the left-front pair ($y_{eb}^t = y_{eb}^l$), or the rightmost feature point when stitching the front-right pair ($y_{eb}^t = y_{eb}^r$).

2.2 Acne Detection on Stitched Images

After obtaining the stitched face image, the next step is to detect acne. However, detecting facial acne presents challenges due to its small size especially on the stitched image and its subtle features among different categories of acne. Meanwhile, recent advancements in object detection, such as Yolo series frameworks (e.g., YoloV5) [5], have been applied successfully. We improve YoloV5 in two ways: **1)** We increase the number of data samples through mosaic data augmentation [13], where new images are generated by combining the original image with three randomly cropped images. This approach enhances the use of high-resolution face images and increases the prominence of small acne lesions. **2)** In order to handle the tiny acne detection, we adjust the loss function to αIoU [12], which improves the accuracy of acne localization and speeds up model convergence. The αIoU loss function is defined as follows:

$$L_{\alpha CIoU} = 1 - \left(IoU^\alpha - \frac{\rho^{2\alpha}(C_{pred}, C_{gt})}{C^{2\alpha}} - (\beta v)^\alpha \right), \qquad (2)$$

where IoU is the intersection over union between the prediction box (Box pred) and ground truth box (Box GT), ρ is the distance between the centers of the two boxes, C is the diagonal length of the minimum enclosing box, and βv measures aspect ratio similarity. α is an exponent to balance the function which is generally set to 3. We refer to this improved version of YoloV5 for acne detection as YoloV5-Acne.

3 Experiments and Analysis

In this Section, we first introduce the dataset and evaluation metrics, and then visualize the effect of image stitching. Moreover, multiple methods are compared to show the effectiveness of acne detection and counting. At last, the effectiveness of our method on reducing redundant acne detection and the significance of our result to the prognosis evaluation on acne treatment are provided.

ACNE-Shanghai Dataset. In order to validate the proposed method, we collected a dataset consisting of three-view face images of 103 patients from Shanghai Skin Disease Hospital. Each patient was required to take images from the left, front, and right faces. In total, there are 309 images, each sized 3456 × 5184. To label the facial acne lesions, six professional doctors divided into three groups marked each acne with its category (i.e., comedos, papules, pustules, and nodules), shape (i.e., a rectangle to minimally contain the acne), and location (i.e., the coordinate of the rectangle's center) using Labelme [14]. All labeled results were double-checked by a supervising doctor. To obtain the ground truth (i.e., gt) for stitched images, we used the previously obtained homography matrix to transform the coordinates of the ground-truth labels from the un-stitched images to the stitched images.

ACNE04 [10] Dataset. ACNE04 is a publicly available dataset for facial acne severity classification task. It contains 1,457 images with 18,983 annotations of bounding boxes to identify the existences of each acne. Thus, for each image, only the total number of acne as well as the acne severity (In total there are four levels of severity provided according to the number of acne detected.) is provided. Compared to ACNE-Shanghai, only the classification task about the severity of acne can be performed on ACNE04 by counting the number of acne. The category of each acne and the its locations cannot be evaluated on this dataset. But it is introduced to further validate the robustness of our acne detection method.

Evaluation Metrics. To assess the acne counting capability of our method, we employ two evaluation metrics: Mean Absolute Error (**MAE**) and relative Mean Squared Error (**rMSE**). To evaluate the capability of our method in various categories of acne detection, we adopt the Mean Average Precision (**mAP**) [22] as the evaluation metric. While the Average Precision (AP) is computed as the area under the Precision-Recall curve for each class. We choose $mAP_{0.5}$ considering that the Intersection over Union (**IoU**) between the prediction region and the ground truth is greater than the threshold (i.e. 0.5) can be treated as a True Positive. **Accuracy** is used to measure the performance of acne grading result.

Effect of face Image Stitching. To validate the proposed method on face image feature matching, we simple compare it with traditional methods based on SIFT feature points and Harris feature points. However, these traditional methods perform poorly when handling images with large-angle parallax and are ineffective in multi-view stitching.

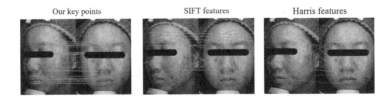

Fig. 4. Comparisons of the feature matching results among our method (left), SIFT features (middle) and harries features (right). (Color figure online)

The visual results of different stitching methods are illustrated in Fig. 4, where green lines represent the results of feature point matching. Clearly, the feature matching effectiveness of traditional methods such as SIFT and Harris corner detection [3] is poor, while the proposed method accurately matches facial points. Experimental results confirm that the method used in this paper can adapt to significant changes in the viewing angle of skin microscope images, showcasing a clear advantage over traditional methods.

Table 1. Comparison of acne counting performance on ACNE-Shanghai and ACNE04 datasets by MAE and rMSE. Here, mAP is adopted only on ACNE-Shanghai for multiple categories of object detection evaluation, while Accuracy is used to evaluate the acne severity grading performance on ACNE04 dataset.

Method	ACNE-Shanghai			ACNE04		
	MAE↓	rMSE↓	mAP$_{0.5}$↑	MAE↓	rMSE↓	Accuracy↑
F-RCNN [17]	–	–	–	6.70	11.51	73.97
RefineDet [18]	–	–	–	5.82	10.14	72.09
LDL [11]	10.94	14.30	–	2.93	5.42	**84.11**
RCNN [23]	4.72	6.12	0.11	–	–	–
YoloV3 [19]	24.40	28.64	0.30	6.69	11.35	63.70
YoloV5 [5]	5.68	7.33	0.36	3.70	7.33	77.49
YoloV5-Acne	**3.09**	**4.28**	**0.40**	**2.91**	**4.79**	83.41

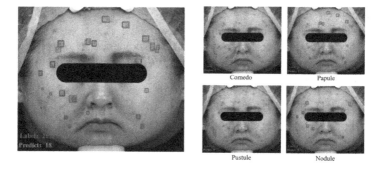

Fig. 5. Visualization results of the acne detection and counting results on the stitched ACNE-Shanghai dataset (Red: the labeled ground truth, Blue: the predicted result). Left: the overall detection results for all kinds of acne. Right: the detection results for four categories of acne. (Color figure online)

Effect of Acne Detection and Counting. To validate the effectiveness of the proposed framework on acne detection as well as the acne number counting, we employed a five-fold cross-validation technique to ensure the robustness and reliability of our acne detection framework, conducting a comparative analysis by evaluating multiple methods alongside our YoloV5-Acne using our stitched ACNE-Shanghai dataset and the public ACNE04 dataset, as outlined in Table 1. In this context, $mAP_{0.5}$ provides a metric for assessing the accuracy of object detection. Additionally, MAE and rMSE are employed to gauge the accuracy of counting. Most methods find it challenging to balance accuracy in both detection and counting aspects. Clearly, our method demonstrates superior counting and detection capabilities, establishing its effectiveness. The successful implementation of robust acne object detection and counting methodologies within dermatological practices has yielded significant impacts on both diagnostic preci-

Table 2. Various categories of acne counting by YoloV5-Acne on the stitched ACNE-Shanghai dataset.

	Comedo	Papule	Pustule	Nodule	Total
MAE↓	1.59	2.67	0.87	1.03	3.09
rMSE↓	2.49	3.47	1.34	1.47	4.28

sion and patient outcomes. Moreover, the visualization of acne detection results on the stitched ACNE-Shanghai dataset is shown in Fig. 5. From this figure, we can see that no more duplicated acne are detected for each individual patient (Table 2).

Effect of Reducing Redundant Acne Detection. To illustrate that duplicated acne constitutes a significant portion of the dataset and cannot be neglected during assessment, 20 subjects were randomly selected to calculate the statistics of acne repetition, as shown in Fig. 6. Specifically, for each patient (i.e., subject), the number of human-labeled acne in the three-view face images is referred to as *total num*, while the number of human-labeled acne on the stitched face image is referred to as *real num*. The gap between these two values indicates the repetition rate. The average repetition rate among the 20 subjects is 24.98%, which is substantial enough to affect the acne assessment results. To validate the

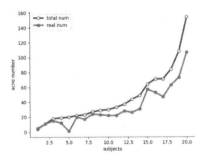

Fig. 6. Comparison between the number of acne in the three face images (total num) and that in the stitched images (real num). The gap between the two curves indicates the duplicated acne rate for each subject.

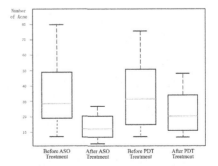

Fig. 7. Comparison of acne quantities before and after treatment for two patient groups.

effectiveness of the proposed method in detecting duplicate acne, we compared the detection results on un-stitched and stitched images using YoloV5-Acne.

Prognosis Assessment by Acne Detection Results. Different from directly grading the severity from the number of acne detected, we grade the severity of acne from its quantity and category simultaneously considering that the severity of acne generally follows: *nodules $\geq$ Pustules $\geq$ papules $\geq$ comedones*. By this way, dermatologists at the Shanghai Skin Disease Hospital developed treatment plans tailored to patients' acne severity. Of the total patients, 51 underwent Isotretinoin (ISO) treatment, while 72 received Photodynamic Therapy (PDT) [20,21]. After treatment, in order to evaluate which treatment gets better effect, the proposed acne detection method is then applied to detect acne on pre-treatment and post-treatment facial images respectively for each patient. The acne counting results for the pre-treatment and post-treatment for the 123 patients is illustrated in the box plot in Fig. 7. It clearly shows a reduction in maximum, minimum, and median acne quantities.

4 Conclusion

In this paper, we proposed a novel redundancy-free acne detection framework by performing acne detection on the stitched multi-view face images. For multi-view face image stitching, we propose a method based on HRNet to successfully solve the problem of the large view angle change problems among face images. For acne detection, YoloV5-Acne is proposed to accurately detect various sizes of acne onto a high-resolution stitched image for each individual patient. Validations on a realistic dataset collected from the collaborated hospital and a publicly available dataset demonstrate the effectiveness of the proposed method. Future research will focus on comparing with some complex stitching and detection methods, and also including testing on a large dataset.

References

1. Hao, J., Li, C.: Acne vulgaris is a special clinical type of pellagra. Am. J. Clin. Exp. Med. **9**(6), 204–208 (2021)
2. Hazarika, N., Archana, M.: The psychosocial impact of acne vulgaris. Indian J. Dermatol. **61**(5), 515–520 (2016)
3. Harris, C., Stephens, M.: A combined corner and edge detector. In: Proceedings of the Alvey Vision Conference, pp. 147–151 (1988)
4. Sun, K., Zhao, Y., Jiang, B., et al.: High-resolution representations for labeling pixels and regions. preprintarXiv:1904.04514 (2019)
5. Redmon, J., Divvala, S., Girshick, R., et al.: You only look once: Unified, real-time object detection. In: Proceedings of the IEEE conference on Computer Vision and Pattern Recognition, pp. 779–788 (2016)
6. Yang, Y., Guo, L., Wu, Q., et al.: Construction and evaluation of a deep learning model for assessing acne vulgaris using clinical images. Dermatol. Therapy **11**(4), 1239–1248 (2021)

7. Zhang, J., Wang, C., Liu, S., et al.: Content-aware unsupervised deep homography estimation. *preprint*arXiv: 1909.05983, (2020)
8. Huynh, Q.T., Nguyen, P.H., Le, H.X., et al.: Automatic acne object detection and acne severity grading using smartphone images and artificial intelligence. Diagnostics **12**, 1879 (2022)
9. Wang, J., Wang, C., Wang, Z., et al.: A novel automatic acne detection and severity quantification scheme using deep learning. Biomed. Sig. Process. Control **84** (2023)
10. Hayashi, N., Akamatsu, H., Kawashima, M.: Establishment of grading criteria for acne severity. J. Dermatol. **35**(5), 255–260 (2008)
11. Wu, X., Wen, N., Liang, J., et al.: Joint acne image grading and counting via label distribution learning. In: Proceedings of the IEEE/CVF International Conference on Computer Vision, pp. 10641–10650 (2019)
12. He, J., Erfani, S., Ma, X., et al.: Alpha-IoU: a family of power intersection over union losses for bounding box regression. Adv. Neural. Inf. Process. Syst. **34**, 20230–20242 (2021)
13. Bochkovskiy, A., Wang, C.Y., Liao, H.Y.M. Yolov4: optimal speed and accuracy of object detection. preprint arXiv:2004.10934 (2020)
14. Torralba, A., Russell, B.C., Yuen, J.: Labelme: online image annotation and applications. Proc. IEEE **98**(8), 1467–1484 (2010)
15. Witkowski, J.A., Parish, L.C.: The assessment of acne: an evaluation of grading and lesion counting in the measurement of acne. Clin. Dermatol. **22**(5), 394–397 (2004)
16. Luo, Y., Li, Z., Luo, S. Research on camera calibration method based on Homography matrix. In: Proceedings of the IEEE Information Technology, Networking, Electronic and Automation Control Conference, pp. 464–468 (2023)
17. Ren, S., He, K., Girshick, R., et al. Faster R-CNN: towards real-time object detection with region proposal networks. In: IEEE Transactions on Pattern Analysis and Machine Intelligence, pp. 1137–1149 (2017)
18. Zhang, S., Wen, L., Bian, X., Lei, Z., et al.: Single-shot refinement neural network for object detection. In: Proceedings of the IEEE Conference on Computer Vision and Pattern Recognition, pp. 4203–4212 (2018)
19. Redmon, J., Farhadi, A.: YOLOv3: an incremental improvement. arXiv preprint arXiv:1804.02767 (2018)
20. Dong, Y., Zhou, G., Chen, J., et al. A new LED device used for photodynamic therapy in treatment of moderate to severe acne vulgaris. Photodiagnosis Photodynamics Ther. 188–195 (2016)
21. Zhang, J., Wang, C., Liu, S., et al.: Content-aware unsupervised deep homography estimation. preprint arXiv:1909.05983 (2020)
22. Girshick, R., Donahue, J., Darrell, T.: Region-based convolutional networks for accurate object detection and segmentation. IEEE Trans. Pattern Anal. Mach. Intell. **38**(1), 142–158 (2016)
23. Girshick, R., Donahue, J., Darrell, T, Malik, J.: Rich feature hierarchies for accurate object detection and semantic segmentation. In: Proceedings of the IEEE Conference on Computer Vision and Pattern Recognition, pp. 580–587 (2014)
24. Prokhorov, K., Kalinin, A.A.: Improving acne image grading with label distribution smoothing. IEEE Int. Sympos. Biomed. Imaging (ISBI) **2024**, 1–5 (2024)

A New Device Placement Approach with Dual Graph Mamba Networks and Proximal Policy Optimization

Meng Han[1], Yan Zeng[1(✉)], Hao Shu[2], Xiaofei Lu[3], Jilin Zhang[1], Yongjian Ren[1], and Wangli Hao[2]

[1] Economic Development Zone, Hangzhou Dianzi University, No. 1158, No. 2 Street, Baiyang Street, Hangzhou, Zhejiang, China
yz@hdu.edu.cn
[2] Shanxi Agricultural University, No.1, Mingxian South Road, Jinzhong, Shanxi, China
[3] Hangzhou Tiankuan Technology Co., Ltd., 15th Floor, Building 1, 252 Wantang Road, Xihu District, Zhejiang 1501, China

Abstract. In recent years, foundation models have achieved state-of-the-art performance, but the computational capabilities of GPUs have not kept up, resulting in bottlenecks. While data parallelism, model parallelism, and pipeline parallelism are common in large-scale clusters, automatic parallelism within a single node remains NP-hard. Reinforcement learning methods show promise but are limited by the capabilities of their modules. To tackle these challenges, we propose a novel device placement method (PPO-DGMA) that combines Proximal Policy Optimization (PPO) with a new dual-branch graph embedding approach(DGMA), featuring a decoupled graph neural network (DGM) and a Graph MAMBA (GMA) branch. The PPO algorithm captures subtle environmental changes for enhanced learning and adaptability. However, our dual-branch structure have the capability to improve the extraction of computation graph's features, and the GMA branch enhances long-distance dependencies, while the DGM branch generates robust node representations by integrating node features, topology graphs, and a semantic graph constructed based on operator attributes. Experimental results show that our method has enhanced execution time by an average of 16.58% over Placeto, 9.65% over GraphSAGE, 3.59% over P-GNN, and 2.81% over CP-GNNAK, while achieving significant reductions in computational time-32.53× faster than Placeto, 37.18× faster than GraphSAGE, 31.34× faster than P-GNN, and 8.68× faster than CP-GNNAK.

Keywords: Graph Neural Networks · Distributed Machine Learning · Automatic Parallel

1 Introduction

In recent years, the rapid scaling of AI models in Natural Language Processing(NLP), Computer Vision(CV), and Recommendation System(RS) has outpaced computing device storage growth, prompting the need for multi-device collaboration and innovative training strategies. Automatic parallelization technology offers a solution by generating parallel strategies that accelerate model execution across multiple devices, allowing researchers to focus on innovation and optimization. Key approaches for parallel training include data parallelism [12] and model parallelism [13]. Data parallelism involves splitting training data across devices while replicating the entire model, enabling independent processing and periodic synchronization of parameters for global consistency. In contrast, model parallelism is tailored for ultra-large models, dividing the model into parts that are flexibly deployed across devices. This approach, also known as "device placement," allows each device to focus on a specific segment, collaborating effectively to complete training tasks.

The efficiency of parallel policy search and policy performance relies heavily on reinforcement learning capabilities and effective representation of computational graph features. While methods like ColocRL [7], HDP [6], and REGAL [9] utilize the Reinforce [16] algorithm to accelerate parallel policy search, they struggle with sample inefficiency, prolonging training cycles. In contrast, Placeto [1], GraphSAGE [8], and P-GNN [8] leverage Markov Decision Processes (MDPs) to optimize device configurations by capturing state transition probabilities and rewards, but they require the Markov property, limiting their effectiveness. Moreover, strategies like Post [2], Spolight [3], GDP [20], Mars [5], and Trinity [18] use Proximal Policy Optimization (PPO) to enhance sampling efficiency and stabilize training through importance sampling. On the other hand, Placeto [1] and TRINITY [18] fail to capture critical node positional information in their embeddings, which can obscure important features and hinder operator placement strategies. GraphSAGE [8] attempts to address this by aggregating local neighbor information via random sampling, but this may lead to loss of essential data as model complexity grows. P-GNN [8] and Aware [19] integrate positional encodings into node representations, yet their reliance on distant nodes limits the effectiveness of Message Passing Neural Networks (MPNNs), impeding comprehensive feature representation and reflecting the complexities of computational graphs. Importantly, existing graph-based device placement methods [1,8,18], and [19] overlook the influence of semantic information, which could further enhance model parallel performance and overall efficiency.

Inspired by the development of Graph Neural Networks [14] [15]. In this paper, we proposes a novel device placement method, PPO-DGMA, which leverages PPO [11] and a dual-branch structural graph embedding approach to enhance strategy performance and search speed, as shown in Fig. 1. The first core idea is to incorporate a loss layer in each node of the computation graph within the reinforcement learning module, enabling more precise adjustments based on environmental feedback. This improves learning efficiency, stability, and exploration capability. The other core idea is to leverage a dual-branch graph embedding module to enhance the computational graph encoding capability. The GMA branch enhances long-range dependencies through a node selection mechanism, boosting graph embedding expressiveness. The DGM branch generates robust node representations by utilizing node attributes, topological features, and semantic information from operator attributes, providing supplementary insights for the topological graph. The final computation graph features result from concatenating and fusing outputs from both branches.

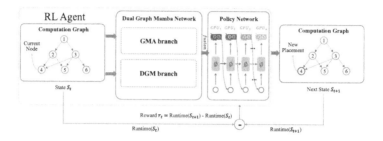

Fig. 1. The overall process of PPO-DGMA for device placement.

Experimental results indicate that the PPO-DGMA method outperforms existing approaches in device placement. Below are the main contribution in the paper.

(1) We propose the dual-branch graph embedding model DGMA, which enhances node representation by constructing a semantic graph from attribute features and improves long-range dependencies through a selection mechanism, thereby increasing the expressiveness of graph embeddings.
(2) We introduce an innovative reinforcement learning method, Proximal Policy Optimization (PPO), which effectively addresses the step-size sensitivity issue by limiting the probability ratio between the new and old policies, and maximizing an objective function that combines the truncated probability ratio and reward value, thus gently adjusting policy generation and parameter updates.
(3) Experimental validation confirms the superior performance of the PPO-DGMA method, supporting efficient training of AI models.

2 Methods

A neural network can be represented as a graph $G(N, E)$, where N is the set of computational nodes (operators) and $n_i \in N$ denotes an individual operator (e.g., matrix multiplication, convolution). The edges E indicate directed data dependencies between nodes, with $e_{i,j} \in E$ representing the dependency from operator n_i to operator n_j. Given a set of devices $D = \{d_1, d_2, d_3, ..., d_m\}$, the goal is to find a mapping strategy π_i that assigns each operator n_i to a device d_i. The objective is to identify the optimal device placement strategy π^* that minimizes the execution time $T(G, \pi^*)$ for the computation graph G.

Our research builds on the advanced device placement scheme called Placeto [1], which follows a standardized process. This process takes a neural network computation graph G and a random device placement strategy π_i as inputs. It iterates through all nodes $n_i \in N$, executing the following steps: First, it uses a graph embedding model, DGMA, to obtain a graph embedding X_G. For each selected node n_i, it applies RL-based(PPO) policy π_i to assign n_i to a device d_i. Then, it updates the reinforcement learning policy π_i based on the execution time $T(G, \pi_i)$, considering the new placement of n_i and previous placements as the reward function. Thus, the algorithm uses execution time as feedback, continuously refining placement strategies for subsequent nodes. Next, we will focus on the introduction of the dual-branch graph embedding framework DGMA (as shown in Fig. 2).

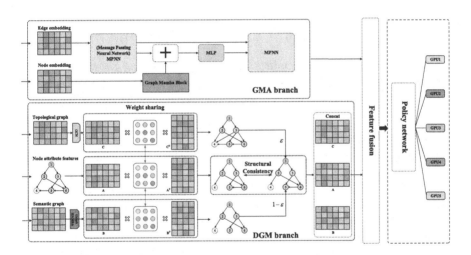

Fig. 2. The framework of DGMA graph embedding, including GMA branch and DGM branch.

2.1 GMA Branch

The Mamba architecture excels in sequential processing, but its complexity limits direct application in graph processing. To address this, we integrate Mamba into Graph Neural Networks (GNN), proposing Graph Mamba Networks (GMA).

We enhance GMA by incorporating position and graph encoding techniques from GCN. This injects structural and locational information into the initial graph encoding $X \in \mathbb{R}^{N \times F}$. The symmetric normalization of the adjacency matrix $\hat{A} = A + I$ helps maintain feature distribution when multiplied by the feature matrix, as shown in Eq. 1:

$$Y_0 = \sigma(D^{-1/2}\hat{A}D^{-1/2}XW). \tag{1}$$

where A is the adjacency matrix, I is the identity matrix, and $\hat{D}$ is the diagonal matrix of node degrees induced by A and I. σ denotes the nonlinear ReLU function. Equations 2 and 3 introduce learnable weight matrices W_B, W_C, W_Δ for encoding, leading to projection parameters B, C and discretization step size Δ. Y_t denotes the feature representation updated at step t. This transforms continuous states into discrete ones, allowing better adaptation in deep learning. The evolution parameter $\bar{A}$ and projection parameter $\bar{B}$ are computed as:

$$B = W_B Y_t, \quad C = W_C Y_t, \quad \Delta = Softplus(W_\Delta, Y_t) \tag{2}$$

$$\bar{A} = \exp(\Delta A), \quad \bar{B} = (\Delta A)^{-1}(\exp(\Delta A) - I)\Delta B \tag{3}$$

In Eq. 4, we utilize Mamba for selective state management (SSM), updating the hidden state:

$$Y_{t+1} = SSM(\bar{A}, \bar{B}, C)Y_t = C\bar{A}Y_{t-1} + \bar{B}Y_t \tag{4}$$

Finally, the output graph code X' is obtained by fusing Y_T with the initial graph encoding Y_0:

$$X' = Combine(Y_T, Y_0) \tag{5}$$

2.2 DGM Branch

We developed our DGM module based on the DGNN framework, retaining its advantages of capturing node attribute encoding and graph structure topology through parallel learning. Additionally, we introduced Graph Mamba(GMA) to further enhance the expressiveness of node encodings. DGM iteratively updates these graph embedding to explore their complementarity, facilitating mutual learning and correction.

First, $X \in \mathbb{R}^{N \times F}$ is encoded as the initial graph, where F is the feature dimension of the nodes in Eq. 6, $A \in \mathbb{R}^{N \times F}$ is used as Node Attribute features, encoded according to random initialized graphs of the same size generated by X. Moreover, X obtains semantic feature embedding and topological structure embedding $B \in \mathbb{R}^{N \times F}$ and $C \in \mathbb{R}^{N \times F}$ through GMA and GCN.

$$t_{A_1} = AW_s A^T, t_{B_1} = BW_s B^T, t_{C_1} = CW_s C^T \tag{6}$$

$$t_{A_2} = AW_s + FW_s^T, t_{B_2} = BW_s + H_f W_s^T, t_{C_2} = CW_s + HW_s^T \tag{7}$$

In the Eqs. 6 and 7, the learnable shared reconstruction factor matrix W_s is introduced to reconstruct the potential correlation between the adjacency matrices. At the same time, in order to ensure the symmetry of the reconstructed matrix, the constraint $W_s = WW^T$.

$$t_M = -t_N = t_{A_1} - \epsilon t_{B_1} - (1-\epsilon) t_{C_1} \tag{8}$$

In Eqs. 8, the error $t_M \in \mathbb{R}^{N \times F}$ and $t_N \in \mathbb{R}^{N \times F}$ between A,B and C is minimized by the reconstruction factor W_s. Where ϵ is the balance coefficient of the semantic feature B and the topology structure C, which balances the weight relationship between the two, and the value ranges from 0–1.

$$A = Mean\,(B+C) - L_3 \sigma \,(t_M t_{A_2}) \tag{9}$$

$$B = L_1 B - \epsilon\,(L_3/L_1)\,\sigma\,(t_N t_{B_2}) \tag{10}$$

$$C = L_2 C - (1-\epsilon)\,(L_3/L_2)\,\sigma\,(t_N t_{C_2}) \tag{11}$$

$$P = Concat\,(A, B, C) \tag{12}$$

In Eqs. 9, 10, 11, 12, σ is the sigmoid function of the nonlinear transformation. L_1, L_2 and L_3 are hyperparameters used to balance the contributions of the three graph encodings (A,B,C) in DGNN. After T iterations, the final A, B, C are concatenated to construct a comprehensive graph encoding P.

3 Experiments and Analysis

This section outlines the experimental setup and results aimed at enhancing device placement performance.

3.1 Experimental Setup

We conducted experiments using Placeto's simulator [1], keeping settings consistent while varying graph embedding and RL components. All tests were performed on a server with 8 T P100 GPUs (12GB each), an Intel(R) Xeon(R) E5-2650 CPU, and 128GB of RAM. We evaluated model performance using execution time, which measures the duration for training a neural network after applying a device placement policy, and computation time, which assesses the time needed to find the optimal policy. Three synthetic datasets were utilized: ptb, cifar10, and nmt, with graph sizes averaging 190, 300, and 500 nodes, respectively, achieved through node fusion [7]. Experiments varied the number of devices (three, five, and eight) across 17 randomly selected graphs per dataset, each consisting of 20 training epochs, adhering to the P-GNN protocol [8].

Table 1. The execution time improvements of various models on different numbers of GPUs across three distinct datasets, compared to Placeto [1].

Methods	Datasets	3 GPUs		5 GPUs		8 GPUs	
		Exe_time(s)	Impro(%)	Exe_time(s)	Impro(%)	Exe_time(s)	Impro(%)
Placeto	ptb	5.6592	-	5.5747	-	5.5134	-
GraphSAGE		5.4236	4.16%	5.3365	4.27%	5.3529	2.91%
P-GNN		5.0996	9.89%	4.9096	11.93%	4.7946	13.04%
CP-GNNAK		5.0184	11.32%	4.8480	13.04%	**4.7237**	**14.32%**
PPO-DGMA		**5.0175**	**11.34%**	**4.8355**	**13.26%**	4.7395	14.04%
Placeto	cifar10	2.1061	-	2.1649	-	1.9609	-
GraphSAGE		1.9349	7.70%	1.8109	16.35%	1.7763	9.46%
P-GNN		1.7682	16.04%	1.6525	23.67%	1.5991	18.45%
CP-GNNAK		1.7443	17.18%	1.6552	23.54%	1.5970	18.56%
PPO-DGMA		**1.6424**	**22.02%**	**1.5305**	**29.30%**	**1.5115**	**22.92%**
Placeto	nmt	2.3136	-	2.0969	-	2.1234	-
GraphSAGE		2.0981	9.31%	2.0392	2.75%	2.0071	5.48%
P-GNN		2.0518	11.32%	1.9869	5.24%	1.9682	7.31%
CP-GNNAK		2.0437	11.67%	1.9719	5.96%	1.9466	8.32%
PPO-DGMA		**2.0016**	**13.49%**	**1.8869**	**10.02%**	**1.8511**	**12.82%**

3.2 Evaluating the Superiority of the Proposed PPO-DGMA Frameworks

To validate the superiority of our proposed PPO-DGMA method, we compared it with classic device placement methods, including Placeto [1], GraphSAGE [8], P-GNN [8], and CP-GNNAK [4]. We utilized execution time and computation time to assess the performance of different models under three different GPU counts. The results are presented in Tables 1 and 2.

Execution Time. In Table 1, we evaluate the performance of various models using the Placeto benchmark [1] across three datasets: ptb, cifar10, and nmt. With three GPUs, GraphSAGE saw execution time reductions of 4.16%, 7.70%, and 9.31%; P-GNN achieved 9.89%, 16.04%, and 11.32%; and CP-GNNAK improved by 11.32%, 17.18%, and 11.67%. Our proposed PPO-DGMA outperformed all, with reductions of 11.34%, 22.02%, and 13.49%. With five GPUs, GraphSAGE's reductions were 4.27%, 16.35%, and 2.75%, while P-GNN showed 11.93%, 23.67%, and 5.24%. CP-GNNAK improved by 13.04%, 23.54%, and 5.96%. PPO-DGMA excelled with 13.26%, 29.30%, and 10.02%. With eight GPUs, GraphSAGE reduced by 2.91%, 9.46%, and 5.48%; P-GNN dropped by 13.04%, 18.45%, and 7.31%; and CP-GNNAK performed better with reductions of 14.32%, 18.56%, and 8.32%. Although PPO-DGMA slightly trailed CP-GNNAK on ptb, it excelled on cifar10 and nmt, improving by 22.92% and 12.82%. These results highlight PPO-DGMA's consistent superiority in execution times across different GPU setups and datasets.

Computation Time. The results in Table 2 highlight the efficiency of the PPO-DGMA model across various datasets. For the ptb dataset, using 3 GPUs, Placeto requires 430.89 s, while CP-GNNAK achieves a 2.29× speedup, and

PPO-DGMA a remarkable 50.59×. With 5 GPUs, Placeto's time extends to 459.07 s, where CP-GNNAK realizes a 2.17× speedup, and PPO-DGMA achieves 54.61×. At 8 GPUs, Placeto's computation time is 437.43 s; CP-GNNAK shows a 2.44× reduction, while PPO-DGMA achieves 46.85×. For the cifar10 dataset, Placeto takes 122.70 s with 3 GPUs. CP-GNNAK reduces time by 3.05×, and PPO-DGMA achieves 25.77×. With 5 GPUs, Placeto's time is 117.92 s, with CP-GNNAK showing 2.91× reduction and PPO-DGMA 24.66×. At 8 GPUs, Placeto's time is 119.07 s, with CP-GNNAK maintaining a 2.98× reduction and PPO-DGMA 25.30×. For the nmt dataset, with 3 GPUs, Placeto takes 64.69 s, while CP-GNNAK achieves 2.44× speedup, and PPO-DGMA 21.72×. At 5 GPUs, Placeto's time is 63.50 s, with CP-GNNAK at 2.64× and PPO-DGMA at 21.47×. With 8 GPUs, Placeto's time remains stable at 63.62 s; CP-GNNAK shows a 2.06× reduction, and PPO-DGMA achieves 21.77×.

Table 2. The computation time improvements of various models on three different numbers of GPUs across three distinct datasets.

Methods	Datasets	3 GPUs		5 GPUs		8 GPUs	
		Comp_time(s)	Speedup	Comp_time(s)	Speedup	Comp_time(s)	Speedup
Placeto	ptb	430.89	51.59 ×	459.07	55.61 ×	437.43	47.85 ×
GraphSAGE		478.88	57.33 ×	495.60	60.04×	553.91	60.60 ×
P-GNN		433.76	51.93 ×	448.31	54.31 ×	451.56	49.40 ×
CP-GNNAK		130.97	15.68 ×	144.72	17.53 ×	127.10	13.90 ×
PPO-DGMA		8.26	–	8.26	–	9.14	–
Placeto	cifar10	122.70	26.77 ×	117.92	25.66 ×	119.07	26.30×
GraphSAGE		137.07	29.90 ×	128.70	28.00 ×	136.17	30.07 ×
P-GNN		93.94	20.49 ×	85.82	18.67 ×	90.40	19.97 ×
CP-GNNAK		30.28	6.61 ×	30.17	6.56 ×	29.94	6.61 ×
PPO-DGMA		4.58	–	4.60	–	4.53	–
Placeto	nmt	64.69	22.72 ×	63.50	22.47 ×	63.62	22.77 ×
GraphSAGE		72.16	25.35×	73.07	25.86 ×	73.87	26.43 ×
P-GNN		68.15	23.94 ×	67.10	23.75 ×	79.88	28.58 ×
CP-GNNAK		18.79	6.60 ×	17.44	6.16×	20.80	7.46 ×
PPO-DGMA		2.85	–	2.83	–	2.79	–

3.3 Evaluating the Effectiveness of DGMA Graph Embedding Methods

To enhance the graph embedding module, we developed several methods, including GMA, DGNN, DGM, and the dual-branch approaches DGM-DGM and DGMA, ultimately selecting DGMA. We validated these methods through extensive ablation experiments, as shown in Fig. 3. Notably, the reinforcement learning approach aligns with Placeto. Results indicate that DGMA's dual-branch structure achieves superior execution times, highlighting its effective parallel strategy.

Compared to GMA, DGNN, and DGM, DGMA outperforms individual methods, and the GMA branch in DGMA offers richer computational features than the DGM branch.

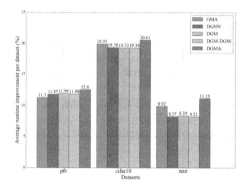

Fig. 3. The average execution time improvements with different graph embedings across three distinct datasets.

3.4 Evaluating the Effectiveness of PPO Methods

To validate our reinforcement learning method for device placement, we implemented a dual-branch graph encoding module. The reinforcement learning in DGMA aligns with Placeto, while PPO-DGMA uses our designed approach. Results in Table 3 and show that PPO-DGMA consistently outperforms DGMA in execution time across various GPU counts and datasets, confirming the effectiveness of our method.

Table 3. The execution time improvements of various models on different numbers of GPUs across three distinct datasets, compared to Placeto [1].

Methods	Datasets	3 GPUs		5 GPUs		8 GPUs	
		Exe_time(s)	Impro(%)	Exe_time(s)	Impro(%)	Exe_time(s)	Impro(%)
DGMA	ptb	5.0563	10.65%	4.9024	12.06%	4.7859	13.20%
PPO-DGMA		**5.0175**	**11.34%**	**4.8355**	**13.26%**	**4.7395**	**14.04%**
DGMA	cifar10	1.7426	17.26%	1.6307	24.68%	1.5707	19.90%
PPO-DGMA		**1.6424**	**22.02%**	**1.5305**	**29.30%**	**1.5115**	**22.92%**
DGMA	nmt	2.0286	12.32%	1.9110	8.87%	1.8610	12.36%
PPO-DGMA		**2.0016**	**13.49%**	**1.8869**	**10.02%**	**1.8511**	**12.82%**

4 Conclusions

This paper proposes a novel device placement method, termed PPO-DGMA, which amalgamates the principles of reinforcement learning with advanced graph embedding techniques. This synergistic integration ensures that PPO-DGMA not only enhances placement efficiency but also provides a better context aware framework for resource distribution, ultimately pushing the boundaries of current device placement methodologies. Its dual-branch graph encoding module extracts robust graph representations through feature fusion, with the DGM branch focusing on topological and semantic embedding, while the GMA branch enhances long-range dependencies via a node selection mechanism. The reinforcement learning module includes a loss layer for fine grained environmental adjustments, allowing for efficient learning of strategies. Our method significantly outperforms traditional device placement models, reducing average execution time by 16.58% and accelerating computation time by 32.53×. Extensive ablation studies confirm its robustness and generalization.

Acknowledgement. This work was supported by the National Key Research and Development Program of China under Grant 2023ZD0120600; the National Natural Science Foundation of China (NSFC) under Grant No.62302133; the Key Research and Development Program of Zhejiang Province under Grant 2024C01104, 2024C01026; the Natural Science Foundation of Zhejiang Province under Grant No.LQ23F020015.

References

1. Addanki, R., Venkatakrishnan, S.B., Gupta, S., Mao, H., Alizadeh, M.: Placeto: learning generalizable device placement algorithms for distributed machine learning. In: Proceedings of the 33rd International Conference on Neural Information Processing Systems, pp. 3981–3991 (2019)
2. Gao, Y., Chen, L., Li, B.: Post: Device placement with cross-entropy minimization and proximal policy optimization. In: Advances in Neural Information Processing Systems, vol. 31 (2018)
3. Gao, Y., Chen, L., Li, B.: Spotlight: Optimizing device placement for training deep neural networks. In: International Conference on Machine Learning, pp. 1676–1684. PMLR (2018)
4. Han, M., Zeng, Y., Zhang, J., Ren, Y., Xue, M., Zhou, M.: A novel device placement approach based on position-aware subgraph neural networks. Neurocomputing **582**, 127501 (2024)
5. Lan, H., Chen, L., Li, B.: Accelerated device placement optimization with contrastive learning. In: Proceedings of the 50th International Conference on Parallel Processing, pp. 1–10 (2021)
6. Mirhoseini, A., Goldie, A., Pham, H., Steiner, B., Le, Q.V., Dean, J.: A hierarchical model for device placement. In: International Conference on Learning Representations (2018)
7. Mirhoseini, A., et al.: Device placement optimization with reinforcement learning. In: International Conference on Machine Learning, pp. 2430–2439. PMLR (2017)

8. Mitropolitsky, M., Abbas, Z., Payberah, A.H.: Graph representation matters in device placement. In: Proceedings of the Workshop on Distributed Infrastructures for Deep Learning, pp. 1–6 (2020)
9. Paliwal, A., et al.: Reinforced genetic algorithm learning for optimizing computation graphs. In: International Conference on Learning Representations (2019)
10. Pham, H., Guan, M., Zoph, B., Le, Q., Dean, J.: Efficient neural architecture search via parameters sharing. In: International Conference on Machine Learning, pp. 4095–4104. PMLR (2018)
11. Schulman, J., Wolski, F., Dhariwal, P., Radford, A., Klimov, O.: Proximal policy optimization algorithms. arXiv preprint arXiv:1707.06347 (2017)
12. Sergeev, A., Del Balso, M.: Horovod: fast and easy distributed deep learning in tensorflow. arXiv preprint arXiv:1802.05799 (2018)
13. Shoeybi, M., Patwary, M., Puri, R., LeGresley, P., Casper, J., Catanzaro, B.: Megatron-lm: training multi-billion parameter language models using model parallelism. arXiv preprint arXiv:1909.08053 (2019)
14. Wang, C., Tsepa, O., Ma, J., Wang, B.: Graph-mamba: Towards long-range graph sequence modeling with selective state spaces. arXiv preprint arXiv:2402.00789 (2024)
15. Wang, J., et al.: Dgnn: decoupled graph neural networks with structural consistency between attribute and graph embedding representations. arXiv preprint arXiv:2401.15584 (2024)
16. Williams, R.J.: Simple statistical gradient-following algorithms for connectionist reinforcement learning. Mach. Learn. **8**, 229–256 (1992)
17. Wu, Y., et al.: Google's neural machine translation system: bridging the gap between human and machine translation. arXiv preprint arXiv:1609.08144 (2016)
18. Zeng, Y., Wu, J., Zhang, J., Ren, Y., Zhang, Y.: Trinity: neural network adaptive distributed parallel training method based on reinforcement learning. Algorithms **15**(4), 108 (2022)
19. Zeng, Y., et al.: Aware: adaptive distributed training with computation, communication and position awareness for deep learning model. In: 2022 IEEE 24th International Conference on High Performance Computing & Communications; 8th International Conference on Data Science & Systems; 20th International Conference on Smart City; 8th International Conference on Dependability in Sensor, Cloud & Big Data Systems & Application (HPCC/DSS/SmartCity/DependSys), pp. 1299–1306. IEEE (2022)
20. Zhou, Y., et al.: Gdp: generalized device placement for dataflow graphs. arXiv preprint arXiv:1910.01578 (2019)

Cross-Generational Contrastive Continual Learning for 3D Point Cloud Semantic Segmentation

Yuan He[1,2], Guyue Hu[5,6(✉)], and Shan Yu[1,3,4]

[1] Laboratory of Brain Atlas and Brain-inspired Intelligence, Institute of Automation, Chinese Academy of Sciences, Beijing, China
heyuan2017@ia.ac.cn, shan.yu@nlpr.ia.ac.cn
[2] School of Artificial Intelligence, University of Chinese Academy of Sciences, Beijing, China
[3] Key Laboratory of Brain Cognition and Brain-inspired Intelligence Technology, Chinese Academy of Sciences, Beijing, China
[4] School of Future Technology, University of Chinese Academy of Sciences, Beijing, China
[5] Anhui Provincial Key Laboratory of Security Artificial Intelligence, Anhui University, Hefei, China
guyue.hu@ahu.edu.cn
[6] School of Artificial Intelligence, Anhui University, Hefei, China

Abstract. Recently, point cloud semantic segmentation technology has made significant progress, boosting the development of autonomous driving, robotic navigation, and urban modeling. However, most current approaches rely on training data of all categories at once. This limitation makes it difficult for models to adapt to dynamic environments, leading to repetitive retraining and high computational costs. To enable continuous learning of new categories by leveraging previous knowledge, and inspired by the brain's ability to learn new knowledge through comparison and association, we introduce a cross-generational contrastive continual learning approach for 3D point cloud semantic segmentation. To mitigate catastrophic forgetting, we contrast representations of old classes and new classes across different generations of encoders. Further, we propose a refined labels guided contrastive loss, which comprehensively accounts for the semantic dependencies between points and leverages previous knowledge. Additionally, we propose a refined label estimation strategy to boost the confidence of all classes while retaining previous knowledge. Extensive experiments on two public 3D point cloud semantic segmentation benchmarks demonstrate the effectiveness of our proposed approach.

Keywords: 3D Point cloud · Continual Learning · Semantic Segmentation

1 Introduction

3D point cloud semantic segmentation [13,23,25] is essential for enabling machines to understand and interact with complex real-world environments, with applications in autonomous driving, robotic navigation, and other fields. Nevertheless, practical deployment demands that models continually learn and adapt to new data while retaining previously acquired knowledge. This presents substantial challenges in enabling point cloud segmentation models to adapt to evolving data and environments.

Currently, while continual learning has been extensively studied in 2D image domain [10,22], its application in 3D point cloud semantic segmentation [30] remains relatively underexplored. In continual learning, the training process is divided into several steps, each handling a set of unseen classes. The most prominent issues are catastrophic forgetting, where learning new classes degrades the performance of previously learned ones, and semantic shift, where representations for old categories lose consistency over time. Moreover, due to the unordered nature of 3D point cloud data and the complexity of scenes, these challenges demand more robust approaches for continual semantic segmentation in point cloud applications. Existing methods in 3D continual learning primarily focus on preserving old knowledge through distillation or replay mechanisms. Most of them are focused on point cloud object classification [9,20,31], with very few addressing 3D point cloud semantic segmentation [30]. Furthermore, [30] primarily focuses on preserving old knowledge but fails to fully capture the relationships between points in the current scene.

To tackle these challenges, we draw inspiration from the brain's natural learning mechanisms. The human brain has a remarkable ability to learn and adapt continuously, by comparing new information with existing knowledge and seamlessly integrating it [8]. For example, when learning something new, the brain relates new knowledge to previously known information and integrates this knowledge within the framework of prior experiences, using past experiences as a scaffold to create connections and draw comparisons. This process accelerates learning and enhances the brain's ability to learn continuously. Inspired by this mechanism, we introduce a supervised contrastive learning strategy into continual learning, which not only integrates new classes into the old model but also enables comparisons between new and old classes, ensuring consistent semantic representations and alleviating catastrophic forgetting.

In this paper, we propose a cross-generational contrastive continual learning (CGC) approach for 3D point cloud semantic segmentation. The CGC framework effectively minimizes catastrophic forgetting and maintains consistent semantic representation across different generations of encoders by leveraging contrastive learning principles, which cluster points of the same category together and push apart points from different categories. Additionally, we propose a refined label-guided contrastive loss for training the model, which helps extract knowledge from previous tasks and leverages the relationships between points to help the model learn more robust representations. To further improve performance, we propose a refined label estimation strategy to generate high-confidence pseudo labels of old classes. Our contributions can be summarized as follows:

- We introduce a cross-generational contrastive continual learning framework for 3D point cloud semantic segmentation to mitigate catastrophic forgetting.
- We introduce a refined labels guided contrastive loss for our learning framework which fully considers the semantic dependencies between points and uses previous knowledge to tackle the semantic shift.
- We propose a refined label estimation strategy to address semantic shift, enhancing the confidence of all category labels (old and new classes) in the current step.
- We demonstrate our approach via extensive experiments and analysis on two publicly available benchmarks, S3DIS [3] and ScanNetv2 [7].

2 Related Works

Continual Learning [1,14,16,28] is characterized as learning from dynamic data distributions, aiming to update the model gradually while receiving new data without forgetting previously learned knowledge. In practice, continuous learning usually faces the problem of forgetting in multi-step learning. Many studies focus on the issue of catastrophic forgetting and can be categorized into the following types: 1) Regularization-based approaches balance the old and new model by adding explicit regularization terms containing weight regularization [16,27] and function regularization [1,24]. 2) Replay-based approaches [4,6,26] help models retain previous knowledge by storing and replaying old data. 3) Optimization-based approaches [17] explicitly design and manipulate the optimization programs to retain the old knowledge. 4) Representation-based approaches exploit the strengths of representations for continual learning, such as self-supervised learning [6,21].

Contrastive Learning aims to learn useful representations by maximizing the similarity between features of similar samples and minimizing the similarity between different samples. This approach has demonstrated strong performance in various domains such as 2D images [12,15] and 3D point cloud analysis [2,18]. Additionally, research has demonstrated that supervised learning can also take advantage of contrastive representation learning by expanding the definition of positive samples using labels [15]. In this work, we apply supervised contrastive representation learning within a continual learning framework.

Continual Semantic Segmentation aims at pixel-level prediction of classes for 2D images or point-level predictions for 3D point clouds, where the old and new classes may appear simultaneously. However, only the new classes are labeled, while the old classes are treated as background. Existing research for continual semantic segmentation is mostly on 2D images. For example, MiB [5] proposed a regular cross-entropy loss and a knowledge distillation to preserve the old knowledge. PLOP [10] and RECALL [22] used pseudo labels generated by the old model to reduce forgetting of old classes. However, research on continual semantic segmentation for point clouds is still quite limited. [30] using pseudo labels and geometry distillation of points to address catastrophic forgetting. However, the relations between points with categories haven't been fully considered. In this work, we introduce a supervised contrastive learning scheme into the continual learning framework, which will help the encoder extract more robust representations. Further, a refined labels guided contrastive loss is proposed to assist in the learning of this continuous learning framework.

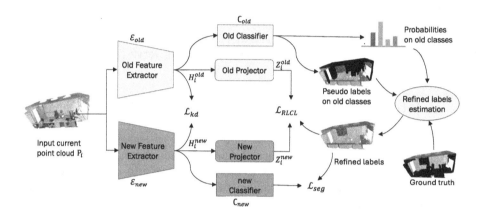

Fig. 1. The overall framework of CGC in the current learning step. The grey blocks denote the network trained on old tasks and these blocks are frozen, while the blue modules correspond to the network trained in the current learning step. (Color figure online)

3 Method

3.1 Overview

Let $\mathcal{D}$ be a point cloud dataset containing a set of (P, Y). Let $P \in \mathcal{R}^{N \times (3+F)}$ represents the input 3D point cloud with xyz coordinates and F-dimensional features where N is the number of points, and $Y \in \mathcal{R}^N$ represents the corresponding point-level labels. In the training paradigm of continual learning, the old classes C_{old} and new classes C_{new} are two disjoint class sets and the model is trained at each learning step using only the current category information. Importantly, the network only has access to the labels of the current category set, considering both previous classes and unseen classes as background. After each training step, the model is evaluated on both previous and current classes, with the expectation that progress will be made on the new classes while maintaining stable performance on the old ones.

Figure 1 shows the overall architecture of our framework CGC, which consists of two parallel streams. The two streams represent the backbone for the current and previous learning steps respectively. Notably, the old feature extractor $\mathcal{E}_{old}$ and the old classifier $\mathcal{C}_{old}$ are frozen in the current learning step. At the current learning step, a batch of point clouds $\{P_i\}(i = 1, 2, ..., B)$ are passed through the old feature extractor $\mathcal{E}_{old}$ and new feature extractor $\mathcal{E}_{new}$ separately to generate their representations $\{h_i^{old}\}$ and $\{h_i^{new}\}$. On the one hand, these representations are fed to the old classifier $\mathcal{C}_{old}$ to produce probability distributions of old classes and new classifier $\mathcal{C}_{new}$ to predict new classes. On the other hand, the representations $\{h_i^q\}$ and $\{h_i^k\}$ are also projected to a metric space with a linear projector $\mathcal{H}_{old}$ and the corresponding momentum projector $\mathcal{H}_{new}$, respectively. The projected and normalized representations $\{z_i^{old}\}$ and $\{z_i^{new}\}$ are used for the following contrastive learning.

3.2 Cross-Generational Contrastive Continual Learning

General and transferable knowledge can effectively help mitigate forgetting for continual learning. [11,15] indicate that knowledge obtained through supervised contrastive learning shows enhanced robustness and transferability. we introduce Cross-generational contrastive continual learning (CGC) to integrate the supervised contrastive learning mechanism into continual learning.

Supervised Contrastive Learning (SCL). SCL aims to push the representation of samples with different classes farther while tightly clustering the representation of samples with the same class. The SCL loss takes the following form:

$$\mathcal{L}_{\text{SCL}} = -\sum_{i \in B} \frac{1}{|A_i|} \sum_{a \in A_i} log \frac{exp(z_i \cdot z_a / \tau)}{\sum_{m \in M_i} exp(z_i \cdot z_m / \tau)} \quad (1)$$

where i is the index of sample in a batch. $\mathcal{M}_i = \{m\}$ represents the contrastive samples set of anchor i. $\mathcal{A}_i = \{a | y_i = y_a\}$ represents the positive set of anchor i. The symbol $\cdot$ denotes dot product, $\tau \in \mathcal{R}^+$ is a temperature hyper-parameter.

CGC. To maximize the benefits of SCL in continual learning, CGC further introduces the Refined Labels Estimation strategy (RLE) and Refined Labels Guided Contrastive Loss (RLCL) to assist CGC's learning, effectively integrating SCL into continual semantic segmentation. On this basis, CGC not only learns more robust and transferable representations with RLCL but also preserves old knowledge more effectively using RLE.

3.3 Refined Labels Estimation

To preserve previous knowledge, we fully exploit implicit information from the old model by using its generated point-level features and pseudo labels. Notably, the old model can only predict the previous categories. The pseudo labels $\hat{Y}_i^{old}$ and label confidences $\hat{S}_i^{old}$ of old classes are obtained by

$$\hat{Y}_{i,n}^{old} = argmax(\mathcal{C}_{old}(\mathcal{E}_{old}(P_{i,n}))), \quad S_{i,n}^{old} = max(softmax(\mathcal{C}_{old}(\mathcal{E}_{old}(P_{i,n})))) \quad (2)$$

where the argmax function converts a classification vector to a classification index and n denotes the index of point in a point cloud sample. However, we can access the real label of the current category, and then mix the groundtruth of new classes and pseudo labeles with label confidences to obtain the refined labels

$$\widetilde{Y}_i = \begin{cases} Y_{i,n}^{new}, & \text{if } Y_{i,n}^{new} \neq 0 \\ \mathbb{1}(S_{i,n}^{old} \geq \gamma) \cdot \hat{Y}_{i,n}^{old}, & \text{otherwise} \end{cases} \quad (3)$$

where γ is a confidence threshold. Then we use these labels for supervised segmentation for new model by segmentation loss given by

$$\mathcal{L}_{\text{seg}} = -\sum_{i \in B} \sum_{c \in \{c_{old} \cup c_{new}\}} \widetilde{Y}_i log(\mathcal{C}_{new}(\mathcal{E}_{new}(P_{i,c}))) \quad (4)$$

3.4 Refined Labels Guided Contrastive Loss

While striving to preserve previous knowledge, we also fully leverage the relationships between points to enhance the model's representation ability for the current classes. To prevent the model from overfitting past tasks, we train the network with only the points of the current class as anchors. Moreover, we use the representations $z_{i,n}^{new}$ generated by the new model as anchor features. Both the point-level features generated by the old model and the new model provide positive and negative samples for the anchor. This approach not only enhances the model's representation capability for the current classes but also helps retain previous knowledge. The loss function $\mathcal{L}_{\text{SCL}}$ used in SCL is primarily designed for classification tasks. For semantic segmentation, we improve $\mathcal{L}_{\text{SCL}}$ by incorporating the refined labels as follows:

$$\mathcal{L}_{\text{RLCL}} = -\sum_{i \in B} \sum_{n \in N} \frac{1}{|A_{i,n}|} \sum_{a \in A_{i,n}} \log \frac{exp(z_{i,n} \cdot z_{i,a}/\tau)}{\sum_{m \in M_{i,n}} exp(z_{i,n} \cdot z_{i,m}/\tau)} \quad (5)$$

where n is the index of points in the point cloud sample belonging to the current categories. $\mathcal{M}_{i,n} = \{m\}$ represents the contrastive samples set of anchor $\{P_{i,n}\}$ containing samples from old classes and current classes. $\mathcal{A}_{i,n} = \{a|y_{i,n} = y_{i,a}\}$ represents the positive samples.

However, the computation cost of the guided contrastive loss is highly correlated with the positive/negative point numbers. Due to the large number of points in point cloud scenes, we can't consider all points as positive and negative samples. Therefore, we must select a subset of samples to serve as positive and negative samples. Point cloud scenes also exhibit imbalanced category distribution. Hence, we can choose a fixed number of points from each category to serve as positive and negative samples. If a category contains fewer points than the fixed number, all of those points will be included.

3.5 Total Training Loss

During training, our CGC network is learned by minimizing the following loss on the training data:

$$\mathcal{L} = \mathcal{L}_{\text{seg}} + \lambda_1 \mathcal{L}_{\text{RLCL}} + \lambda_2 \mathcal{L}_{\text{kd}} \tag{6}$$

where $\mathcal{L}_{\text{kd}} = -\sum_{i \in B} \sum_{c \in c_{new}} \|\boldsymbol{h}_{i,c}^{new} - \boldsymbol{h}_{i,c}^{old}\|^2$ is a distillation loss, helping model prevent catastrophic forgetting. λ_1 and λ_2 are non-negative weights.

Table 1. Experimental comparisons of 3D continual segmentation methods on S3DIS dataset of S1 split. We apply the mIoU (%) as the evaluation metric.

Methods	$C_{new}=5$			$C_{new}=3$			$C_{new}=1$		
	0–7	8–12	all	0–9	10–12	all	0–11	12	all
BT	37.51	–	–	40.57	–	–	46.23	–	–
FT	11.13	50.24	26.67	17.21	55.68	26.19	23.75	5.61	22.37
EWC [16]	23.05	55.17	34.29	29.62	55.37	35.26	25.38	9.93	24.16
LWF [19]	32.64	55.27	41.82	36.89	56.03	41.38	32.24	18.41	31.26
GUC [30]	27.83	55.39	44.56	38.65	57.25	42.58	40.57	19.40	38.92
CGC(Ours)	38.46	57.14	45.23	38.84	58.36	44.10	42.09	24.17	40.23
Joint	38.72	60.35	47.26	44.23	59.87	47.14	48.13	43.08	47.52

Table 2. Experimental comparisons of 3D continual segmentation methods on ScanNetv2 dataset of S1 split. We apply the mIoU (%) as the evaluation metric.

Methods	$C_{new}=5$			$C_{new}=3$			$C_{new}=1$		
	0–14	15–19	all	0–16	17–19	all	0–18	19	all
BT	29.84	–	–	32.05	–	–	31.57	–	–
FT	5.47	35.43	13.17	5.34	42.67	11.35	4.83	8.25	5.17
EWC [16]	15.62	33.90	19.84	9.28	32.85	13.27	12.77	9.24	12.53
LWF [19]	23.71	37.95	27.43	22.93	41.88	25.69	19.74	14.28	19.41
GUC [30]	26.58	35.90	28.76	27.56	40.36	29.94	23.36	13.53	22.89
CGC(Ours)	27.87	37.84	30.15	28.98	41.22	31.06	26.57	20.41	26.19
Joint	31.65	39.28	33.47	31.42	41.27	33.51	33.86	32.58	33.72

4 Experiments

4.1 Experimental Setup

Datasets. Our experiments are performed on two main point cloud datasets: S3DIS [3] and ScanNetv2 [7]. S3DIS dataset has 13 semantic categories, containing 3D RGB point clouds of 6 indoor areas covering 272 rooms. We use the more challenging area 5 as validation and the other areas as training. ScanNetv2 dataset contains 1,513 scans in 707 indoor scenes. Each point is labeled as one of 21 classes (20 semantic classes and unannotated place). Following [30], we apply specific data processing and configuration for point cloud continual semantic segmentation. S1 introduces classes in alphabetical order.

Implementation Details. To facilitate comparison with these point cloud semantic segmentation methods in [30], we also choose DGCNN [29] as the feature extractor though our approach is compatible with most point cloud feature extractors. In the initial training phase, we utilize only the current classes to train the network using label guided contrastive loss and segmention loss. In subsequent learning steps, we then integrate the information of old classes into the network training.

Baselines. We make comparisons with continual semantic segmentation methods which are also compared in [30] and reproduce the results of these methods. Fine-tuning (FT) randomly initializes the new classifier and joins the base model to training. EWC [16] and LWF [19] from classical incremental learning models to 3D point cloud incremental segmentation setting by GUC [30]. GUC is the latest research in the field of continuous semantic segmentation of point clouds, and we will compare our work with it.

4.2 Experimental Results

Table 1 and 2 show the results on S3DIS and ScanNetv2 datasets on S1 split. In the table, "BT" and "FT" represent Base Training and Fine-Tuning respectively. "Joint" denotes Joint Training on all classes at once. At different incremental Settings, we report the mIoU on the old and the new classes after incremental learning. Compared to these baselines, our method outperforms the state-of-the-art with about 0.67–3.3%. From the table, we find that the recognition ability of the new model to the old categories is greatly reduced in "FT". EWC [16] and LWF [19], by introducing strategies for tackling catastrophic forgetting, have greatly reduced the forgetting of old categories compared to the FT. GUC [30] leverages the geometric relationships between points based on specific task characteristics to mitigate forgetting, achieving impressive results. Nevertheless, our method achieves the best performance on the two datasets, which clearly demonstrates the superiority of the proposed CGC.

Ablation Study. We then investigate the contributions of proposed modules in CGC. In Table 3, we obverse that adding RLE and RLCL modules boost the performance. The proposed RLE and RLCL are crucial for mitigating catastrophic forgetting and enhancing knowledge retention. In more detail, the results show that the RLE performs better than the KD, with an improvement on mIoU by 15.58% on S3DIS and by 11.59% on ScanNetv2, respectively. The RLCL further enhances the model's performance, outperforming the RLE by 1.29% on S3DIS and by 1.08% on ScanNetv2, respectively.

Table 3. Ablation study of the proposed method on the S3DIS and ScanNet.

KD	RLE	RLCL	S3DIS			ScanNetv2		
			0–7	8–12	all	0–14	15–19	all
✓			12.83	51.95	28.36	10.82	34.67	17.38
✓	✓		37.27	55.72	43.94	25.83	36.19	28.97
✓	✓	✓	38.46	57.14	45.23	27.87	37.84	30.15

5 Conclusion

In this paper, we propose a cross-generational continual contrastive learning framework (CGC) for 3D point cloud semantic segmentation. By leveraging the relationships between points and the implicit information within the old model, our approach captures the global semantic relationships between new and old features through a contrastive loss and a refined label estimation strategy, significantly mitigating catastrophic forgetting compared to previous techniques. Through extensive experiments, we validate the efficacy of the proposed refined label guided contrastive loss and refined label estimation, demonstrating the superior performance of our method.

Acknowledgments. This work was supported in part by the Natural Science Foundation of Anhui Province (No. 2408085QF201), and in part by the Open Project of Anhui Provincial Key Laboratory of Security Artificial Intelligence (No. SAI2024003).

References

1. Adel, T., Zhao, H., et al.: Continual learning with adaptive weights (claw). arXiv preprint arXiv:1911.09514 (2019)
2. Afham, M., Dissanayake, I., et al.: Crosspoint: self-supervised cross-modal contrastive learning for 3d point cloud understanding. In: Proceedings of the IEEE/CVF Conference on Computer Vision and Pattern Recognition, pp. 9902–9912 (2022)
3. Armeni, I., Sener, O., et al.: 3d semantic parsing of large-scale indoor spaces. In: Proceedings of the IEEE Conference on Computer Vision and Pattern Recognition, pp. 1534–1543 (2016)

4. Ashok, A., Joseph, K., et al.: Class-incremental learning with cross-space clustering and controlled transfer. In: European Conference on Computer Vision, pp. 105–122. Springer (2022)
5. Cermelli, F., Mancini, M., et al.: Modeling the background for incremental learning in semantic segmentation. In: Proceedings of the IEEE/CVF Conference on Computer Vision and Pattern Recognition, pp. 9233–9242 (2020)
6. Cha, H., Lee, J., et al.: Co2l: contrastive continual learning. In: Proceedings of the IEEE/CVF International Conference on Computer Vision, pp. 9516–9525 (2021)
7. Dai, A., Chang, A.X., et al.: Scannet: richly-annotated 3d reconstructions of indoor scenes. In: Proceedings of the IEEE Conference on Computer Vision and Pattern Recognition, pp. 5828–5839 (2017)
8. Ding, Y., Wang, Y., et al.: A simplified plasticity model based on synaptic tagging and capture theory: Simplified stc. Front. Comput. Neurosci. 798418 (2022)
9. Dong, J., Cong, Y., et al.: I3dol: incremental 3d object learning without catastrophic forgetting. In: Proceedings of the AAAI Conference on Artificial Intelligence, pp. 6066–6074 (2021)
10. Douillard, A., Chen, Y., et al.: Plop: learning without forgetting for continual semantic segmentation. In: Proceedings of the IEEE/CVF Conference on Computer Vision and Pattern Recognition, pp. 4040–4050 (2021)
11. Gunel, B., Du, J., et al.: Supervised contrastive learning for pre-trained language model fine-tuning. arXiv preprint arXiv:2011.01403 (2020)
12. He, K., Fan, H., et al.: Momentum contrast for unsupervised visual representation learning. In: Proceedings of the IEEE/CVF Conference on Computer Vision and Pattern Recognition, pp. 9729–9738 (2020)
13. He, Y., Hu, G., Yu, S.: Hard-soft pseudo labels guided semi-supervised learning for point cloud classification. IEEE Sig. Process. Lett. (2024)
14. Hu, G., Cui, B., Yu, S.: Joint learning in the spatio-temporal and frequency domains for skeleton-based action recognition. IEEE Trans. Multimedia 2207–2220 (2019)
15. Khosla, P., Teterwak, P., et al.: Supervised contrastive learning. Adv. Neural. Inf. Process. Syst. **33**, 18661–18673 (2020)
16. Kirkpatrick, J., Pascanu, R., et al.: Overcoming Catastrophic Forgetting in Neural Networks. In: Proceedings of the National Academy of Sciences, pp. 3521–3526 (2017)
17. Kong, Y., Liu, L., et al.: Balancing stability and plasticity through advanced null space in continual learning. In: European Conference on Computer Vision, pp. 219–236. Springer (2022)
18. Li, X., Chen, J., et al.: Tothepoint: efficient contrastive learning of 3d point clouds via recycling. In: Proceedings of the IEEE/CVF Conference on Computer Vision and Pattern Recognition, pp. 21781–21790 (2023)
19. Li, Z., Hoiem, D.: Learning without forgetting. IEEE Trans. Pattern Anal. Machi. Intell. 2935–2947 (2017)
20. Liu, Y., Cong, Y., et al.: L3doc: lifelong 3d object classification. IEEE Trans. Image Process. 7486–7498 (2021)
21. Madaan, D., Yoon, J., et al.: Representational continuity for unsupervised continual learning. arXiv preprint arXiv:2110.06976 (2021)
22. Maracani, A., Michieli, U., et al.: Recall: replay-based continual learning in semantic segmentation. In: Proceedings of the IEEE/CVF International Conference on Computer Vision, pp. 7026–7035 (2021)
23. Nguyen, A., Le, B.: 3d point cloud segmentation: a survey. In: 2013 6th IEEE conference on robotics, automation and mechatronics, pp. 225–230. IEEE (2013)

24. Nguyen, C.V., Li, Y., et al.: Variational continual learning. arXiv preprint arXiv:1710.10628 (2017)
25. Qi, C.R., Yi, L., et al.: Pointnet++: deep hierarchical feature learning on point sets in a metric space. Im: Advances in Neural Information Processing Systems (2017)
26. Rebuffi, S.A., Kolesnikov, A., et al.: icarl: incremental classifier and representation learning. In: Proceedings of the IEEE Conference on Computer Vision and Pattern Recognition, pp. 2001–2010 (2017)
27. Ritter, H., Botev, A., et al.: Online structured laplace approximations for overcoming catastrophic forgetting. In: Advances in Neural Information Processing Systems (2018)
28. Wang, L., Zhang, X., et al.: A comprehensive survey of continual learning: theory, method and application. IEEE Trans. Ppattern Anal. Mmach. intell (2024)
29. Wang, Y., Sun, Y., et al.: Dynamic graph cnn for learning on point clouds. ACM Trans. Graph. 1–12 (2019)
30. Yang, Y., Hayat, M., et al.: Geometry and uncertainty-aware 3d point cloud class-incremental semantic segmentation. In: Proceedings of the IEEE/CVF Conference on Computer Vision and Pattern Recognition, pp. 21759–21768 (2023)
31. Zamorski, M., Stypułkowski, M., et al: Continual learning on 3d point clouds with random compressed rehearsal. Comput. Vis. Image Understand. 103621 (2023)

TGAM-SR: A Sequential Recommendation Model for Long and Short-Term Interests Based on TCN-GRU and Attention Mechanism

Jiajing Zhang[1], Zhiya Shen[1], Jinlan Chen[2(✉)], and Qilang Li[3]

[1] School of Electronics and Information Engineering, Anhui Jianzhu University, Anhui Hefei, China
[2] School of Mechanical and Electrical Engineering, Anhui Jianzhu University, Anhui Hefei 230301, China
jlchen@163.com
[3] School of Mathematics and Physics, Anhui Jianzhu University, Anhui Hefei 230301, China

Abstract. Sequential recommendation can establish user behavior sequences based on the historical interaction records between users and items to dynamically model user preferences. In sequential recommendation, accurate modeling of both the long and short-term interests of users is the key to accurate recommendations. To address the issue of existing models' inability to effectively capture both long and short-term interests, this paper proposed a Temporal Convolutional Network and Gated Recurrent Unit-based Sequential Recommendation Model with Attention Mechanism (TGAM-SR) to address existing models' inability to effectively capture both long and short-term interests. Firstly, TGAM-SR uses a TCN and attention mechanism to model long-term interests, fully exploring user long-term interest information. At the same time, it uses a low-complexity GRU and multi-head self-attention mechanism to model short-term interests. Lastly, the attention mechanism calculates the weights of both long and short-term interests, and a gated fusion module is employed to integrate these interests. Experimental results show that compared to popular and newer sequential recommendation models, TGAM-SR exhibits superior performance in metrics such as hit rate and recall rate.

Keywords: sequential recommendation · temporal convolutional network · attention mechanism · long-term and short-term interest · gate recurrent unit

1 Introduction

In recent years, with the rapid development of the Internet and the exponential growth of information, recommendation systems have played an increasingly important role in fields such as news [1], e-commerce [2], and videos [3]. Most websites use item-based Collaborative Filtering (CF) models as their primary recommendation models [4]. To address issues such as sparse data in CF, user generated content methods are integrated with CF to expand the scope of information extraction [5]. However, recommendation systems based on content and collaborative filtering cannot accurately capture users' dynamic interests and the dependencies between different interactions.

Sequential recommendation models, through learning the temporal patterns of user behavior, can effectively capture changes in users' interests and provide increasingly precise personalized recommendations. Nevertheless, as time passes, users' interests may undergo alterations, resulting in notable disparities between current and earlier sessions. Researchers categorize user behavior into short-term and long-term interest actions. The interest exhibited in the current session might be influenced by users' longstanding habits and preferences. Therefore, disentangling short-term and long-term interests and modeling them individually can lead to a more accurate prediction of user behavior. In terms of integrating long and short-term interests, Ying [6] and Li [7] used a simple method of concatenating long and short-term interests to fuse long and short-term interest recommendations.

However, in practice, customer shopping needs are diverse and rich, and their long-term interests are also complex and diverse, directly concatenating long and short-term interests cannot fully reveal long-term interests. Therefore, Lv [8] proposed the SDM model, which uses a gate fusion module to merge long and short-term interest preference features. Nevertheless, while this model utilizes an attention mechanism for extracting long-term interest features, it falls short in reflecting the varying importance of long and short-term interest features. Additionally, the model employs a LSTM for short-term interest feature extraction, adding unnecessary complexity.

Addressing the above issues, this paper proposes a long and short-term interest sequential recommendation model, TGAM-SR, which incorporates a wide receptive field TCN in the long-term interest modeling to enhance the feature extraction capability of long-term interest, and replaces the LSTM in the short-term interest modeling with a lightweight GRU, and incorporates an attention mechanism in the fusion of long and short-term interests to improve the accuracy of recommendations and reduce the complexity of the model. In addition, in terms of long-term and short-term fusion, an attention mechanism is added to strengthen the fusion module's ability to allocate weights to long and short-term interests.

2 Related Work

Currently, research on session-based sequential recommendation can be categorized into three directions: traditional method-based sequential recommendation, deep learning-based sequential recommendation, and long-term and short-term interest-based sequential models.

2.1 Traditional Sequential Recommendation

Traditional methods for sequential recommendation primarily include Sequential Pattern Mining-based recommendation methods, Markov Chain-based recommendation methods, and Matrix Factorization-based recommendation methods. Yap [9] proposed a method based on Sequential Pattern Mining by learning the importance of knowledge of users' sequences. However, this method generated a lot of redundancy during Mining, leading to increased system time and space costs. To address the issues, Feng [10] embedded Markov Chains into Euclidean space and calculated the transition probabilities between items using Euclidean distance. The Matrix Factorization (MF) model [11] decomposes the user-item implicit feedback into user and item vectors, resolving the problem of significant differences in calculation results when increasing or decreasing a dimension in the algorithm proposed by Feng. However, these early models only consider low-order interactions of latent factors and have issues in handling long sequence dependencies.

2.2 Sequential Recommendation Based on Deep Learing

In response to the limitations of traditional models in handling long sequence dependencies, deep learning-based sequential models have been proposed. The DSSM model [12] utilizes deep neural networks and its unique dual-tower structure to embed users and items. However, DSSM has no sensitivity to contextual information and continuous variables. YouTubeDNN [13] uses deep neural networks to incorporate continuous and categorical variables into the model. Since YouTubeDNN uses only one vector to represent user interests, it cannot capture users' diverse interests. Li [14] proposed the MIND model, which uses multiple vectors to represent a user's various interests. As YouTubeDNN cannot model dynamic features, which are essential for sequential recommendation systems to capture the evolving interests of users over time, Hidasi [15] proposed GRU4Rec, which uses the RNN to model the dependencies between items in user behavior sequences dynamically. However, RNN's recurrent structure has inherent weaknesses in handling spatial and image related problems. Tang [16] proposed the Convolutional Sequence Embedding Recommendation Model, which treats the embedding matrix of N previously interacted items by a user as an "image" and uses horizontal and vertical convolutional kernels to capture sequential patterns. Methods based on RNN or CNN cannot capture the hierarchical nature of user interests within and across sessions, and CNN-based methods have high memory consumption. You [17] proposed HierTCN, which combines TCN with GRU. In the TCN, dilated causal convolutions are used to model long-term behavior, to extend the receptive field of behavioral features, and to construct longer adequate memory with fewer parameters. This approach combines the convolutional of CNN with the sequential modeling of RNN, making it more suitable for long-sequence modeling.

The models mentioned above, only focus on short-term or long-term user interests without distinguishing between them, making it difficult to accurately address the wide range of user interests and interaction behaviors in real-world scenarios.

2.3 Sequential Recommendation Model Based on Long and Short-Term Interests

Feng [18] proposed the long and short-term interests of the Multi-Neural Network Hybrid Dynamic Recommendation Model, which uses both RNN and Feedforward Neural Network (FNN) to model users' short-term interests and long-term preferences. Ji [19] proposed a multi-hop time-aware attention memory network that captures long-term preferences and short-term preferences. Although the above mentioned studies consider both long and short-term features, they simply concatenate the long and short-term interest features for prediction, without accurately describing the correlation between these features. Lv [8] proposed the SDM model, which uses a gate fusion module to combine long-term preferences with current needs, revealing information related to short-term interests in long-term interests more accurately. However, using only the gate fusion module cannot distinguish the degree of impact of long-term and short-term interests. In addition, the SDM model has issues with effectively capturing long-term interest from historical information features and with high model complexity and low training efficiency in extracting short-term interest features.

3 Problem Description and Symbolism

In this paper, U represents the set of users, and I represents the set of items. The objective is to predict whether user u will interact with item i at time t. The method for session segmentation is inspired by the SDM [8]. The specific method is as follows.

- The system first records the products with the same session ID.
- Adjacent interactions with a less than 10 min time difference are input into a session.
- The maximum length of a session is set to 50 (i.e. no more than 50 items). When the session length exceeds 50, a new session will be started.

The short-term behavior is represented by the recent session, namely $S^u = [i_1^u, ..., i_t^u, ..., i_m^u]$, where m is the sequence length. The long-term behavior is represented by the sessions within the past seven days, excluding the latest session, denoted as L^u. Feature extraction is carried out on the S^u and L^u to match the features of users and items to recommend products. $s_t^u \in R^{d \times 1}$ Represents the short-term behavior at time t, while $p^u \in R^{d \times 1}$ represents the long-term behavior up until time t. The fusion of s_t^u and p^u generates the complete user behavior at time t, denoted by $o_t^u \in R^{d \times 1}$. The score z_i is calculated using the user's behavior o_t^u at time t and the item $v_i \in R^{d \times 1}$, as shown in formula (1).

$$z_i = \text{score}(o_t^u, v_i) = o_t^{uT} v_i \tag{1}$$

The goal of the model is to predict the first N candidates for the user to click on item i_{t+1}^u at the next moment based on this score.

4 TGAM-SR Model

The frame diagram of the TGAM-SR model is presented in Fig. 1, which comprises three distinct parts. These include the long-term interest modeling component utilizing Attention Mechanism and TCN, the short-term interest modeling component leveraging GRU and Multi-head Self-attention Mechanism, as well as the fusion component of long-term and short-term interests utilizing Attention Mechanism and a gated fusion unit.

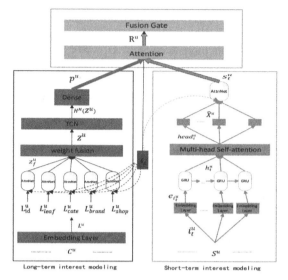

Fig. 1. Model diagram

4.1 Long-Term Interest Modeling Part

4.1.1 Embedding Layer

From a long-term perspective, the accumulated interests of users in different dimensions will have an impact on the current behavior. Therefore, the L^u obtained after the embedding of long-term behavior C^u comes from different feature scales. $L^u = \{L_f^u | f \in F\}$ consists of multiple subsets: L_{id}^u, L_{leaf}^u (leaf Category), L_{cate}^u (first level category), L_{shop}^u (shop), L_{brand}^u (brand). For each subset, different users typically show a variety of preferences. We use the user module to mine personalized information, such as the user's age, gender, etc., and embed the representation as $e_u \in R^{d \times 1}$.

4.1.2 Attention Mechanism Layer

Combining L^u with user feature e_u, the attention algorithm is used to calculate the score, equivalent to extracting the user's favourite store, brand and other information, and finally obtaining the representation vector. The calculation is shown in formula (2–3).

$$\alpha_k = \frac{\exp(g_k^{u^T} e_u)}{\sum_{k=1}^{|L_f^u|} \exp(g_k^{u^T} e_u)} \quad (2)$$

$$z_f^u = \sum_{k=1}^{|L_f^u|} \alpha_k g_k^u \quad (3)$$

where $g_k^u \in R^{d \times 1}$ represents the dense vector, and $g_k^{u^T}$ is the transpose vector of g_k^u.

The representation α_k of each feature is obtained by using the attention mechanism, after α_k is weighted and fused to get z_f^u, multiple features are spliced, and z^u is input for the attention mechanism layer, as shown in formula (4).

$$z^u = \text{concat}\left(\{z^u_f | f \in F\}\right) \tag{4}$$

4.1.3 TCN Layer

The spliced long-term feature z^u is sent into the TCN Layer. In the TCN, the extended causal convolution is used to model the long-term behavior sequence, change the size of the expansion coefficient, and enhance the sensitivity field with fewer parameters. The convolution kernel of a TCN is defined as $f = [W_0, ..., W_k, ..., W_{K-1}] \in \mathbf{R}^{d_z \times K}$. The time convolution of z^u in s time step s is shown in formula (5).

$$F(z^u) = \text{ReLU}\left(\sum_{j=0}^{K-1} w_j \odot z^u_{s-j \times d}\right) \tag{5}$$

where j is the size of the convolution kernel and $d \in [1,2,4,...,2^n]$ is the expansion coefficient of the TCN.

To alleviate gradient disappearance caused by the superposition of multi-convolutional layers, the residual structure is used. The m-th layer of residual is shown in formula (6).

$$H^m(z^u) = F^m(z^u) + z^u \tag{6}$$

$F^m(z^u)$ is the output of the m-th convolutional layer, and $H^m(z^u)$ is the output of the last layer of the convolutional layer.

4.1.4 Deep Neural Networks Layer

The output $H^M(z^u)$ of the last TCN layer is fed into the deep neural network, as formula (7).

$$p^u = \tanh(W^P z^u + b) \tag{7}$$

where W^P is the coefficient matrix, b is the bias, p^u is the eigenvector of long-term interest.

4.2 Short-Term Interest Modeling Part

4.2.1 Embedding Layer

After adding side information such as category, store, and brand to the last session $S^u = [i^u_1, ..., i^u_t, ..., i^u_m]$, the final embed of the short-term as $[e_{i^u_1}, ..., e_{i^u_t}]$ is obtained.

4.2.2 GRU Layer

We use GRU as circulating cells for recommendations, described in formula (8)–(11).

$$r^u_t = \sigma\left(W_r \cdot [h^u_{t-1}, e_{i^u_t}]\right) \tag{8}$$

$$z_t^u = \sigma(W_z \cdot [h_{t-1}^u, e_{i_t^u}]) \quad (9)$$

$$\widetilde{h}_t^u = \tanh(W_{\widetilde{h}} \cdot [r_t^u * h_{t-1}^u, e_{i_t^u}]) \quad (10)$$

$$h_t^u = (1 - z_t^u) * h_{t-1}^u + z_t^u * \widetilde{h}_t^u \quad (11)$$

where W_r, W_z, $W_{\widetilde{h}}$ respectively represents the weight matrix to be learned, r_t^u represents the reset gate, z_t^u represents the update gate.

GRU encodes the short-term interaction sequence of u as a output vector $h_t^u \in R^{d \times 1}$ at time t, h_t^u represents the sequence preference at t and is the output of the GRU layer.

4.2.3 Multi-head Self-attention Layer

Users often click on unrelated items randomly, resulting in the noise in short-term interests. Therefore, we mitigate the impact of irrelevant activities using self-attention. In addition, users typically have multiple points of interest within a session. We employ a multi-head attention mechanism to capture representations from different subspaces and positions.

Multi-head self-attention takes the hidden output aggregation h_t^u of multiple time points of the GRU as input, denoted as $X^u = [h_1^u, ..., h_t^u]$, as shown in formula (12–13).

$$head_i^u = \text{Attention}(W_i^Q X^u, W_i^K X^u, W_i^V X^u) = W_i^V X^u \text{softmax}(Q_i^{u^T} K_i^u) \quad (12)$$

$$\widehat{X}^u = \text{MultiHead}(X^u) = W^O \text{concat}(head_1^u, ...head_h^u) \quad (13)$$

where, $W^O \in R^{d \times hd_k}$ represents the weight of the output linear transformation, h is the number of heads, $d_k = \frac{1}{h}d$, $W_i^Q, W_i^K, W_i^V \in R^{d_k \times d}$ indicates the query, keyword, and value respectively, $Q_i^u = W_i^Q X^u$, $K_i^u = W_i^K X^u$, $\widehat{X}^u$ indicates the output of the GRU.

4.2.4 User Attention Mechanism Layer

Similar to the modeling of long-term, the output $\widehat{X}^u = [\widehat{h}_1^u, ..., \widehat{h}_t^u]$ of multi-head self-attention and the personal information e_u are taken as the input, the user's attention is added, weighted and summed to get the final vector s_t^u of short-term interest, as formula (14–15).

$$\alpha_k = \frac{\exp(\widehat{h}_k^{u^T} e_u)}{\sum_{k=1}^{t} \exp(\widehat{h}_k^{u^T} e_u)} \quad (14)$$

$$s_t^u = \sum_{k=1}^{t} \alpha_k \widehat{h}_k^u \quad (15)$$

where $\widehat{h}_k^u$ indicates the output at time k, e_u indicates the user's personal information.

4.3 Long-Term and Short-Term Interest Fusion Layer

SDM [8] and other models directly splice, add or gate fusion to integrate the two interests. However, only a small part of the behaviors in the long-term sequence is related to the current behaviors, direct fusion lead to an excessive proportion of long-term behaviors, it is difficult to find the behaviors associated to short-term behaviors in the long-term behaviors, and the prediction results are biased. Based on SDM, this paper adds the attention mechanism to set the weight of long and short-term user interests dynamically.

4.3.1 Attention Mechanism Layer

The attention mechanism formula is shown in (16-17).

$$\alpha_k = \frac{\exp(s_t^u p^u)}{\sum_{k=1}^{t} \exp(s_t^u p^u)} \quad (16)$$

$$R^u = \sum_{k=1}^{t} \alpha_k s_t^u \quad (17)$$

where, input s_t^u represents the final vector of short-term interest, p^u represents the final vector of long-term interest, and R^u represents the output vector of the attention mechanism.

4.3.2 Gate Fusion Layer

R^u Is input into a gating mechanism to determines the degree of contribution of short - and long-term interest at time t. Short-term and long-term weights are fused in formula (18–19).

$$G_t^u = \text{sigmoid}(W^1 e_u + W^2 R^u + b) \quad (18)$$

$$o_t^u = (1 - G_t^u) \odot p^u + G_t^u s_t^u \quad (19)$$

where, e_u represents user personalized information, s_t^u represents the short-term interest final vector, p^u represents the long-term interest final vector, and $\odot$ represents element multiplication. Finally, the predicted behavior output of user u is o_t^u.

5 Analysis of Experiment Results

To verify the effect of the TGAM-SR, the experimental are carried out on two data sets. The relevant recommendation model such as DSSM [12], YouTubebeDNN [13], MIND [14], SDM [8] are selected to compare with TGAM-SR.

5.1 Datasets

The following two datasets are selected for experimental verification, as shown in Table 1.

Table 1. ·Dataset information

Dataset	Users	Articles/Items	Average user clicks
MovieLens 1M	6040	3416	163.5
News recommendation dataset	20000	48383	29.3028

(1) MovieLens 1M dataset: This paper uses MovieLens 1M movie rating. A total of 6,000 users watched 4,000 movies with 1 million ratings.

(2) Recommendation System Competition news recommendation dataset: This data set is from the real data of the industry. There are 8 multi-gigabytes of data, a total of 3 files: user portrait, article portrait, click log, the number of users more than 1 million, more than 60 million clicks. Due to the large size of the data set, this paper samples the data set, selects a relatively regular data set, which has a total of more than 1 million clicks of 20,000 users.

5.2 Evaluation Metrics

To accurately evaluate the performance of the model, HitRate@K and Recall@K are used, where K is most interested list returned to the user in the final recommendation result.

5.3 Analysis of Experimental Results

Shown in Table 2 and 3, the TGAM-SR model achieves optimal performance in all indexes.

Table 2. The model comparison with respect to the Hit Rate across two datasets

Models	MovieLens 1M		News recommendation dataset	
	HitRate @50	HitRate @150	HitRate @50	HitRate @150
DSSM	27.21%	50.44%	0.046%	0.107%
Youtube DNN	27.98%	49.51%	0.048%	0.111%
MIND	26.12%	50.77%	0.045%	0.105%
SDM	42.52%	61.24%	0.075%	0.160%
TGAM-SR	43.74%	62.32%	**0.17%**	**0.46%**

5.3.1 Analysis of HitRate Results for the MovieLens 1M Dataset

Considering the index of HitRate@50, the hitrate of TGAM-SR model is 43.74%. Compared with the YoutubeDNN, our model exhibits hitrate improvements of 15.76%.

Table 3. The model comparison based on the Recall for the MovieLens1M dataset

Models	Recall@20	Recall@30
DSSM	15.99%	21.14%
YoutubeDNN	14.45%	19.48%
MIND	14.80%	19.66%
SDM	22.56%	30.89%
TGAM-SR	**24.53%**	**31.64%**

Against the DSSM, our model's hitrate increases by 16.53%. Relative to the MIND, our model's hitrate rises by 17.62%. Compared with the SDM model, which similarly considers both long-term and short-term interests, our model's hitrate increases by 1.22%.

Considering the index of HitRate@150, the hitrate of TGAM-SR is 62.32%. Against the DSSM, our model's hitrate rises by 11.88%. Relative to the YoutubeDNN, our model's hitrate increases by 12.81%. In comparison with the MIND, our model's hitrate improves by 11.55%. Finally, compared with the SDM model, our model's hitrate increases by 1.08%.

The TGAM-SR model is capable of simultaneously modeling long-term and short-term interests, while accounting for the differing levels of importance these interests have on the final recommendation outcome.

5.3.2 Analysis of HitRate Results for the News Recommendation Dataset

To present the improvement in model performance in a more intuitive manner, the following text utilizes relative increase (decrease) rates. Considering the index of HitRate@50, the hitrate of TGAM-SR is 0.17%. Against the DSSM model, our model's hitrate shows an 269% improvement. Relative to the YoutubeDNN, our model's hitrate demonstrates an 254% improvement. In comparison with the MIND, our model's hitrate exhibits an 278% improvement. Compared with the SDM, our model's hitrate indicates an 126% improvement.

Considering the index of HitRate@150, the hitrate of TGAM-SR is 0.46%. Relative to the baseline SDM, our model's hitrate represents an 187% improvement. Compared with DSSM model, YoutubeDNN model, and MIND model, our model's hitrate shows varying degrees of improvement. Overall, this model is insensitive to the data set, and its recommendation effect can be improved on multiple data sets.

News is more serious, and users may be relatively less interested, with fewer clicks and comments. They are also less to express their true thoughts when posting comments. Most users are more interested in movies, so there are not too many restrictions on commenting on movies. There are more commenting users, and the comments they post are relatively authentic. This can be seen from the average hit rate that movie clicks are relatively active. These situations may result in lower hit rates for various algorithms on news datasets. The TGAM-SR shows better performance improvement on news datasets, possibly due to lower hit rates in other datasets and relatively longer news comments.

The TGAM-SR enhances the extraction of long-term features. We chose the news dataset to demonstrate that the TGAM-SR not only performs better on datasets with higher hit rates, but also performs better on datasets with lower hit rates.

5.3.3 Analysis of Recall Results for the MoviLens1M Dataset

Considering the index of Recall@20, the recall rate of TGAM-SR is 24.53%. Against the DSSM, our model's recall rate shows an 8.54% improvement. Relative to the YoutubeDNN, our model's recall rate demonstrates a 10.08% improvement. In comparison with the MIND, our model's recall rate exhibits a 9.73% improvement. Compared with the SDM model, our model's recall rate indicates a 1.97% improvement.

Considering the index of Recall@30, the recall rate of TGAM-SR is 31.64%. Against the DSSM, our model's recall rate shows a 10.5% improvement. Relative to the YoutubeDNN, our model's recall rate demonstrates a 12.16% improvement. In comparison with the MIND, our model's recall rate exhibits an 11.98% improvement. Compared with the SDM, our model's recall rate indicates a 0.75% improvement.

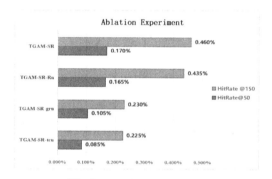

Fig. 2. Ablation Experiment

5.4 Ablation Experiment

To verify the validity of each module in TGAM-SR, ablation experiments are conducted on the news recommendation data set, and the results are shown in the Fig. 2. TGAM-SR-tcn is for TGAM-SR to remove the TCN part in the user's long-term interest model, and TGAM-SR-gru is for TGAM-SR to replace the GRU part in the user's short-term interest model with LSTM. TGAM-SR-Ru is a variant of TGAM-SR that removes the attention mechanism component responsible for fusing long-term and short-term interests.

Considering the index of HitRate@50, the hitrate of the TGAM-SR-tcn is 0.085%, and the hitrate is 0.17%, and the model result is reduced by 50%, indicating that the TCN module has a significant effect on capturing long history information in long-term interest. The hitrate of the TGAM-SR-gru is 0.105%, which is 38% lower than the hitrate of 0.17% without replacement, indicating that the extraction performance of GRU is better under the condition of shorter sequence. The hitrate of the TGAM-SR-Ru is

0.165%, which is 3% lower than the 0.17% hitrate without the fusion module, indicating that the attention mechanism can help the gating mechanism better assign weight to long and short-term interest. Considering the index of HitRate@150, The hitrate of the TGAM-SR-tcn, TGAM-SR-gru, TGAM-SR-Ru also decrease by a similar magnitude respectively.

6 Conclusions

This paper proposes the TGAM-SR based on multiple neural networks and attention mechanisms to capture a user's dynamic preferences by combining short-term and long-term interests. We used the TCN to enhance the ability of the attention mechanism to extract users' long-term interests, and used the GRU with relatively low complexity to capture short-term interest sequences. In the long and short-term interest fusion part, we added the attention mechanism to make the fusion result more accurate and focus on the critical information of the model as a whole. We tested data sets with different domains and different sparsity, and find that the TGAM-SR has better performance. Ablation experiments also show that the TCN, gated cycle unit and the attention mechanism of the long and short-term fusion part of the model have positive effects on the performance of the model. Due to the problem of time and effort, the work can also consider the use of better neural networks and user comments in the future to accurately model the long and short-term interests of users.

Acknowledgments. This work was supported in part by the National Key Research and Development Program Project with Grant No. 2023YFC3205705, the Provincial Key Project of Natural Science Research in Anhui Province with Grant No. 2022AH050252, the Provincial Quality Engineering Project of Anhui Province with Grant No. 2021xnfzxm019.

References

1. Ding, Y., et al.: Popularity prediction with semantic retrieval for news recommendation. Expert Syst. Appl. 247, 123308 (2024). https://doi.org/10.1016/j.eswa.2024.12330
2. Sahu, S., Satapathy, S.: Leveraging modified social group optimization for enhanced e-commerce recommendation systems. J. Sci. Ind. Res. **83**(3), 274–281 (2024)
3. Quan, Y., et al.: Hierarchical multi-modal attention network for time-sync comment video recommendation. ACM Trans. Inf. Syst. **42**(2), 44 (2024). https://doi.org/10.1145/3617826
4. Rendle, S.: Factorization machines. In: Proceedings of the 10th IEEE International conference on data mining. Danvers: Institute of Electrical and Electronics Engineers Inc. (2010). https://doi.org/10.1109/ICDM.2010.127
5. Zhang, J., et al.: Blending recommendation algorithm based on tensor decompositions and deep learning. J. Nanjing Univ. (Nat. Sci.) **55**(06), 952–959 (2019). https://doi.org/10.13232/j.cnki.jnju.2019.06.008
6. Ying, H., et al.: Sequential recommender system based on hierarchical attention networks. In: Proceedings of the 27th International Joint Conference on Artificial Intelligence, International Joint Conferences on Artificial Intelligence, Stockholm (2018)

7. Li, Z., et al.: Learning from history and present: next-item recommendation via discriminatively exploiting user behaviors. In: Proceedings of the 24th ACM SIGKDD International Conference on Knowledge Discovery and Data Mining. Association for Computing Machinery, New York (2018). https://doi.org/10.1145/3219819.3220014
8. Lv, F., et al.: SDM: Sequential deep matching model for online large-scale recommender system. In: Proceedings of the 28th ACM International Conference on Information and Knowledge Management, Association for Computing Machinery, New York (2018).https://doi.org/10.1145/3357384.3357818
9. Yap, G., Li, X., Philip, S.: Effective next-items recommendation via personalized sequential pattern mining. In: Proceedings of the17th International Conference on Database Systems for Advanced Applications. Springer, Busan (2012). https://doi.org/10.1007/978-3-642-29035-0_4
10. Feng, S., et al.: Personalized ranking metric embedding for next new poi recommendation. In: Proceedings of the 24th International Joint Conference on Artificial Intelligence. International Joint Conferences on Artificial Intelligence, Buenos Aires (2015)
11. Cen, Y., et al.: Controllable multi-interest framework for recommendation. In: Proceedings of the 26th ACM SIGKDD International Conference on Knowledge Discovery and Data Mining, Association for Computing Machinery, New York (2020). https://doi.org/10.1145/3394486.3403344
12. Huang, P., et al.: Learning deep structured semantic models for web search using click through data. In: Proceedings of the 22nd ACM international conference on Conference on information & knowledge management. Association for Computing Machinery, New York (2013). https://doi.org/10.1145/2505515.2505665
13. Covington, P., Adams, J., Sargin, E.: Deep neural networks for youtube recommendations. In: Proceedings of the 10th ACM Conference on Recommender Systems. Association for Computing Machinery, New York (2016). https://doi.org/10.1145/2959100.2959190
14. Li, C., et al.: Multi-interest network with dynamic routing for recommendation at tmall. In: Proceedings of the 28th ACM International Conference on Information and Knowledge Management. Association for Computing Machinery, New York (2019). https://doi.org/10.1145/3357384.3357814
15. Hidasi, B., Karatzoglou, A.: Recurrent neural networks with top-k gains for session-based recommendations. In: Proceedings of the 27th ACM International Conference on Information and Knowledge Management. Association for Computing Machinery, New York (2017).https://doi.org/10.1145/3269206.3271761
16. Tang, J., Ke, W.: Personalized top-N sequential recommendation via convolutional sequence embedding. In: Proceedings of the 11th ACM International Conference on Web Search and Data Mining. Association for Computing Machinery, New York (2018). https://doi.org/10.1145/3159652.3159656
17. You, J., et al.: Hierarchical temporal convolutional networks for dynamic recommender systems. In: 2019 World Wide Web Conference,Association for Computing Machinery, New York (2019)
18. Feng, Y., et al.: MN -HDRM: long short interest multiple neural network hybrid dynamic recommendation model. J. Comput. Sci. **42**(1), 16–28 (2019). https://doi.org/10.11897/SP.J.1016.2019.00016
19. Ji, W., et al.: Sequential recommender via time-aware attentive memory network. In: Proceedings of the 29th ACM International Conference on Information and Knowledge Management. Association for Computing Machinery, New York (2020). https://doi.org/10.1145/3340531.3411869

Investigating ChatGPT Translation Hallucination from an Embodied-Cognitive Translatology Perspective

Hui Jiao[1], Xinwei Li[2], Jonathan Ding[3], and Xiaojun Zhang[1](✉)

[1] Xi'an Jiaotong-Liverpool University, Suzhou, Jiangsu, China
Xiaojun.Zhang01@xjtlu.edu.cn
[2] Southeast University, Nanjing, Jiangsu, China
[3] Western Academy of Beijing, Beijing, China

Abstract. ChatGPT faces significant challenges in realising the dynamic construction of reality, cognition and language emphasised in Embodied-cognitive translatology. This study first examines the translation application of ChatGPT, defines hallucination phenomena, and reviews related research. It then explores the hallucination phenomena in ChatGPT's translation tasks, categorizing them into semantic, pragmatic, syntactic, and cultural hallucinations. To quantify these manifestations, we manually annotate ChatGPT's English-Chinese translations on the Flores-101 dataset, providing statistical insights into the frequency and severity of each type of hallucination. This analysis offers clarity on the prevalence of various hallucinations and suggests directions for improvement. Finally, this study summarises the current research on hallucination mitigation strategies in a targeted way, including the improvement of corpus quality, model architecture development, and prompt engineering techniques, offering novel insights for advancing machine translation technology towards greater intelligence.

Keywords: Embodied-cognitive translatology · ChatGPT · Translation · Hallucination

1 Introduction

ChatGPT, evolving from the GPT (Generative Pre-Trained Transformer) series, showcases improved domain adaptation and context sensitivity in translation tasks, surpassing Neural machine translation (NMT) systems. Starting with GPT-1, leveraging unsupervised learning [23], the series saw enhancements with GPT-2 through multi-task learning and increased model dimensions [24]. GPT-3 marked a significant leap in few-shot learning capabilities by integrating meta-learning with contextual learning [3]. The series culminated with GPT-3.5 and GPT-4, the latter being multimodal, both achieving, or exceeding, human-like linguistic task performance, particularly in translation [1].

Although ChatGPT has demonstrated its superior ability to perform direct multilingual translation, in practice, the so-called translation hallucination phenomenon may occur due to factors including errors during model training and data noise [26]. According to [20], hallucination is defined as the perception of entities or events that do not exist in reality. In general, hallucinations in Natural Language Generation (NLG) tasks can be classified into two main types: intrinsic and extrinsic hallucinations [19]. Intrinsic hallucinations misalign with the source content, while extrinsic ones produce content unverifiable from the source. In Machine Translation (MT), hallucinations are notably rare and hard to detect in clean data setups due to the task's closed nature [10]. [26] subdivided the hallucination phenomenon in MT into two main categories: hallucinations under perturbation and natural hallucinations. [11] further classified natural hallucinations into oscillatory hallucinations and largely fluent detached hallucinations, with the former referring to incomplete translations containing incorrectly repeated words and phrases, and the latter to translations that are completely unrelated to the source text.

However, there are several limitations to existing research on hallucination definitions and classifications. Firstly, these definitions and classifications tend to be specific to NLG tasks, often overlooking the nuances of translation tasks. Secondly, current classifications, such as the intrinsic and extrinsic dichotomy, may oversimplify the spectrum of translation hallucinations. These approaches also tend to emphasize the superficial agreement or discrepancy between the output and source content, neglecting the cognitive deviation between the machine-generated content and human expectations. Finally, in contrast to more flexible LLMs, traditional MT systems usually rely on predefined rules or statistical models that generate translations in a more controlled way and depend on the quality and coverage of the training data. Therefore, existing definitions and classifications of MT hallucinations are not sufficient to fully describe the translation hallucination phenomenon in LLMs.

In this paper, we aim to redefine and classify translation hallucinations in ChatGPT through the lens of Embodied-cognitive translatology. We will thoroughly analyze the linguistic causes behind these hallucinations and summarise mitigation strategies. This study provides insights for model improvement by identifying cognitive biases in LLM translations, optimizing training strategies and generative algorithms. Additionally, understanding the generation of translation hallucinations can guide more effective prompt design, enhancing translation accuracy.

2 Overview of Embodied-Cognitive Translatology

Embodied-cognitive translatology, emerging from the interdisciplinary integration of cognitive science and experiential philosophy, critiques traditional transformational generative grammar, highlighting the role of experience in language acquisition. [16] initially challenged existing linguistic theories, paving the way for [7] to outline cognitive linguistics, focusing on concepts like categorization, metaphor, and construction grammar, all grounded in the principle of embodied cognition. This principle posits that cognitive processes are deeply intertwined with bodily experiences and the physical world, influencing language generation and usage. [37] then fused this with Marxist materialism and humanism, creating Embodied Cognitive Linguistics, which argues that human cognition and language are shaped by physical interaction and experience. Extending this, [40] introduced Embodied-cognitive translatology, viewing translation as an interplay of embodiment and cognition, thus shifting the focus towards the translator's subjective experience and cognitive framework in reinterpreting and reconstructing the source text's meaning. This approach marks a significant shift in translation studies, emphasizing the translator's role in the translation process.

According to [37], the core principle of Embodied Cognitive Linguistics is "reality (embodiment)-cognition-language", and all languages are products of "embodiment" and "cognition". Translation work needs to be based on the comparison and transformation of "embodiment" and "cognition". Cognition is grounded in the real world, and language arises from bodily experiences of reality, shaped by cognitive processes. Thus, these elements are intertwined, forming the core principles of this approach and guiding the cognitive path in translation [39]. Translators must emphasize the interactive and dynamic transformation of these elements. They should understand the objective world presented in the source text and "creatively imitate" it in the target language, considering cultural and cognitive differences. Different translators' varied understandings lead to different creative imitations and translations. This embodied cognition activity, based on multiple interactions, necessitates using various embodied cognitive modalities (e.g., sensory perception, conceptualization, cognitive modeling, implicit metaphors and conceptual integration) to map meanings into the target language, constructing and articulating them through creative imitation [40].

Embodied-cognitive translatology offers a novel perspective for examining cognitive mechanisms in translation. This approach not only enriches theoretical understanding but also aids in developing translation tools that align with the physical and cognitive traits of translators.

3 Types and Analysis of ChatGPT Translation Hallucination Phenomena

From a cognitive perspective, LLMs' hallucinations are responses that do not reflect human cognition and facts. [36]. Combined with the theory of Embodied-cognitive translatology, this paper defines LLMs' hallucination as the error or deviation in understanding and reconstructing the meaning of the original text by LLMs in the absence of its internal cognitive framework and experience. This hallucination phenomenon usually manifests itself in the form of deviation of the translated text from the expected semantics, context or facts of the source text, which are rooted in the LLMs' lack of authentic bodily experience and human cognitive complexity.

3.1 Translation Hallucination Categories

Different from the current research on translation hallucinations of LLMs focusing on the internal model architecture, this paper classifies ChatGPT's hallucinations in translation tasks into four types from the perspective of translatology.

Syntactic Hallucinations. Syntactic hallucinations in translation introduce grammatical structures that deviate from target language norms or the source text's original meaning. Dynamic Grammar Theory posits that language comprehension and output are procedural, with grammar emerging from real-time meaning construction based on linguistic and contextual cues [15]. Thus, syntactic hallucinations arise from deviations in this procedural meaning construction, leading to the misuse of tenses, moods, voices, or other grammatical elements, distorting the intended message.

Semantic Hallucinations. Semantic hallucinations in translation are not merely word-level errors but deep deviations in concept construction and understanding, occurring when a model inaccurately captures the source text's meaning or misconstrues words' meanings [29]. These misunderstandings span concrete words and abstract concepts. For one, ChatGPT sometimes makes mistakes when translating words that involve specific sensory experiences or physical behaviours, reflecting its inadequate understanding of the embodied conceptualisation of human beings. For another, Conceptual Metaphor Theory posits that metaphors facilitate understanding one concept through another via cross-domain mappings, rooted in embodied experiences [14]. This nuanced metaphorical meaning often eludes ChatGPT, resulting in a semantic hallucination.

Pragmatic Hallucinations. As [17] highlights, pragmatics delves into language use, emphasizing discourse intentions, the significance of context, and communicators' expectations. In translation, ChatGPT often encounters pragmatic hallucinations, struggling to render the original text's implicit meanings and communicative intentions. This includes failing to capture non-obvious information reliant on context-specific knowledge, leading to a loss of the nuanced tone intended by the original author. Additionally, ChatGPT may not accurately grasp and convey the author's real intention, a challenge underscored from an Embodied-cognitive translatology perspective, which posits translation as a cognitive process demanding deep understanding and appropriate expression of the source text's intended meaning in the target language.

Cultural Hallucinations. From the perspective of Embodied-cognitive translatology, the cultural hallucination phenomenon refers to ChatGPT's inability to accurately understand or convey cultural nuances, idioms, proverbs, humour, and other culturally specific references from the source language to the target language during the translation process. Embodied cognition suggests that our understanding of language is deeply rooted in our bodily and cultural experiences [38]. Thus, when ChatGPT attempts to translate culturally specific expressions without having a cultural experience, it may result in a translation that has a very different meaning from the source text.

3.2 Empirical Analysis

Based on the above four types of translation hallucinations, this section will delve into the translation performance of ChatGPT on the Flores-101 [9] dataset. We will identify, classify and quantify the hallucinations that occur during English-Chinese translation, and analyse the frequency and severity of these hallucinations. This empirical study will help to gain insight into the underlying causes of translation hallucinations and provide a reference for future researchers to develop effective translation mitigation strategies.

Dataset. The Flores-101 dataset [9] is a large multilingual benchmark dedicated to the evaluation of MT models. It consists of 1012 sentences extracted from the English Wikipedia and carefully translated into 101 languages by professional linguists through a rigorously controlled process. Automated and manual quality checks were also used to ensure the accuracy of the translations. Notably, all translations are multilingual aligned, providing a comprehensive testbed for evaluating the model's translation capabilities. In this study, we focus on English-Chinese language pairs as it provides rich material for exploring linguistic nuances, semantic complexities, and cultural specificities that can lead to translation hallucinations. The dataset covers a variety of topics and language domains, ensuring robust analyses in different contexts.

However, we acknowledge the limitations of using a dataset primarily derived from English Wikipedia. While Wikipedia covers a wide range of topics, it may not fully capture the linguistic complexity and diversity of real-world translation tasks. Due to length constraints, we have not expanded the dataset for present study. In future studies, we plan to include more varied and challenging texts, so as to enhance the generalisability of our results.

Methodology. The methodology of this empirical study comprises three main steps: translation, manual labeling, and quantitative analysis. ChatGPT was tasked with translating the Flores-101 dataset from English to Chinese using a standardized prompt: "Please translate the source text into Chinese." This was done under controlled conditions to ensure consistency. Subsequently, three translation experts reviewed the translations to identify hallucinations, classifying them into syntactic, semantic, pragmatic, and cultural categories according to criteria outlined in Sect. 3.1. To ensure the objectivity of this annotation work, we adopted the annotating method used by [35], asking two experts to conduct the initial labeling, while a third to decide the final annotation for the case of divergence for the first two experts. Each hallucination type was recorded and statistically analyzed to assess its frequency and severity, including calculating the percentage of each type relative to the total translations reviewed.

Findings. According to our statistics (Table 1), despite the absence of hallucination phenomenon in most of the translation tasks (no hallucinations in 744 instances), ChatGPT still has significant difficulties in processing complex semantic and pragmatic information.

Table 1. Frequency of Hallucination Types in ChatGPT's English-Chinese Translation on the Flores-101 Dataset annotated by three expert translators

Hallucination Type	Number of Instances	Percentage of Total Instances
Syntactic Hallucination	49	4.84%
Semantic Hallucination	118	11.66%
Pragmatic Hallucination	80	7.91%
Cultural Hallucination	21	2.08%
No Hallucination	744	73.51%

The high incidence of semantic hallucinations in ChatGPT highlights significant challenges in accurately capturing or interpreting the meaning of source texts. These hallucinations, rooted in deep-seated biases, underscore limitations in understanding nuanced meanings, particularly with abstract concepts and specific words. Pragmatic hallucinations, the second most common type (7.91%), reveal the model's difficulty in grasping implicit meanings, social cues, and communicative intentions, reflecting the complexity of human communication beyond lexical or grammatical accuracy. Syntactic hallucinations (4.84%) indicate that while ChatGPT generally generates accurate grammatical structures, its application of these rules in specific contexts remains flawed. Cultural hallucinations, though least frequent (2.08%), point to deficiencies in capturing and translating cultural nuances. This low incidence might not signal a minor issue but rather the inherent complexity of cultural elements, which are difficult to quantify and may be underrepresented in datasets.

These findings illustrate the limitations of current LLMs within the Embodied-cognitive translatology paradigm. Despite advances in simulating human language processing, ChatGPT struggles to fully comprehend and translate complex semantic and pragmatic information, as well as convey meaning in cross-cultural contexts.

4 Mitigation Strategies

Building on the above analysis, we will summarize the current research on hallucination mitigation strategies and explore effective methods to reduce translation hallucination, focusing on coping strategies for the four types: syntactic, semantic, pragmatic, and cultural.

4.1 Enhancement of Corpus Quality and Diversity

To address semantic and cultural hallucinations in translation, enhancing the quality and diversity of the corpus is fundamental. Mitigating misinformation and bias can be achieved through high-quality factual data collection and data cleansing to reduce bias [24]. High-quality data collection has proven effective in improving the factual accuracy of LLMs [30], thereby reducing semantic hallucinations. However, this becomes more challenging with larger datasets. To combat repetition and social bias, de-duplication strategies and selecting diverse training data are essential [12]. Data augmentation allows for the inclusion of varied texts, enhancing the model's generalization and reducing translation hallucinations from corpus bias. This approach also aids the model in accurately translating abstract and culturally specific concepts, thereby reducing semantic and cultural hallucinations.

4.2 Development of Models

For addressing pragmatic hallucinations and syntactic hallucinations, developing mitigation strategies in models is particularly crucial. By introducing new decoding strategies such as Context-Aware Decoding (CAD) [27], Decoding by Contrasting Layers (DoLa) [5], and Inference-Time Intervention (ITI) [18], models can pay more attention to the original text's semantic fidelity and pragmatic intentions during translation, effectively reducing pragmatic hallucinations caused by overlooking implicit meanings or communicative intentions. Additionally, knowledge graph methods enrich models with background knowledge, improving their understanding of complex sentence structures and cultural contexts, thereby reducing syntactic and cultural hallucinations. Tools like FLEEK [2] identify verifiable facts and query knowledge graphs and the web for evidence, helping to minimize cultural translation errors. Supervised Fine-Tuning (SFT) [32] with specific data also enhances translation performance. Knowledge Injection and Teacher-Student Approaches also offer promising solutions. For instance, [6] utilized detailed Q&A sessions from stronger models to guide weaker ones, while [33] proposed Refusal-Aware Instruction Tuning (R-Tuning) to identify and reject questions beyond the model's reliable answering capabilities. These methods collectively offer effective strategies for enhancing the performance of LLMs in translation tasks.

4.3 Prompt Engineering

Prompt engineering techniques are crucial for addressing translation hallucinations caused by models' knowledge limitations. Retrieval-enhanced generation (RAG) [28] mitigates these issues by incorporating external knowledge sources, providing additional context to reduce hallucinations. One-time retrieval [25] integrates external information into the model input, reducing pragmatic and semantic hallucinations. Iterative retrieval [22,31] continuously updates the model with new knowledge, addressing cultural hallucinations. Post-hoc retrieval

[8,34] refines translations by revising outputs, enhancing output accuracy. Moreover, self-improvement methods through feedback and reasoning, such as ChatProtect [21] and Self-Reflection [13], can effectively improve the consistency and accuracy of the model output, which helps to identify and correct syntactic and semantic errors in translation. Furthermore, prompt tuning strategies can improve the model's performance on specific tasks by adjusting the prompts. For example, [4] proposes UPRISE, a universal prompt retriever that automatically generates prompts for zero-sample tasks and demonstrates cross-task and cross-model generalisability.

In summary, while current research doesn't directly address ChatGPT's translation hallucinations, it suggests solutions through improving corpus quality and diversity, refining model architecture, and employing advanced prompting techniques. Combining these innovations with an exploration of embodied cognition mechanisms in translation can enhance translation services, moving towards greater humanity and intelligence.

5 Conclusion

This paper explores translation hallucinations in ChatGPT from the perspective of Embodied-cognitive translatology, categorizing them into semantic, pragmatic, syntactic, and cultural hallucinations. We propose mitigation strategies for each, focusing on improving corpus quality and diversity, refining model architectures, utilizing knowledge graphs, employing prompt engineering, and emphasizing supervised fine-tuning for specific tasks. Despite advancements by LLMs in multilingual translation, limitations remain due to the lack of embodied experience and real-world cognition. Future translation technologies should aim not only to optimize algorithms and models but also to simulate and integrate human cognitive and cultural knowledge. This study provides a new perspective on translation hallucinations in ChatGPT and offers valuable insights for enhancing translation technology and quality.

Acknowledgments. This work was supported by the Open Project of Anhui Provincial Key Laboratory of Multimodal Cognitive Computation, Anhui University (Grant No. MMC202414), Guangdong Provincial Key Laboratory of Novel Security Intelligence Technologies (Grant No. 2022B1212010005), and XJTLU Research Development Funding (Grant No. RDF-22-01-053).

References

1. Achiam, J., et al.: Gpt-4 technical report. arXiv preprint arXiv:2303.08774 (2023)
2. Bayat, F.F., et al.: Fleek: factual error detection and correction with evidence retrieved from external knowledge. arXiv preprint arXiv:2310.17119 (2023)
3. Brown, T., et al.: Language models are few-shot learners. Adv. Neural. Inf. Process. Syst. **33**, 1877–1901 (2020)

4. Cheng, D., et al.: Uprise: universal prompt retrieval for improving zero-shot evaluation. arXiv preprint arXiv:2303.08518 (2023)
5. Chuang, Y.S., Xie, Y., Luo, H., Kim, Y., Glass, J., He, P.: Dola: decoding by contrasting layers improves factuality in large language models. arXiv preprint arXiv:2309.03883 (2023)
6. Elaraby, M., Lu, M., Dunn, J., Zhang, X., Wang, Y., Liu, S.: Halo: Estimation and reduction of hallucinations in open-source weak large language models. arXiv preprint arXiv:2308.11764 (2023)
7. Evans, V., Green, M.: Cognitive Linguistics: An Introduction. Routledge, New York (2018)
8. Gao, L., et al.: RARR: researching and revising what language models say, using language models. In: 61st Annual Meeting of the Association for Computational Linguistics, pp. 16477–16508. Association for Computational Linguistics, July 2023
9. Goyal, N., et al.: The flores-101 evaluation benchmark for low-resource and multilingual machine translation. Trans. Assoc. Comput. Linguist. **10**, 522–538 (2022)
10. Guerreiro, N.M., et al.: Hallucinations in large multilingual translation models. Trans. Assoc. Computat. Linguist. **11**, 1500–1517 (2023)
11. Guerreiro, N.M., Voita, E., Martins, A.F.: Looking for a needle in a haystack: a comprehensive study of hallucinations in neural machine translation. arXiv preprint arXiv:2208.05309 (2022)
12. Gyawali, B., Anastasiou, L., Knoth, P.: Deduplication of scholarly documents using locality sensitive hashing and word embeddings (2020)
13. Ji, Z., Yu, T., Xu, Y., Lee, N., Ishii, E., Fung, P.: Towards mitigating hallucination in large language models via self-reflection. arXiv preprint arXiv:2310.06271 (2023)
14. Johnson, M., Lakoff, G.: Metaphors We Live By. University of Chicago Press, Chicago (1980)
15. Kempson, R., Meyer-Viol, W., Gabbay, D.: Dynamic syntax. Pronouns-Grammar Rep. **137**, (2001)
16. Lakoff, G., Johnson, M., Sowa, J.F.: Review of philosophy in the flesh: the embodied mind and its challenge to western thought. Comput. Linguist. **25**(4), 631–634 (1999)
17. Levinson, S.C.: Pragmatics, 2nd edn. Cambridge University Press, Cambridge (1983)
18. Li, K., Patel, O., Viégas, F., Pfister, H., Wattenberg, M.: Inference-time intervention: eliciting truthful answers from a language model. In: Advances in Neural Information Processing Systems, vol. 36 (2024)
19. Li, W., Wu, W., Chen, M., Liu, J., Xiao, X., Wu, H.: Faithfulness in natural language generation: a systematic survey of analysis, evaluation, and optimization methods. arXiv preprint arXiv:2203.05227 (2022)
20. Macpherson, F., Platchias, D. (eds.).: Hallucination: Philosophy and Psychology. MIT Press, Cambridge, MA (2013)
21. Mündler, N., He, J., Jenko, S., Vechev, M.: Self-contradictory hallucinations of large language models: Evaluation, detection and mitigation. arXiv preprint arXiv:2305.15852 (2023)
22. Press, O., Zhang, M., Min, S., Schmidt, L., Smith, N.A., Lewis, M.: Measuring and narrowing the compositionality gap in language models. In: Findings of the Association for Computational Linguistics: EMNLP 2023, pp. 5687–5711 (2023)
23. Radford, A., Narasimhan, K., Salimans, T., Sutskever, I.: Improving language understanding by generative pre-training (2018)
24. Radford, A., Wu, J., Child, R., Luan, D., Amodei, D., Sutskever, I.: Language models are unsupervised multitask learners. OpenAI Blog **1**(8), 9 (2019)

25. Ram, O., et al.: In-context retrieval-augmented language models. Trans. Assoc. Comput. Linguist. **11**, 1316–1331 (2023)
26. Raunak, V., Menezes, A., & Junczys-Dowmunt, M.: The curious case of hallucinations in neural machine translation. arXiv preprint arXiv:2104.06683 (2021)
27. Shi, W., Han, X., Lewis, M., Tsvetkov, Y., Zettlemoyer, L., Yih, S.W.T.: Trusting your evidence: Hallucinate less with context-aware decoding. arXiv preprint arXiv:2305.14739 (2023)
28. Shuster, K., Poff, S., Chen, M., Kiela, D., Weston, J.: Retrieval augmentation reduces hallucination in conversation. In: Findings of the Association for Computational Linguistics: EMNLP 2021, pp. 3784–3803 (2021)
29. Talmy, L.: Toward a Cognitive Semantics: concept structuring Systems, vol. 1. 2nd edn. MIT Press, Cambridge, MA (2000)
30. Touvron, H., et al.: Llama: open and efficient foundation language models. arXiv preprint arXiv:2302.13971 (2023)
31. Trivedi, H., Balasubramanian, N., Khot, T., Sabharwal, A.: Interleaving retrieval with chain-of-thought reasoning for knowledge-intensive multi-step questions. In: 61st Annual Meeting of the Association for Computational Linguistics, pp. 10014–10037, July 2023 (2023)
32. Wang, Y., et al.: Self-instruct: aligning language models with self-generated instructions. In: The 61st Annual Meeting of the Association for Computational Linguistics, July 2023 (2023)
33. Wang, Y., et al.: Self-instruct: aligning language models with self-generated instructions. In: The 61st Annual Meeting of the Association for Computational Linguistics, July 2023 (2023)
34. Zhao, R., Li, X., Joty, S., Qin, C., Bing, L.: Verify-and-edit: a knowledge-enhanced chain-of-thought framework. In: 61st Annual Meeting of the Association for Computational Linguistics, pp. 5823–5840, July 2023, (2023)
35. Zhu, J., Li, H., Liu, T., Zhou, Y., Zhang, J., Zong, C.: MSMO: multimodal summarization with multimodal output. In: Proceedings of the 2018 Conference on Empirical Methods in Natural Language Processing, pp. 4154–4164 (2018)
36. Yao, J.Y., Ning, K.P., Liu, Z.H., Ning, M.N., Yuan, L.: Llm lies: hallucinations are not bugs, but features as adversarial examples. arXiv preprint arXiv:2310.01469 (2023)
37. 王寅: 后现代哲学视野下的体认语言学. 外国语文 **2014**(06), 61–67 (2014)
38. 王寅: 基于认知语言学的翻译过程新观. 中国翻译 **2017**(06), 5–10+17+129 (2017)
39. 王寅: 体认翻译学视野下的'映射'与'创仿'. 中国外语 **2020**(05), 37–44 (2020). https://doi.org/10.13564/j.cnki.issn.1672-9382.2020.05.006
40. 王寅: 体认翻译学的理论建构与实践应用. 中国翻译 **2021**(03), 43–49+191 (2021)

A Study on Chinese Acronym Prediction Based on Contextual Thematic Consistency

Wan Tao[1], Xiaoran Wang[1], and Qiang Zhang[2](✉)

[1] School of Computer and Information, Anhui Polytechnic University, Wuhu, China
taowan@ahpu.edu.cn
[2] School of Literature, Huaiyin Normal University, Huaian, China
zhangqiang_dh@163.com

Abstract. Common forms of lexicalization, acronyms are characterized by their concise structure, precise meaning, and retention of the original intent. They can effectively condense text and convey key information, playing a crucial role in various natural language processing tasks such as information retrieval and entity linking. However, the structure of Chinese acronyms is complex and diverse, and some acronyms are polysemous, which means they may represent different entities in different contexts. Most existing methods primarily rely on the entities themselves and rarely incorporate contextual information when studying Chinese acronyms, thereby overlooking the rich semantics within the context. Therefore, this paper first enhances a large Chinese acronym dataset with contextual information. Then, it proposes a new generation-evaluation framework for Chinese acronym prediction. This framework consists of a generation model that generates multiple candidate acronyms and an evaluation model based on thematic consistency. Experimental results on the public dataset demonstrate the effectiveness of our method, achieving improvements of 3.5 and 1.6 in Hit@1 and Hit@3, respectively.

Keywords: Chinese Abbreviation Prediction · BERTopic · Sequence Generation Model · Comparative Evaluation · Natural Language Processing

1 Introduction

In daily language interactions, complex vocabulary often leads to inefficient communication, which conflicts with the fast-paced demands of modern society [1]. To adapt to this pace, communication must follow established linguistic norms. Consequently, vocabulary that is concise, accurately conveys meaning, and retains the original intent is gaining recognition, commonly referred to as acronyms. In the context of neologisms, acronyms are widely used in natural language processing due to their brevity and effectiveness in conveying meaning. For example, in information extraction and named entity recognition, identifying and interpreting acronyms helps to accurately extract key entity information, supporting applications like information retrieval, entity linking, knowledge graph construction, and entity relationship analysis.

This paper primarily focuses on acronym prediction, aiming to anticipate abbreviated forms of entity names. Most current research centers on English initialisms, such as abbreviating "Natural Language Processing" to "NLP." In contrast, Chinese acronyms exhibit a broader range of forms. Table 1 lists examples of Chinese acronyms, categorized into three main types: morpheme-based, headword-based, and mixed-method constructions [2, 3].

Table 1. Examples of Chinese Acronym Construction Methods.

Type	Full form	Abbreviation
Morpheme-based	北京大学(Peking University)	北大
	奥林匹克运动(Olympic Games)	奥运
Headword-based	清华大学(Tsinghua University)	清华
	上海迪士尼(Shanghai Disneyland)	迪士尼
Mixed-method	中国科学院(Chinese Academy of Sciences)	中科院
	中央电视台(China Central Television)	中央台

Most Chinese acronym prediction methods currently rely on binary classification in sequence labeling, where labels are assigned to each character to indicate whether it should be kept or removed. However, these methods often overlook the coherence across characters and the broader context, focusing solely on individual characters. In practice, an acronym is heavily influenced by its surrounding context, such as adjacent vocabulary, phrase structure, and semantic relationships, which traditional models tend to neglect.

To address this, we propose a novel two-stage framework: candidate generation and contextual thematic consistency evaluation. In the first stage, we generate high-quality candidate abbreviations, and in the second, we evaluate and select the abbreviation that best aligns with the context and is both clear and representative of the intended meaning. The highest-scoring option is chosen as the final output. Experimental results confirm the effectiveness of this approach.

2 Related Works

In the early exploration of Chinese acronym prediction, researchers viewed it as a sequence labeling problem and used sequence labeling models. Sun and Yang et al. [5–7] developed feature functions specifically tailored for this task and employed Conditional Random Fields (CRF) to label the sequences. Chen [8] enhanced the model by integrating first-order logic, which captures long-range and global dependencies that CRF models typically struggle to capture. Additionally, Zhang [9] investigated methods that combined search engine results with manual adjustments to the ranking of CRF predictions.

In recent years, with the rise of neural network technology, the advantages of automatic feature extraction offered by neural networks have been increasingly applied to Chinese acronym prediction tasks, replacing traditional manual feature design methods. For instance, Zhang [10] proposed a novel neural network architecture to address the issue of generated acronyms not conforming to traditional Chinese acronym conventions in sequence labeling tasks. This structure combines Recurrent Neural Networks (RNN) with a mechanism that determines whether a given character sequence can form a word, completing the Chinese acronym generation task.

Regarding the aforementioned studies, Ma [11] highlighted that sequence labeling methods struggle to fully capture the overall meaning of sentences and face limitations in handling label dependencies. Neural network-based models often rely on transition matrices and often overlook label dependencies. To address these issues, Wang [12] was the first to frame Chinese acronym prediction as a sequence generation problem, proposing a multi-layer pre-trained model that integrates character, word, and concept-level embeddings, thereby enhancing prediction accuracy. Cao [13] introduced a context-enhanced Transformer model with an acronym recovery strategy, predicting acronym sequences through an iterative decoding process, which alternates between acronym generation and recovery operations. Tong [14] took a novel approach by evaluating acronyms within their textual context, establishing new criteria for the accuracy and practicality of predictions. These methods enrich research pathways for Chinese acronym prediction and suggest new directions for future studies.

Despite the varying degrees of performance improvement brought by the aforementioned methods, the unique characteristics of the Chinese language make acronym forms complex and diverse. Existing research has not sufficiently captured the intrinsic diversity of acronyms, which limits improvements in model generalization capabilities. Additionally, an acronym may carry different meanings across different contexts, yet current research often overlooks this diversity in acronym meanings, which may adversely affect the accuracy of Chinese acronym prediction.

3 Research Content

3.1 Problem Definition

The core objective of the Chinese acronym prediction task is to accurately predict the possible abbreviations for the complete form of a source entity. Specifically, consider a source entity represented by a complete name consisting of m characters, denoted as $X = [x_1, x_2, ..., x_m]$, while also considering the textual context information related to this entity, denoted as D. In the Chinese acronym prediction task, the goal is to predict an effective abbreviation sequence $Y = [y_1, y_2, \ldots, y_n]$, where each character in Y is contained within the complete name X. This problem can be formalized as:

$$Y = [y_1, y_2, ..., y_n] = f(x, d) \tag{1}$$

3.2 Research Framework

This paper proposes a Chinese acronym prediction method based on contextual thematic consistency. The model framework is shown in Fig. 1, which consists of two parts: the candidate generation phase and the context-enhanced evaluation phase.

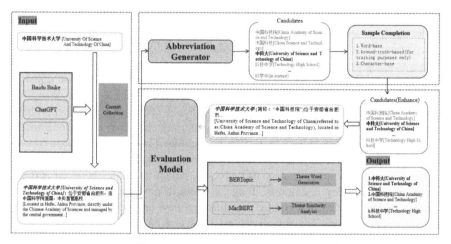

Fig. 1. Overall Architecture Diagram.

Preparation Phase. In this phase, contextual information D is collected for each source entity X to enhance the model's understanding of its complete semantics. The contextual information primarily comes from Baidu Baike and ChatGPT.

Candidate Generation Phase. In this phase, the semantic information of the text is obtained using the BARTTransformer CPT pre-trained model to generate candidate acronyms. Specifically, given a full form X and its related contextual information D, the pair (x, d) is input into the model, which is trained to generate k candidate acronyms $\hat{Y} = \{\hat{y}_1, \hat{y}_2, ..., \hat{y}_k\}$. In addition, a set of heuristic rules is used to pre-screen the candidate acronyms to produce an enhanced set of candidates.

Contextual Thematic Consistency Evaluation Phase. In this phase, the rationality of each candidate acronym is evaluated from the perspective of thematic consistency. Specifically, given (x, d) and $\hat{Y} = \{\hat{y}_1, \hat{y}_2, ..., \hat{y}_k\}$, the full name X in the original entity context D is first replaced with each candidate acronym $\hat{Y}$, yielding the contextual information $d_k \in D$ for k candidate acronyms. Then, a topic model is used to perform topic analysis on the original context $\hat{D}$ and each $\hat{d}_k \in \hat{D}$, resulting in f topics $T = \{t_1, t_2, ..., t_f\}$ and $\hat{T} = \{\hat{t}_1, \hat{t}_2, ..., \hat{t}_f\}$. Next, a contrastive learning-based evaluation model is trained to encode both T and each $\hat{T}$, and cosine similarity between them is used as the score for $\hat{D}$. In this way, the evaluation model assigns higher scores to candidate acronyms that have greater thematic consistency with the original entity. Finally, the candidate acronym with the highest score from $\hat{Y}$ is selected as the final output.

4 Research Methodology

4.1 Dataset Context Expansion

In this study, we utilize the Chinese acronym dataset introduced by Wang et al. [12]. While the original dataset lacks contextual information about entities, we enhance it by incorporating relevant contextual details from Baidu Baike and ChatGPT. The context expansion process was automated using web scraping techniques, and the specific steps are as follows:

Baidu Baike Search. For each entity pair (X, Y), we programmatically call the Baidu Baike API to search for relevant information. If successful, the paragraph summary of x is extracted as its contextual information.

Fallback to ChatGPT. If no relevant information for the entity is found on Baidu Baike, we use GPT-3.5 to supplement the contextual data. A carefully designed prompt is used to gather context by querying the GPT-3.5 API.

Contextual Information Filtering. Considering that excessive contextual information could increase the processing burden on the evaluation model, and that the first two sentences of a summary are usually sufficient to describe the full form of an entity, this study only collects the first two sentences of each entity's full description as its contextual information. Additionally, to prevent data leakage, any sentences containing actual acronyms were removed from the context.

By combining these methods, we aimed to build an authoritative and enriched set of contextual information, providing a solid data foundation for subsequent research and analysis.

4.2 Candidate Generation Phase

The candidate generation phase focuses on producing multiple high-quality candidate acronyms using the Chinese BARTTransformer CPT [15] model, a pre-trained language model designed for Chinese NLU and NLG tasks. CPT features an asymmetrical Transformer architecture with a shared encoder (S-Enc), a shallow encoder for understanding (U-Dec), and a shallow decoder for generation (G-Dec). These components are pretrained through masked language modeling (MLM) and denoising autoencoding (DAE) tasks, making CPT highly effective for sequence generation.

As the number of candidate acronyms k gradually increases, characters that do not belong to the subsequence of the original phrase may appear among the candidates. These candidates clearly cannot serve as valid abbreviations and therefore need to be filtered out. To supplement the k candidate acronyms, this study employs a series of heuristic rules [14] to construct complex negative samples, resulting in a final set of k enhanced candidate acronyms.

Word-based. In Chinese, words are the smallest semantic units, and removing certain words from X does not affect the overall fluency of the text. Therefore, by deleting a few words from the full form X, we can generate negative samples that maintain fluency but differ in meaning from the original entity.

Ground-truth-based(Used Only for Training). During the training phase, real data is available, and candidate acronyms that closely resemble real data can be treated as negative samples.

Character-based. Candidates that are lexically similar to the original phrase X can also serve as negative samples, as they retain some of the characters from the original phrase.

After filtering out unreasonable candidates and constructing complex negative samples using heuristic rules, we obtain the final set of k enhanced candidate acronyms, which will be used in the subsequent contextual thematic consistency evaluation phase.

4.3 Contextual Thematic Consistency Evaluation Phase

The second phase, the contextual enhancement evaluation, assesses the accuracy of candidate acronyms within their contexts and consists of two sub-stages: thematic word generation and thematic consistency evaluation.

Thematic Word Generation. Topic modeling is a powerful unsupervised tool for uncovering common themes and latent information. In this phase, we use the BERTopic model [16] to perform topic analysis on each instance of $d_k \in D$ and $\hat{d}_k \in \hat{D}$. BERTopic excels at capturing deep semantic information, especially in long or complex sentences, with three main steps: constructing text embeddings, reducing dimensionality and clustering embeddings, and representing topics.

Text Embedding Construction. In this stage, candidate acronyms are embedded into vector space using BERTopic, allowing for semantic comparisons between them.

Dimensionality Reduction and Clustering. To address the fuzziness of spatial locality in high-dimensional vector spaces, BERTopic employs the UMAP algorithm for dimensionality reduction, which preserves both local and global features, enhancing the effectiveness of clustering. The HDBSCAN algorithm is then applied to cluster the reduced vectors, effectively identifying and excluding noise and outliers.

Topic Discovery and Representation. Based on the sentence vector clusters obtained through clustering, the c-TF-IDF algorithm is used to represent topics by extracting representative topic words. The c-TF-IDF improves upon the traditional TF-IDF algorithm by considering the importance of words within specific categories (i.e., clusters). By selecting the most representative words from each cluster, the algorithm forms a clearer representation of the topics' themes.

Thematic Consistency Evaluation. The goal of the evaluation model is to ensure that valid candidate acronyms receive higher scores compared to invalid candidates, based on the original and candidate contextual information. To achieve this, topic consistency evaluation is introduced. Given $(t, \hat{t})$, the pre-trained Chinese Transformer encoder MacBERT [17] is used to encode both T and $\hat{T}$. The embeddings of the[CLS]tokens from the output are used as their representations, as shown in the following equations:

$$t = MacBERT(T)[CLS] \qquad (2)$$

$$\hat{t}_f = MacBERT(\hat{T}_f)[CLS] \qquad (3)$$

where t represents the embedding of T, and $\hat{t}_f$ represents the embedding of $\hat{T}_f$. The topic words generated from the original context, t, are treated as positive samples, denoted as $\hat{t}^+$, where $\hat{t}^+ \in \{\hat{t}_f\}_{f=1}^e$. The evaluation model is then trained using the contrastive loss described in Eq. (4), where $sim(\cdot, \cdot)$ denotes the cosine similarity, and τ represents the temperature hyperparameter:

$$Leval = -\log \frac{e^{sim(t,\hat{t}^+)/\tau}}{\sum_{f=1}^{e} e^{sim(t,\hat{t}_f)/\tau}} \qquad (4)$$

During inference, the evaluation model uses the cosine similarity between the embedding of T and each candidate embedding $\hat{T}$ as the score for $\hat{T}$, which also serves as the score for the corresponding candidate acronym $\hat{Y}$. Finally, the acronym $\hat{Y}$ with the highest score is selected from the set of candidates Y.

5 Experiments

5.1 Experimental Setup and Evaluation Metrics

In the candidate generation phase, we fine-tune the model using the Adam W optimizer with training parameters: batch size of 128, learning rate of 2e-5, and 10 epochs. In the evaluation phase, the parameters are adjusted to a batch size of 4, learning rate of 4e-5, and 10 epochs. To prevent overfitting and speed up convergence, Dropout is applied, and the Adam W optimizer minimizes model loss with parameter normalization after each batch to improve performance.

For accurate model performance evaluation, we use Hit as the metric. A higher hit rate signifies closer predictions to the true abbreviations for each entity:

Hit@1: The proportion of correctly predicted acronyms in the test set.

Hit@3: Proportion of correctly predicted acronyms among the top-3 predictions in the test set.

5.2 Determining the Number of Candidate Acronyms

As the number of candidate abbreviations increases, the time and space overhead for training the evaluation model significantly rises. To balance hit rate and resource consumption, this study experiments to find the optimal number of candidate acronyms, as shown in Fig. 2.

From Fig. 2, it is evident that as k increases, the hit rate also rises. However, the benefits of adding more candidates diminish when k rises from 8 to 15. Considering the computational complexity and resource consumption, setting k to 8 achieves good performance with a lower burden. Therefore, we select $k = 8$ as the optimal number of candidate acronyms.

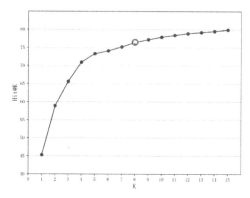

Fig. 2. Hit@K results of the generation model.

5.3 Comparative Methods and Result Analysis

Comparative Experiments. To evaluate the proposed model's effectiveness, this study compares it with several baseline models on the same dataset. The baseline models include sequence labeling models such as CRF, RNN-RADD, Bi-LSTM + CRF, and BERT + Bi-LSTM + CRF, as well as sequence generation models like Attention-based Seq2seq, Transformer, CPT, and Wang et al. [12].

Table 2. The overall performance of different methods.

Model	Hit@1	Hit@3
CRF	39.4	50.6
RNN-RADD	42.5	65.4
Bi-LSTM + CRF	43.4	66.6
BERT + Bi-LSTM + CRF	44.9	68.8
Attention-based Seq2seq	32.8	47.6
Transformer	34.6	48.5
Wang et al. [12]	47.6	71.6
CPT-base	45.3	65.6
Our Method	51.1	73.2

The overall results of all comparison methods are shown in Table 2. We conclude the following: (1) Our method clearly outperforms the baseline models, with improvements of 3.5 and 1.6 in Hit@1 and Hit@3, respectively. (2) When we introduce context thematic consistency evaluation, the performance improves significantly, demonstrating the effectiveness of the evaluation phase.

Manual Evaluation. To enable a fairer and more accurate comparison, this study further considers the following metrics, which are computed based on a manual evaluation of 1000 randomly selected samples from the test set:

CC: Acronyms that exhibit the characteristics of an abbreviation, effectively representing the original form of the entity but are not the only annotated acronym in the dataset, are denoted as CC.

CE: Abbreviations that possess a normal linguistic structure but do not effectively convey the original semantics of the entity are denoted as CE.

EE: Abbreviations that fail to effectively express the original semantics of the entity and exhibit abnormal linguistic structures are denoted as EE.

Examples of these three types (CC, CE, and EE) are provided in Table 3.

Table 3. Four Types of Acronym Prediction.

Type	Full Form	Correct	Predicted
Hit	清华大学	清华	清华
CC	华中师范大学	华师	华中师大
CE	数学期望	期望	数学
EE	上海迪士尼	迪士尼	上尼

Table 4. Manual Sampling Results.

Judges	Hit	R_{CC}	R_{CE}	R_{EE}
1	45.7%	29.3%	17.8%	7.2%
2	46.3%	28.6%	18.5%	6.6%
3	45.5%	29.7%	17.3%	7.5%
Average	45.8%	29.2%	17.9%	7.1%

We selected three students to conduct a manual evaluation on a randomly extracted sample of 1,000 entries, with the results shown in Table 4. The Hit and R_{cc} values are relatively high, indicating that the method successfully predicted abbreviations in over half of the samples, further validating its effectiveness.

6 Conclusion

This paper primarily focuses on Chinese abbreviation prediction and innovatively proposes a method for evaluating abbreviation quality. Unlike traditional abbreviation prediction methods, we first define the Chinese abbreviation prediction problem as a sequence generation task, effectively overcoming the inherent limitations of sequence labeling methods. Furthermore, we introduce contextual thematic consistency to assess the quality of abbreviations, aiming to accurately select those that best reflect the original entity's meaning. In addition, we enrich a large-scale Chinese abbreviation dataset through context collection, providing a solid data foundation for model training and validation, and validate the effectiveness of the proposed method through a series of extensive experiments.

References

1. Li, G.: A Comparative Study on Syntactic and Semantic Features of Chinese Verb Abbreviations and Their Original Forms. Master's Thesis, Huazhong Normal University (2017)
2. Niu, X.: Research and Standardization of Modern Chinese Abbreviations. Master's Thesis, Hebei Normal University (2004)
3. Wu, Z., Zheng, J.: Research on methods for automatic recognition of modern Chinese abbreviations. Comput. Eng. Des. **28**(16), 4052–4054 (2007)
4. Ma, D., Li, S., Wu, F., Xie, X., Wang, H.: Exploring sequence-to-sequence learning in aspect term extraction. In: Proceedings of the 57th Conference of the Association for Computational Linguistics(ACL 2019), Florence, Italy, July 28-August 2 2019, Volume 1: Long Papers, pp. 3538–3547 (2019)
5. Sun, X., Li, W., Meng, F., et al.: Generalized abbreviation prediction with negative full forms and its application on improving Chinese web search. In: Proceedings of the Sixth International Joint Conference on Natural Language Processing, pp. 641–647 (2013)
6. Yang, D., Pan, Y.C., Furui, S.: Vocabulary expansion through automatic abbreviation generation for Chinese voice search. Comput. Speech Lang. **26**(5), 321–335 (2012)
7. Yang, D., Pan, Y., Furui, S.: Automatic Chinese abbreviation generation using conditional random field. In: Proceedings of Human Language Technologies: The 2009 Annual Conference of the North American Chapter of the Association for Computational Linguistics, Companion Volume: Short Papers, pp. 273–276 (2009)
8. Chen, H., Zhang, Q., Qian, J., et al.: Chinese named entity abbreviation generation using first-order logic. In: Proceedings of the Sixth International Joint Conference on Natural Language Processing, pp. 320–328 (2013)
9. Zhang, L., Li, S., Wang, H., et al.: Constructing Chinese abbreviation dictionary: a stacked approach. In: Proceedings of COLING 2012, pp. 3055–3070 (2012)
10. Jiao, Y., Wang, H.: A method for abbreviation prediction based on machine learning and search engine validation. In: Chinese Information Processing Society of China (eds.) Frontier Progress in Computational Linguistics Research in China (2009–2011), p. 6. Tsinghua University Press, Beijing (2011)
11. Zhang, Q., Qian, J., Guo, Y., et al.: Generating abbreviations for Chinese named entities using recurrent neural network with dynamic dictionary. In: Proceedings of the 2016 Conference on Empirical Methods in Natural Language Processing, pp. 721–730 (2016)
12. Ma, D., Li, S., Wu, F., et al.: Exploring sequence-to-sequence learning in aspect term extraction. In: Proceedings of the 57th Annual Meeting of the Association for Computational Linguistics, pp. 3538–3547 (2019)
13. Wang, C., Liu, J., Zhuang, T., et al.: A sequence-to-sequence model for large-scale Chinese abbreviation database construction. In: Proceedings of the Fifteenth ACM International Conference on Web Search and Data Mining, pp. 1063–1071 (2022)
14. Tong, H., Xie, C., Liang, J., et al.: A context-enhanced generate-then-evaluate framework for Chinese abbreviation prediction. In: Proceedings of the 31st ACM International Conference on Information & Knowledge Management, pp. 1945–1954 (2022)

15. Shao, Y., et al.: CPT: a pre-trained unbalanced transformer for both Chinese language understanding and generation. arXiv preprint arXiv:2109.05729 (2021)
16. Grootendorst, M.: BERTopic: neural topic modeling with a class-based tf-idf procedure. arXiv preprint arXiv:2203.05794 (2022)
17. Cui, Y., Che, W., Liu, T., Qin, B., Wang, S., Hu, G.: Revisiting pre-trained models for Chinese natural language processing. In: Findings of the Association for Computational Linguistics: EMNLP 2020, pp. 657–668 (2020)

Learning Supportive Two-Stream Network for Audio-Visual Segmentation

Hongfan Jiang, Tianyang Xu, Xuefeng Zhu, and Xiaojun Wu[✉]

School of Artificial Intelligence and Computer Science, Jiangnan University, Wuxi, China
wu_Xiaojun@jiangnan.edu.cn

Abstract. Audio Visual Segmentation is an emerging problem in multi-modality video analysis. Sound signals in the task enable to help the network segment the vocal object in the video. However, how to effectively utilize sound signals for segmentation remains an open challenge. We introduce an innovative segmentation network in this manuscript, referred to as Two-stream Network for Audio Visual Segmentation. The network comprises two distinct streams, namely Multi-Modality Stream and Video-Feature Stream, which respectively provide the model with fusion features and deep visual features. In particular, Multi-Modality Stream is responsible for fusing features from two modalities to generate fusion features, while Video-Feature Stream employs self-attention to extract deeper visual features. Then, Two output features are combined by Cross-Modality Fusion Module, which enables to adaptively keep the balance between two different streams based on the influence extent of noise in the audio. In our experiments on AVSBench, our method surpasses several current methods, showcasing advanced performance while utilizing the same backbone.

Keywords: Audio Visual Segmentation · Multi-modality · Convolutional Neural Network

1 Introduction

A standard semantic segmentation, using the visual modality as the sole input, assigns each pixel in an image or video to one of the pre-defined semantic categories [5,6,17,18]. Although semantic segmentation has been developed for many years, the task is still limited by the requirement of pre-defined categories, which restricts trained models to specific scenarios. In contrast, Audio Visual Segmentation (AVS) considers both videos and additional corresponding audio signals as inputs. The objective of AVS is to densely predict the correspondence between each pixel and the provided audio input, thereby generating a mask specifically highlighting the vocal object(s), which focuses on segmenting image regions based on the audio understanding [1,14,20,21]. Compared to conventional semantic segmentation, AVS is a more practical problem. In principle, AVS is not constrained by pre-defined semantic categories, developing its

(a) high-quality sound signal (b) low-quality sound signal

Fig. 1. The sound signal corresponding to the left image (a) is a piece of high-quality sound of bird vocalization, where the region of interest aligns closely with the bird. Conversely, the sound signal corresponding to the right image (b) represents a low-quality sound of helicopter flight. The sound quality is compromised with less relevance, leading to the interference of the predicted interesting region, resulting in inaccurate localization of the helicopter.

application to a wide range of scenes. Furthermore, by incorporating additional audio signals, AVS excels in handling complex scenes, providing more accurate and comprehensive segmentation results. Therefore, AVS, as an open-set problem, holds great potential for various applications in computer vision and multi-modal understanding.

As illustrated in Fig. 1, when the audio signal carries an excessive amount of noise, the fusion feature of audio and visual modalities is contaminated, leading to an insufficient provision of discriminative information for segmentation.

To tackle the aforementioned challenge, we put forward the idea of developing an original supportive two-stream network for AVS (TsAVS-Net). The difference between traditional single module for fusion and two-stream network structure is depicted in Fig. 2. Our TsAVS-Net is comprised of two streams: the Multi-Modality Stream (MM Stream) and the Video-Feature Stream (VF Stream), obtaining fusion features and deep visual features, respectively. Specifically, the MM Stream is equipped with co-attention between audio and vision to highlight the region of interest in the images by the audio, while the VF Stream utilizes an attention-based scheme for deeper semantic features. Temporal modeling is also employed in the VF Stream to extract the sequential information to incorporate into TsAVS-Net. Additionally, we design a Cross-Modality Fusion Module (CMF) that adaptively combines the output features from two streams. CMF is capable of dynamically adjusting the weights between the fusion features and complementary deep visual features based on the influence extent of noise in the audio. We conduct an evaluation of TsAVS-Net on benchmark datasets to verify the effectiveness of our model. Thorough experimental outcomes confirm that TsAVS-Net is more effective for AVS, via learning supportive multi-modal clues.

2 Related Work

2.1 Audio Visual Segmentation

Audio Visual Segmentation is an emerging task of computer vision proposed by [21]. This task necessitates the network to predict categories of pixels and generate masks in a dense manner in the video of the vocal object(s) involved in the

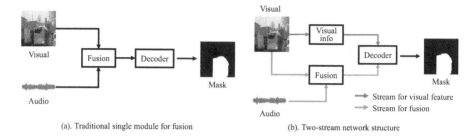

Fig. 2. The difference between the framework of existing methods and that of the proposed method. The traditional structure performs direct fusion for interaction between two modalities, while the proposed structure designs two streams for adaptive multmodal segmentation.

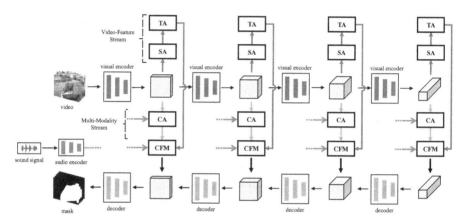

Fig. 3. The overall structure of TsAVS-Net. TsAVS-Net includes a Multi-Modality Stream (MM Stream), a Video-Feature Stream (VF Stream), and a Cross Modality Fusion Module (CMF). MM Stream is to generate fusion features from audio and visual pieces. VF Stream is to extract deeper features for visual semantics. CFM is to adaptively fuse features from two streams.

audio. In particular, the network for AVS [1,12,14,15] aims at fusing audio and visual modalities. Zhang et al. [20] focuses on the audio modality. Though reshaping the audio feature, they acquire the attention matrix between two modalities to fuse visual and audio features. On this basis, Darrell et al. [1] focuses on the fusion between the two modalities. The researchers use mutual information analysis to realize the interaction of video and audio modalities, generating the masks. Afterwards, researchers aim to obtain more accurate segmentation results by optimizing the network structure. Rouditchenko et al. [14] divides the model into different networks. The audio synthesizer network is designed for the interaction of two modalities. The image analysis network is for image segmentation while the audio analysis network is for audio source separation.

2.2 Audio-Visual Dataset

We choose AVSBench [21] for training, validation, and testing. AVSBench includes two different subsets. The one is single-source called S4 in short and the other is multi-source called MS3 in short. In the S4 task, there is only one category present in the audio signal, hence requiring the segmentation of a single category. Conversely, in the MS3 task, the audio signal comprises multiple categories, necessitating multi-target segmentation. Each video was edited into five seconds, and takes one frame per second. So each sample in the AVSBench contains five images and one five-second audio clip. The S4 subset contains 4932 videos in 23 categories. To each video in the S4 training set, the annotations are only performed on the first frame. The frames of the rest of AVSBench are annotated (including MS3). Therefore, S4 is a semi-supervised task, while MS3 is a fully-supervised task, which contains 424 videos. The training data accounts for 70% of the total dataset, while validation and test data both accounts for 15%.

3 Methodology

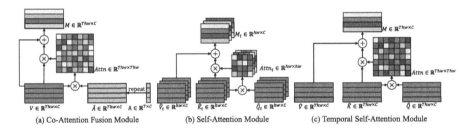

Fig. 4. The structure of Co-Attention Fusion Module (a), Self-Attention Module (b), and Temporal Self-Attention Module.

We present an original network for AVS: Two-stream Network for Audio-Visual Segmentation (TsAVS-Net) as shown in Fig. 3, which enables us to complete both S4 and MS3 tasks.

3.1 Overview

A VGGlike-Net [16] pretrained on AudioSet [3] is equipped as the audio encoder, while resNet-50 [7] is equipped as the visual encoder to generate hierarchical visual features. Panoptic-FPN [9], which serves as the decoder, is equipped to seamlessly integrate hierarchical features. These features are then activated by sigmoid function to generate the final masks. We employ the binary cross entropy function [8] and Kullback-Leibler divergence [4] to ensure effective supervision for TsAVS-Net. The main loss for supervision of the final masks in both S4 and MS3 is the binary cross entropy loss. Kullback-Leibler divergence is only equipped in MS3 for supervision of the audio.

3.2 Multi-Modality Stream

Multi-Modality Stream (MM Stream) is equipped for fusion between audio and vision. The core of this stream is Co-Attention Fusion Module (CA), whose structure is shown in Fig. 4 (a). FC layer and repeat are equipped to get the audio feature $\hat{A}$, whose size is the same with the visual feature F_i. Utilize the attention mechanism to capture the interdependencies between disparate features. Subsequently, we employ the obtained attention weights $attn_{CA}$ to weight the corresponding elements of F_i, resulting in the generation of the output Z_i for the MM Stream. The computation of Z_i involves the following step:

$$attn_{CA} = \frac{\mu_1(V_i) \cdot \theta_1(\hat{A})}{N} \qquad (1)$$

$$Z_i = V_i + \phi_1(attn_{CA} \cdot \psi_1(V_i)) \qquad (2)$$

where μ_1, θ_1, ϕ_1 and ψ_1 are 1×1 convolutions; $V_i, \hat{A}, Z_i \in \mathbb{R}^{Th_iw_i \times C}$; $N = Th_iw_i$ is a normalization factor; $attn_{CA} \in \mathbb{R}^{Th_iw_i \times Th_iw_i}$ is the attention matrix.

3.3 Video-Feature Stream

We design Self-Attention Module (SA) and Temporal Self-Attention Module (TA) in the Video-Feature Stream (VF Stream). Figure 4 (b) and (c) present the structures of them. SA focuses on one single image, while TA focuses on the whole video. SA is set before TA. Considering the memory of the model, we only equip TA modules in two deeper stages.

About SA, we resize the visual feature of the i-th stage $V_i \in \mathbb{R}^{T \times h_iw_i \times C}$. Through self attention, SA module multiplies the visual feature and self attention matrix $attn_SA$ of the images to get the SA output $Out_{SA} \in \mathbb{R}^{T \times h_iw_i \times C}$ of the video which includes T frames. Out_{SA} can be computed as,

$$attn_{SA} = \frac{\mu_2(V_i) \cdot \theta_2(V_i)}{N'} \qquad (3)$$

$$Out_{SA} = V_i + \phi_2(attn_{SA} \cdot \psi_2(V_i)) \qquad (4)$$

where μ_2, θ_2, ϕ_2 and ψ_2 are 1×1 convolutions; $N' = h_iw_i$ is a normalization factor; $attn_{SA} \in \mathbb{R}^{T \times h_iw_i \times h_iw_i}$ is the attention matrix of the video.

In TA, to get multi-frame temporal features, we put dimension T into h_iw_i to implement interaction across frames of the same video, taking $V_i \in \mathbb{R}^{Th_iw_i \times C}$ as the input of TA. Then through self-attention mechanism to get Out_{TA}, which can be computed as,

$$attn_{TA} = \frac{\mu_3(V_i) \cdot \theta_3(V_i)}{N} \qquad (5)$$

$$Out_{TA} = V_i + \phi_3(attn_{TA} \cdot \psi_3(V_i)) \qquad (6)$$

where μ_3, θ_3, ϕ_3 and ψ_3 are 1×1 convolutions; $Out_{TA} \in \mathbb{R}^{Th_iw_i \times C}$; $N = Th_iw_i$ is a normalization factor; $attn_{TA} \in \mathbb{R}^{Th_iw_i \times Th_iw_i}$ is the attention matrix.

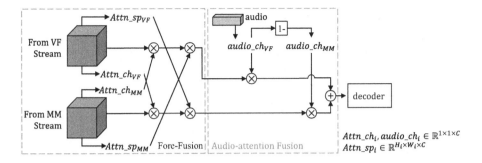

Fig. 5. The structure of Cross Modality Fusion Module, including fore-fusion and audio-attention fusion.

3.4 Cross Modality Fusion

To keep balance between the outputs of two streams adaptively, we design Cross Modality Fusion Module (CMF) for fusion between two features. As shown in Fig. 5, CMF has two different parts, fore-fusion and audio-attention fusion. In fore-fusion, we use channel and spatial attention to reduce distance between two different features. In audio-attention fusion, because we believe that VF Stream and MM Stream are complementary, A_z obtained after processing the audio A through the FC layer and sigmoid function is equipped as the channel attention of VF Stream, while all-ones vector minus A_z is the channel attention of the other. At last, add them to get the fusion feature D_i as the output, which is computed as,

$$\hat{M}_i = [M_i \cdot \varphi_1(Z_i)] \cdot \gamma_1(Z_i) \tag{7}$$

$$\hat{Z}_i = [Z_i \cdot \varphi_2(M_i)] \cdot \gamma_2(M_i) \tag{8}$$

$$D_i = \hat{M}_i \cdot FC(A) + \hat{Z}_i \cdot [1 - FC(A)] \tag{9}$$

where φ_1, φ_2, γ_1 and γ_2 are combination of convolution and pooling; FC is full connection layer; $\varphi_1(Z_i), \varphi_2(M_i) \in \mathbb{R}^{1 \times 1 \times C}$ are channel attention; $\gamma_1(Z_i), \gamma_2(M_i) \in \mathbb{R}^{h_i \times w_i \times 1}$ are spatial attention; $\hat{M}_i, \hat{Z}_i, D_i \in \mathbb{R}^{h_i \times w_i \times C}$.

4 Experiments

We evaluate TsAVS-Net objectively by the results of qualitative experiments. A comprehensive range of ablation experiments is conducted to assess the impact of various network structures and modules in our study.

4.1 Qualitative Experiments

Performance Comparison. Jaccard index [2] and F-score are equipped for evaluating. MJ and MF represent the average measurements of the entire dataset. $Mscore$ is the average of the two evaluation indicators.

The results of experiments are presented in the Table 1. Among the networks of comparison, AVS [21] are the network focused on AVS. It also adopts ResNet-50 [7] as the visual encoder, which is our primary object of comparison. Due to the presence of VF Stream and CMF, TsAVS-Net exhibits an effective response to various audio qualities, thereby enhancing segmentation accuracy. Due to the less number of network for AVS, we select some networks that perform similar tasks such as Sound Source Localization (SSL), Video Object Segmentation (VOS) and Salient Object Detection (SOD) for comparison. MSSL [13],3DS [10] and iGAN [11] is the network for SSL, VOS and SOD respectively. Without a dedicated audio input and fusion module, the gap with our network is predictable.

4.2 Ablation Experiments

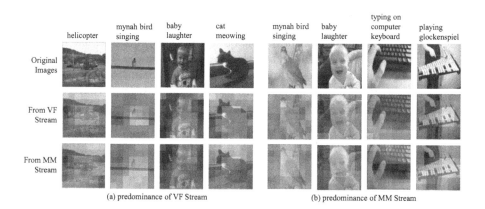

Fig. 6. The interesting regions of VF Stream and MM Stream in the $4th$ stage. The left part (a) shows the case that interesting regions of VF Stream is more accurate than MM Stream. The right part (b) shows the case that MM Stream is more effective.

About Structure of Video-Feature Stream We conduct an experiment to investigate the optimal marshalling sequence of various modules in VF Stream. The experimental results, presented in Table 2, illustrate the outcomes of the ablation study for S4 conducted on the SA and TA modules (respectively represented by S and T). In TsAVS-Net, we employ the ST structure for VF Stream, wherein one SA module is followed by one TA module. By comparing the TS

Table 1. Comparison with other related direction networks on AVSbench for S4 and S3.

Methods	S4			MS3		
	MJ	MF	$Mscore$	MJ	MF	$Mscore$
MSSL [13]	44.90	66.31	55.60	26.17	36.31	31.24
3DC [10]	57.11	75.93	66.52	36.94	50.32	43.63
iGAN [11]	61.63	77.83	69.73	42.92	54.46	48.69
AVS [19]	72.46	82.45	77.45	45.34	56.15	50.74
Ours	**74.41**	**84.49**	**79.45**	**47.41**	**57.96**	**52.69**

and ST structures, the advantages of employing the ST structure are confirmed. The significantly lower $Mscore$ of the SS and TT structures compared to the ST structure provides evidence for the necessity of the SA and TA modules in VF Stream. Furthermore, we conduct a comparison between the S structure and the SST structure. Both of these structures are found to be inferior to the ST structure. We believe that a lower number of layers may result in insufficient feature extraction, while an excessive number of layers leads to overfitting. These conclusions also hold true for MS3.

Table 2. Ablation Study About the structrue of Video-Feature Stream.

Structure	S4			MS3		
	MJ	MF	$Mscore$	MJ	MF	$Mscore$
SST	73.28	83.29	78.29	46.72	54.06	50.39
T	73.13	83.57	78.35	45.15	55.87	50.51
TT	73.68	84.01	78.85	46.66	57.36	52.01
SS	73.28	83.83	78.56	45.12	55.20	50.16
TS	73.57	83.71	78.64	46.76	57.04	51.90
ST	**74.41**	**84.49**	**79.45**	**47.41**	**57.96**	**52.69**

Feature Visualization. In order to highlight the complementary relationship between the two streams, we visualize the output features of two streams. Visual feature maps are presented in Fig. 6. Specifically, we select the respective streams at the deepest stage, summing the features along the channel dimension. Brighter colors in the one-dimensional images indicate higher attention from the corresponding stream.

As depicted in Fig. 6 (a), VF Stream generates higher quality features in most cases. Furthermore, these features provided more accurate localization compared to MM Stream, which exhibited coarser interesting regions, resulting from the noise presented in the audio modality. However, we observed two specific cases where the features of MM Stream outperformed those of VF Stream, as shown in Fig. 6 (b). The first case occurs when the vocal object occupies most of the space in the image. In such instances, MM Stream focuses on the entire target, while VF Stream concentrates only on a portion of the target. Such as the case of mynah bird singing, MM Stream successfully attends to the entire bird, whereas VF Stream predominantly focuses on the beak. The second case is that the primary object in the image do not align with the vocal object. In this case, MM Stream correctly focuses on the object that is producing the sound. For instance, in the example of typing on a computer keyboard as shown in Fig. 6 (b), the vocal object is the keyboard, while the primary object is the hand. MM Stream correctly directs attention towards the keyboard, whereas VF Stream primarily attends to the hands. These findings confirm the complementary relationship between VF Stream and MM Stream, demonstrating the necessity of two streams in our network.

5 Conclusion

We design a Two-stream network for Audio-Visual Segmentation. The Video-Feature Stream within the network serves as a reliable source of discriminative features for segmentation when the audio quality is poor. The Cross Modality Fusion module dynamically combines the two modalities, allowing the features from both streams to complement each other. We conduct a comparative analysis of TsAVS-Net and other advanced models for AVS on the AVSBench. Through this evaluation, we successfully validated the excellence of TsAVS-Net. For subsequent research, we believe that the proposed Two-stream architecture can serve as a reference for multi-modal segmentation tasks.

References

1. Darrell, T., Fisher Iii, J.W., Viola, P.: Audio-visual segmentation and the cocktail party effect. In: Advances in Multimodal Interfaces-ICMI 2000: Third International Conference Beijing, China, 14–16 October 2000, Proceedings, pp. 32–40. Springer (2001)
2. Everingham, M., Van Gool, L., Williams, C.K., Winn, J., Zisserman, A.: The pascal visual object classes (voc) challenge. Int. J. Comput. Vis. **88**, 303–308 (2009)
3. Gemmeke, J.F., et al.: Audio set: an ontology and human-labeled dataset for audio events. In: 2017 IEEE International Conference on Acoustics, Speech and Signal Processing (ICASSP), pp. 776–780. IEEE (2017)
4. Georgiou, T.T., Lindquist, A.: Kullback-leibler approximation of spectral density functions. IEEE Trans. Inf. Theory **49**(11), 2910–2917 (2003)
5. Guo, Y., Liu, Y., Georgiou, T., Lew, M.S.: A review of semantic segmentation using deep neural networks. Int. J. Multimed. Inf. Retr. **7**, 87–93 (2018)
6. Hao, S., Zhou, Y., Guo, Y.: A brief survey on semantic segmentation with deep learning. Neurocomputing **406**, 302–321 (2020)

7. He, K., Zhang, X., Ren, S., Sun, J.: Deep residual learning for image recognition. In: Proceedings of the IEEE Conference on Computer Vision and Pattern Recognition, pp. 770–778 (2016)
8. Jadon, S.: A survey of loss functions for semantic segmentation. In: 2020 IEEE Conference on Computational Intelligence in Bioinformatics and Computational Biology (CIBCB), pp. 1–7. IEEE (2020)
9. Kirillov, A., Girshick, R., He, K., Dollár, P.: Panoptic feature pyramid networks. In: Proceedings of the IEEE/CVF Conference on Computer Vision and Pattern Recognition, pp. 6399–6408 (2019)
10. Mahadevan, S., Athar, A., Ošep, A., Hennen, S., Leal-Taixé, L., Leibe, B.: Making a case for 3d convolutions for object segmentation in videos. arXiv preprint arXiv:2008.11516 (2020)
11. Mao, Y., et al.: Transformer transforms salient object detection and camouflaged object detection. arXiv preprint arXiv:2104.10127 (2021)
12. Owens, A., Efros, A.A.: Audio-visual scene analysis with self-supervised multisensory features. In: Proceedings of the European Conference on Computer Vision (ECCV), pp. 631–648 (2018)
13. Qian, R., Hu, D., Dinkel, H., Wu, M., Xu, N., Lin, W.: Multiple sound sources localization from coarse to fine. In: Computer Vision–ECCV 2020: 16th European Conference, Glasgow, UK, 23–28 August 2020, Proceedings, Part XX 16, pp. 292–308. Springer (2020)
14. Rouditchenko, A., Zhao, H., Gan, C., McDermott, J., Torralba, A.: Self-supervised audio-visual co-segmentation. In: ICASSP 2019-2019 IEEE International Conference on Acoustics, Speech and Signal Processing (ICASSP), pp. 2357–2361. IEEE (2019)
15. Sidiropoulos, P., Mezaris, V., Kompatsiaris, I., Meinedo, H., Bugalho, M., Trancoso, I.: Temporal video segmentation to scenes using high-level audiovisual features. IEEE Trans. Circuits Syst. Video Technol. **21**(8), 1163–1177 (2011)
16. Simonyan, K., Zisserman, A.: Very deep convolutional networks for large-scale image recognition. arXiv preprint arXiv:1409.1556 (2014)
17. Strudel, R., Garcia, R., Laptev, I., Schmid, C.: Segmenter: transformer for semantic segmentation. In: Proceedings of the IEEE/CVF International Conference on Computer Vision, pp. 7262–7272 (2021)
18. Thoma, M.: A survey of semantic segmentation. arXiv preprint arXiv:1602.06541 (2016)
19. Tian, Y., Shi, J., Li, B., Duan, Z., Xu, C.: Audio-visual event localization in unconstrained videos. In: Proceedings of the European Conference on Computer Vision (ECCV), pp. 247–263 (2018)
20. Zhang, T., Kuo, C.C.J.: Audio content analysis for online audiovisual data segmentation and classification. IEEE Trans. Speech Audio Process. **9**(4), 441–457 (2001)
21. Zhou, J., et al.: Audio–visual segmentation. In: Computer Vision–ECCV 2022: 17th European Conference, Tel Aviv, Israel, 23–27 October 2022, Proceedings, Part XXXVII, pp. 386–403. Springer (2022)

Multi-exposure Driven Stable Diffusion for Shadow Removal

Zheng Yan, Wenhao Tan, and Linbo Wang[✉]

School of Computer Science and Technology, Anhui University, Hefei, Anhui, China
wanglb@ahu.edu.cn

Abstract. Shadow removal has long been a challenging task in the field of computer vision. Despite the significant progress since the advent of deep learning, existing approaches still fall short of robustness in generating high-quality shadow-free images due to the scarcity of training data and the complexity of shadow images. In this paper, we approach the task as an image-denoising process and present a deep neural network based on the ControlNet-driven stable diffusion model, whose rich prior knowledge compensates for the data shortage and better facilitates the task modeling of shadow removal. Moreover, we simulate multi-exposure shadow images as conditional inputs to regularize the denoising process during training. An intensity modulation block is also integrated to boost the intensity of the recovered scene. Experiments on two benchmark datasets, ISTD+ and SRD, witness the superior performance of the proposed approach.

Keywords: Shadow Removal · Stable Diffusion · Multi-exposure · ControlNet

1 Introduction

Shadow is a common natural phenomenon, which usually refers to a dark area generated when light is obstructed by objects, creating dark areas. While shadows can limit human perception, they also pose challenges for various visual tasks, including object detection, tracking, and semantic segmentation.

Recent approaches usually design various deep models [1–4] for the shadow removal task. Despite the significant progress achieved, it is still very challenging to generate a satisfactory shadow-free image from a given shadow image. The difficulty can arise from several aspects: 1) Shadows often exhibit arbitrary shapes and complex structures, making them difficult to distinguish from non-shadow regions, particularly in intricate environments. 2) The degradation caused by shadows is spatially non-uniform, varying in intensity and coverage across the image. 3) Ensuring consistent illumination between shadow and non-shadow areas during shadow removal is challenging, often leading to noticeable discrepancies.

We argue that these challenges can partially be attributed to the lack of sufficient training data for the modeling of the shadow and non-shadow context. Effective context encoding may better distinguish between shadow and non-shadow features, which further facilitates the shadow-free image recovering process. The advent of large-scale models can partially alleviate the issue of training data scarcity. In particular, existing large-scale models are usually pre-trained on some large-scale datasets, which contain rich prior knowledge of natural scenes and objects and thus provide a good foundation for modeling specialized tasks. In particular, the shadow regions are typically profiled with general object shapes, filled with arbitrary ground textures while shaded in gray. All these features can be described by the general knowledge contained in large models, whose applications to the shadow removal task remain largely unexplored.

To this end, we approach shadow removal as a denoising process and propose a network architecture based on the stable diffusion model with conditional inputs served by ControlNet [5]. Since the original ControlNet is tailored for general image generation and denoising, we incorporate shadow-removal-specific inputs to better guide the diffusion-based de-shadowing process.

To do so, we first observe that the appearance of shadow regions can vary with the exposure level, and shadows in high-exposure conditions look more prominent. Feeding such knowledge to the diffusion model may help it to better understand the trait of shadow, and further facilitate the modeling of shadow removal. Therefore, we seek to generate multi-exposure shadow images from the raw input by simulating various lighting conditions. Here, a linear exposure model is hypothesized with key parameters controlling the exposure level. Those parameters are then estimated by assuming that the shadow effect can be diminished at the proper exposure level, whereby the appearance of the shadow region can be recovered perfectly as in the shadow-free image. Nevertheless, as only the shadow image is available at the inference stage, we thus further construct the convolutional neural network (CNN) to estimate the exposure parameters. For training the CNN model, we take pairs of shadow and shadow-free images and evaluate the ground truth parameters by linear square regression in the shadow regions. Once trained, we can easily apply the CNN model to evaluate the exposure parameters for a new shadow image, and further manipulate the image to obtain multi-exposure shadow images using the linear exposure model with the estimated parameters. These images are then paired with the original shadow image and shadow mask and fed to the ControlNet to drive the de-shadowing process.

By adapting the ControlNet with the multi-exposure shadow inputs, we find that the generated shadow-free images often exhibit inconsistent intensity between the shadow and non-shadow areas. Note that the multi-exposure shadow input is integrated into the diffusion model at the de-shadowing phase of the training stage, while the shadow noise is added at the sampling process, independent of the shadow image. Therefore, the diffusion model may fail to sense the environmental intensity of the raw input shadow image. This can lead to an overall intensity gap between the predicted noise map and the ground-truth, and further cause the intensity inconsistency between the shadow and non-shadow regions. To address this issue, we further introduce a specialized intensity encoder to estimate an environmental intensity scale and bias term in the non-shadow region. Those factors are then used to modulate the intensity of the output noise map and boost noise prediction, whereby further enhances the de-shadowed outputs. Note that the exposure estimator is trained independently, while the environmental intensity predictor is optimized along with the diffusion model.

The pre-trained stable diffusion model provides rich general knowledge, while the multi-exposure shadow inputs feed the shadow knowledge to customize the diffusion model for the de-shadow task. Moreover, the integration of the environmental intensity encoder further modulates the predicted shadow noise map, which helps improve the quality of shadow-free image generation. Equipped with all these merits, the proposed model achieves consistently robust shadow removal, which is verified by extensive experiments on two benchmark datasets, SRD and ISTD+. To summarize,

- We introduce a deep neural network, leveraging the pre-trained stable diffusion model, to harness the potential prior knowledge embedded in large-scale models for shadow removal.
- We incorporate multi-exposure shadow input generation and environmental intensity prediction to better regulate the diffusion process.
- Experiments on two benchmark datasets, ISTD+ and SRD, demonstrate the superior performance of our proposed method.

2 Related Works

2.1 Deep Learning Based Shadow Removal

Recent approaches usually tackle the shadow removal task by designing various deep models. DeshadowNet [4] expands the receptive field to capture more contextual information, enhancing shadow removal by incorporating both global and local features. G2R [6] and BMNet [7] propose that shadow generation and shadow removal are complementary processes, each benefiting the other. SG-ShadowNet [1] transforms the shadow removal problem into an image style transfer task, transferring the style learned from non-shadow regions to shadowed areas. CANet [2] further improves shadow removal by matching patches from non-shadow regions to similar shadowed regions, leveraging cross-image information. More recently, Transformer-based models [3,8] have demonstrated success by exploring the self-attention scheme to effectively capture global dependencies, thereby improving performance of shadow removal.

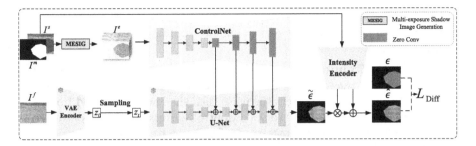

Fig. 1. Pipeline of the proposed method. We adapt the stable diffusion model with ControlNet to the shadow removal task. The diffusion process is guided by multi-exposure shadow inputs as well as environmental intensity components predicted.

2.2 Diffusion Model

In recent years, diffusion models have emerged as powerful tools for image generation, transforming random noise into the desired distribution through a series of transformations. Stable Diffusion [9] pre-trained on large image-text datasets can effectively translate between images and natural language, enabling high-quality image generation based on specific prompts. ControlNet [5], a popular conditional diffusion model, extends the capabilities of Stable Diffusion by using a conditional encoder to generate images based on user-provided constraints or conditions. In this work, we build our model on ControlNet and introduce innovative designs to address the specific challenges of shadow removal.

3 Method

The pipeline of our approach is shown in Fig. 1. We take the shadow as one kind of noise in the image and design a stable diffusion-based model to execute the de-shadowing process. Below, we first briefly introduce the diffusion framework, followed by the details of multi-exposure shadow generation and noise map intensity adaption.

3.1 Diffusion Model

Our diffusion model consists of two phases. Initially, the sampling process recursively adds noises in an expanded shadow region to transform a shadow-free image into a shadow one. Next, the de-shadowing process estimates the shadow noise from the generated shadow image and recovers the shadow-free image.

Sampling Process. The diffusion model is built by extending the ControlNet-based Stable Diffusion framework. During the sampling phase, a shadow-free image I^f is fed into the variational auto-encoder (VAE) to generate initial feature maps z_0. These feature maps are then gradually perturbed with noise to produce $z_{t|t=1,...,T}$ over T steps. Since shadow removal targets to recover shadow regions, noise is added selectively in those areas. This ensures that the denoising model remains focused on the shadow regions, minimizing interference with the non-shadowed parts of the image.

De-shadowing Training Process. The de-shadowing process works in the reverse direction of the sampling process. It also consists of T steps. At the t-th step, a pre-trained U-Net predicts the noise (i.e., shadow) map added from z_{t-1} to z_t, guided by conditions injected via a ControlNet encoder. The ControlNet is responsible for conveying shadow-related information to the U-Net, thus guiding the de-shadowing process. During training, the model is optimized based on the difference between the Gaussian sampling noise map ϵ and the predicted one $\hat{\epsilon}$ in the shadow area, formally,

$$\mathcal{L}_{\text{Diff}} = E_{t,\epsilon \sim \mathcal{N}(0,1)} \parallel I^m \otimes (\epsilon - \hat{\epsilon}) \parallel_2^2, \qquad (1)$$

where I^m denotes the shadow area.

3.2 Multi-exposure Shadow Image Generation

Effective inputs to the ControlNet are the key to the de-shadow training. Note that the main differences between shadow and non-shadow regions lie in their intensity, which visually varies with exposure levels. In high-exposure images, shadows often appear more pronounced, while in low-exposure images, they tend to be more faint. This trait of the shadow can be helpful for the shadow removal task.

To enhance ControlNet's understanding of shadow characteristics, we simulate the appearance of shadows under varying exposure conditions to generate a series of exposure maps. These maps, along with the shadow map and the original shadow image, are then fed into the ControlNet. Given a shadow image I_s, we follow [10] to manipulate its overall intensity using an intensity scale vector $\alpha \in \mathbb{R}^3$ and an intensity shift vector $\beta \in \mathbb{R}^3$, producing an exposure image I^e by

$$I^e = \alpha I^s + \beta, \qquad (2)$$

The key idea for estimating α and β is that the shadow region in the augmented image should better resemble the corresponding region in the shadow-free image I^f. Since the shadow-free image is not available during the testing phase, we design a deep neural network $\phi(\cdot)$ to predict these two parameters from its shadow regions by

$$(\alpha, \beta) = \phi(I^s, I^m) \qquad (3)$$

The specific structure of the exposure estimator network is presented in Fig. 2. To train the model, the ground truth values for α and β are estimated using least squares regression on the shadow regions in I_s and I^f. Once α and β are determined, a median-exposure image I^e is generated by applying these parameters to I_s. Additionally, two more exposure images are created by applying α and β with coefficients 0.8 and 1.2, respectively, to simulate shadow images under different exposure conditions. All of these images, along with the original shadow image and the shadow mask, are concatenated and fed into the ControlNet after a linear projection.

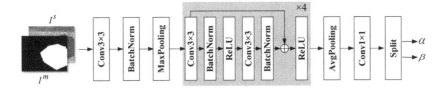

Fig. 2. Architecture of the exposure factor estimator.

3.3 Intensity Adaption and Post-processing

Figure 3 illustrates the architecture of our environmental intensity predictor. The network receives a shadow image I^s and a non-shadow mask $\overline{I^m} = 1 - I^m$ as inputs, producing an intensity scale vector s and a bias vector b as outputs. The core architecture consists of a ResNet-18 backbone [11], followed by a 3×3 convolutional layer to reduce the channel dimension. The final output is obtained through linear transformations and splitting operations. These intensity factors are then applied to modulate the intensity of the predicted noise map $\tilde{\epsilon}$ generated by:

$$\hat{\epsilon} = s \otimes \tilde{\epsilon} + b \tag{4}$$

while the $\hat{\epsilon}$ is the predicted noise map after modulation.

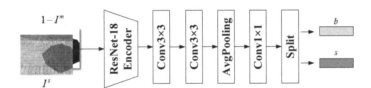

Fig. 3. Architecture of the environmental intensity predictor.

During the inference stage, given a shadow image, we first feed it into the deshadowing U-Net to obtain shadow-free features. These features are then passed through the VAE decoder of the stable diffusion model to produce a shadow-free image. Additionally, we have also designed a post-processing that subtly adjusts the output image to correct color shifts, resulting in the final image.

4 Experiments

4.1 Experimental Settings

Implementation Details. We developed our method using PyTorch 1.12.1, training the model with the Adam optimizer at a constant learning rate of 1e-5 over 50 epochs. All experiments were conducted on a GeForce RTX 3090 Ti GPU. All test results are resized to 256 × 256 for comparisons as in [12].

Datasets. Two benchmark datasets are tested. The Adjusted ISTD (ISTD+) dataset [13] includes 1870 image triplets (shadow images, non-shadow images, and shadow masks), divided into 1330 training triplets and 540 testing triplets. The SRD dataset [4] consists of 2680 training pairs and 408 testing pairs. Due to the lack of ground truth shadow masks, we utilized the publicly available SRD shadow masks provided by DHAN [12] for training and testing.

Evaluation Metrics. We utilize the Root Mean Square Error (RMSE) [14] in the LAB color space, Peak Signal-to-Noise Ratio (PSNR), and Structural Similarity Index (SSIM) to compare the estimated image with ground truths. For RMSE, lower values indicate better performance while PSNR and SSIM tell the opposite.

Table 1. Quantitative comparisons of different methods on the SRD dataset. The best and second-best results are in bold and underlined respectively.

Method	Shadow			Non-Shadow			All		
	PSNR↑	SSIM↑	RMSE↓	PSNR↑	SSIM↑	RMSE↓	PSNR↑	SSIM↑	RMSE↓
Input images	18.96	0.871	36.69	31.47	0.975	4.83	18.19	0.829	14.05
Guo et al. [15] (TPAMI'12)	–	–	29.89	–	–	6.47	–	–	12.60
DeshadowNet [4] (CVPR'17)	–	–	11.78	–	–	4.84	–	–	6.64
DSC [16] (TPAMI'19)	30.65	0.960	8.62	31.94	0.965	4.41	27.76	0.903	5.71
DHAN [12] (AAAI'20)	33.67	0.978	8.94	34.79	0.979	4.80	30.51	0.949	5.67
Fu et al. [10] (CVPR'21)	32.26	0.966	8.55	31.87	0.945	5.74	28.40	0.893	6.50
Jin et al. [17] (ICCV'21)	34.00	0.975	7.70	35.53	0.981	3.65	31.53	0.955	4.65
BMNet [7] (CVPR'22)	35.05	0.981	6.61	36.02	0.982	3.61	31.69	0.956	4.46
SG-ShadowNet [1] (ECCV'22)	–	–	7.53	–	–	2.97	–	–	4.23
ShadowFormer [3] (AAAI'23)	36.91	**0.989**	5.90	36.22	<u>0.989</u>	3.44	32.90	0.958	4.04
TBRNet [18] (TNNLS'23)	–	–	7.69	–	–	4.89	–	–	5.57
Liu et al. [19] (AAAI'24)	<u>36.51</u>	0.983	<u>5.49</u>	<u>37.71</u>	0.986	3.00	<u>33.48</u>	0.967	<u>3.66</u>
Ours	**37.28**	<u>0.985</u>	**4.89**	**39.03**	**0.991**	**2.85**	**34.29**	**0.970**	**3.57**

4.2 Results

Tables 1 and 2 summarize the results on the SRD and ISTD+ datasets. For the SRD dataset, performance comparisons across all metrics are provided, whereas for ISTD+, following [1,7,10], we report RMSE for all methods. As indicated, our approach consistently outperforms competitors across all metrics evaluating the overall quality of the generated images. Specifically, in shadow regions, our method excels in both PSNR and RMSE, while achieving performance comparable to ShadowFormer in SSIM. Figure 4 showcases visual examples from both datasets, demonstrating that our approach generates high-quality de-shadowed images with fewer structural artifacts and less noise compared to other methods. Overall, our method achieves superior performance both quantitatively and qualitatively, underscoring its effectiveness.

Table 2. Quantitative results on the ISTD+ dataset in RMSE. The best and second-best are in bold and underlined respectively.

Method	Shadow	Non-Shadow	All
Input images	40.2	2.6	8.5
Guo et al. [15] (TPAMI'12)	22.0	3.1	6.1
DeshadowNet [4] (CVPR'17)	15.9	6.0	7.6
ST-CGAN [20] (CVPR'18)	13.4	7.7	8.7
ShadowGan [21] (CVPR'19)	12.4	4.0	5.3
SP+M-Net [13] (ECCV'20)	7.9	3.1	3.9
G2R [6] (CVPR'21)	7.3	2.9	3.6
Fu et al. [10] (CVPR'21)	6.5	3.8	4.2
Jin et al. [17] (ICCV'21)	10.3	3.5	4.6
BMNet [7] (CVPR'22)	5.6	2.5	3.0
SG-ShadowNet [1] (ECCV'22)	5.9	2.9	3.4
ShadowFormer [3] (AAAI'23)	<u>5.2</u>	<u>2.3</u>	<u>2.8</u>
TBRNet [18] (TNNLS'23)	6.4	3.3	3.8
Liu et al. [19] (AAAI'24)	5.6	<u>2.3</u>	<u>2.8</u>
Ours	**5.1**	**2.2**	**2.7**

Table 3. Ablation studies on the ISTD+ datasets. All results are reported in RMSE, where lower values indicate better performance.

Case	Multi-exposure Inputs	Intensity Modulation	Shadow	Non-Shadow	All
C1	-	✓	6.1	2.4	2.9
C2	✓	-	5.3	2.3	2.8
C3	✓	✓	5.1	2.2	2.7

4.3 Ablation Studies

To validate different design choices in our approach, we conduct ablation studies using the ISTD+ dataset (Table 3). Specifically, we evaluate the impact of replacing multi-exposure inputs with the original shadow image for the diffusion model (case C1) and the absence of intensity modulation for noise maps during de-shadowing training (case C2). As shown, comparing case C1 with C2, the performance significantly improves when multi-exposure inputs are utilized. Furthermore, intensity modulation enhances the quality of the shadow-free images, as evident when comparing case C2 with C3. Overall, both components are essential for the success of our proposed method.

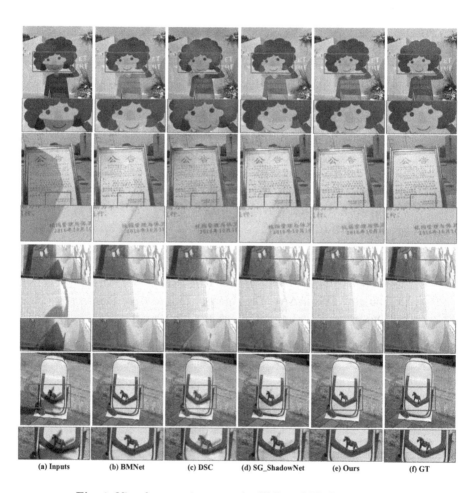

Fig. 4. Visual comparisons on the SRD and ISTD+ datasets.

5 Conclusion

In this paper, we introduce a novel network model for shadow removal based on the conditional stable diffusion model. We generate the multi-exposure shadow images as well as design an environmental intensity encoder to guide the deshadowing process. Our experiments demonstrate that our model achieves superior performance on the ISTD+ and SRD datasets.

References

1. Wan, J., Yin, H., Wu, Z., Wu, X., Liu, Y., Wang, S.: Style-guided shadow removal. In: ECCV, pp. 361–378 (2022). Springer
2. Chen, Z., Long, C., Zhang, L., Xiao, C.: Canet: a context-aware network for shadow removal. In: IEEE ICCV, pp. 4743–4752 (2021)
3. Guo, L., Huang, S., Liu, D., Cheng, H., Wen, B.: Shadowformer: global context helps image shadow removal. arXiv preprint arXiv:2302.01650 (2023)
4. Qu, L., Tian, J., He, S., Tang, Y., Lau, R.W.: Deshadownet: a multi-context embedding deep network for shadow removal. In: IEEE CVPR, pp. 4067–4075 (2017)
5. Zhang, L., Rao, A., Agrawala, M.: Adding conditional control to text-to-image diffusion models. In: IEEE ICCV, pp. 3836–3847 (2023)
6. Liu, Z., Yin, H., Wu, X., Wu, Z., Mi, Y., Wang, S.: From shadow generation to shadow removal. In: IEEE CVPR, pp. 4927–4936 (2021)
7. Zhu, Y., Huang, J., Fu, X., Zhao, F., Sun, Q., Zha, Z.-J.: Bijective mapping network for shadow removal. In: IEEE CVPR, pp. 5627–5636 (2022)
8. Vasluianu, F.-A., et al: Ntire 2024 image shadow removal challenge report. In: IEEE CVPR, pp. 6547–6570 (2024)
9. Fernandez, P., Couairon, G., Jégou, H., Douze, M., Furon, T.: The stable signature: rooting watermarks in latent diffusion models. In: IEEE ICCV, pp. 22466–22477 (2023)
10. Fu, L., et al.: Auto-exposure fusion for single-image shadow removal. In: IEEE CVPR, pp. 10571–10580 (2021)
11. He, K., Zhang, X., Ren, S., Sun, J.: Deep residual learning for image recognition. In: IEEE CVPR, pp. 770–778 (2016)
12. Cun, X., Pun, C.-M., Shi, C.: Towards ghost-free shadow removal via dual hierarchical aggregation network and shadow matting gan. In: AAAI, vol. 34, pp. 10680–10687 (2020)
13. Zhang, L., Yan, Q., Zhu, Y., Zhang, X., Xiao, C.: Effective shadow removal via multi-scale image decomposition. Vis. Comput. **35**, 1091–1104 (2019)
14. Wang, J., Li, X., Yang, J.: Stacked conditional generative adversarial networks for jointly learning shadow detection and shadow removal. In: IEEE CVPR, pp. 1788–1797 (2018)
15. Guo, R., Dai, Q., Hoiem, D.: Paired regions for shadow detection and removal. IEEE TPAMI **35**(12), 2956–2967 (2012)
16. Hu, X., Fu, C.-W., Zhu, L., Qin, J., Heng, P.-A.: Direction-aware spatial context features for shadow detection and removal. IEEE TPAMI **42**(11), 2795–2808 (2019)
17. Jin, Y., Sharma, A., Tan, R.T.: Dc-shadownet: single-image hard and soft shadow removal using unsupervised domain-classifier guided network. In: IEEE ICCV, pp. 5027–5036 (2021)

18. Liu, J., Wang, Q., Fan, H., Tian, J., Tang, Y.: A shadow imaging bilinear model and three-branch residual network for shadow removal. IEEE TNNLS 1–15 (2023)
19. Liu, Y., Ke, Z., Xu, K., Liu, F., Wang, Z., Lau, R.W.: Recasting regional lighting for shadow removal. In: AAAI, vol. 38, pp. 3810–3818 (2024)
20. Zhang, H., Dai, Y., Li, H., Koniusz, P.: Deep stacked hierarchical multi-patch network for image deblurring. In: IEEE CVPR, pp. 5978–5986 (2019)
21. Hu, X., Jiang, Y., Fu, C.-W., Heng, P.-A.: Mask-shadowgan: learning to remove shadows from unpaired data. In: IEEE ICCV, pp. 2472–2481 (2019)

Human Disease Prediction Based on Symptoms Using Novel Machine Learning

Ibukunoluwa Oluwabusayo Efunwoye[1], Mandar Gogate[1], Adeel Hussain[1], Bin Luo[2], Jinchang Ren[3], Fengling Jiang[1], Amir Hussain[1], and Kia Dashtipour[1](✉)

[1] School of Computing, Edinburgh Napier University, Edinburgh EH10 5DT, UK
K.Dashtipour@napier.ac.uk
[2] Anhui University, Hefei, China
[3] National Subsea Centre, Robert Gordon University, Aberdeen AB21 0BH, UK
40584472@live.napier.ac.uk

Abstract. Nowadays, individuals are often preoccupied with their daily lives and may disregard minor illnesses they are experiencing. However, these seemingly insignificant diseases can sometimes escalate into more serious health problems. Therefore, in this paper, we propose a novel approach for the detection of various types of diseases using machine learning algorithms, such as Support Vector Machine (SVM), Naïve Bayes, Multilayer Perceptron (MLP), Convolutional Neural Network (CNN), Long Short-Term Memory (LSTM), and others. The primary objective of this research is to develop a personalized health monitoring system that leverages an individual's medical history and current health status, utilizing advanced machine learning techniques. The main aim of this study is to provide personalized health predictions based on individual's medical and current health information. Through our proposed approach, we aim to create a robust and accurate disease prediction model that can effectively identify a wide range of ailments, including minor diseases. Our approach involves employing diverse machine learning algorithms to capture complex patterns and features in the data, enabling a comprehensive analysis of health conditions. The results of our machine learning experiments demonstrate the effectiveness of our proposed approach in detecting small diseases. By leveraging machine learning algorithms, we are able to provide timely and accurate disease predictions, enabling early identification of potential health issues. This proactive approach to disease detection has the potential to facilitate timely medical intervention, leading to improved health outcomes and overall well-being. Our research contributes to the field of personalized health monitoring and has implications for proactive disease prevention and management strategies. Further studies and validations are warranted to refine and optimize our approach for real-world clinical applications.

Keywords: Machine Learning · Disease prediction · Deep Learning

1 Introduction

In the recent years, to learn from the past and identify meaningful patterns from large and unstructured data, machine learning algorithms utilize statistical, probabilistic, and optimization techniques. These techniques are employed in many applications, including automatic text analysis [7,24], network intrusion detection [8,16], hearing aids [10,18], and others. However, the majority of these applications have been built using supervised machine learning algorithms instead of unsupervised ones. In supervised machine learning, a prediction model is created by learning from a labeled dataset, where the outcomes of unlabeled cases can be predicted based on the identified labels [1,2,4,6,10–14,17,22].

People are facing different types of diseases due to various lifestyle habits [3]. Therefore, it is important to predict diseases in the early stages. Early prediction of diseases can be challenging for doctors. To overcome this issue and make accurate predictions based on different symptoms, machine learning can be used. The use of machine learning for detecting diseases in the early stages has rapidly increased [5,23]. However, there is currently no comprehensive framework available for detecting various types of diseases in the early stages. Therefore, we propose a novel framework based on machine learning and deep learning approaches to detect different types of diseases in the early stages. We utilize a disease prediction dataset consisting of 132 parameters and 42 different types of diseases, such as diabetes, allergies, hepatitis, etc. We employ various machine learning algorithms, such as logistic regression, multilayer perceptron (MLP), K-nearest neighbors (KNN), random forest, support vector machine (SVM), as well as deep learning algorithms, such as convolutional neural network (CNN), 2D-CNN, and long short-term memory (LSTM).

Gavhane et al. [9] proposed an approach for heart disease prediction based on machine learning. However, the multilayer perceptron (MLP) achieved better performance compared to other approaches. The MLP is more efficient and accurate compared to other algorithms. Additionally, the algorithm generates reliable heart disease predictions based on limited input. Traditional approaches for heart disease prediction are not reliable, and Haq et al. [15] proposed a novel hybrid approach to classify healthy individuals and patients with heart disease problems. Seven well-known datasets were used to evaluate the performance of the approach. The machine learning results demonstrate that the approach can easily identify heart disease patients from healthy individuals.

Mohan et al. [19] proposed a method which used to find the most relevant features by applying machine learning techniques resulting to improve the performance of prediction of cardiovascular disease. The model introduced with different types of features and several classification techniques. In addition, they proposed the enhance model with high level of performance in terms of prediction of heart diseases with random forest and linear model.

In addition, Shen et al. [21] introduced MEGMA (Microbial Embedding, Grouping, and Mapping Algorithm) that utilized unsupervised learning techniques to enhance the tasks of disease prediction and identification of key biomarkers. The approach demonstrate that MEGMA can be used to construct unsupervised microbial embeddings, structured multichannel feature maps, and signal-amplified metagenomic feature maps, which in turn can improve the performance of downstream supervised tasks such as disease prediction and key biomarker recognition. Table 1 shows the summary of the latest research on human disease prediction.

Table 1. Summary of the latest research on Human Disease Prediction

Research Method	Methods	Key Findings
Machine learning	SVM, KNN, Random Forest	Predict diease with high accuracy
Data	Electronic health records (EHRs)	Can be used machine learning models to predict diseases
Algorithms	Deep learning	Can used accuracy and performance of machine learning models
Application	Early Diagnosis	Machine learning can be used to improve the diagnosis, treatment, and prevention of diseases

In addition, the machine learning (ML) can be used in different fields such as health, agriculture, text classification. The ML can analyse the large amount of data without learning. There are different types of machine learning such as supervised, semi-supervised and unsupervised machine learning are available. The types of ML are used to analyse the data in more accurate way. However, it is worth to mention that the supervised ML are mostly applied into medical data. Because the medical data is vital and it can related to life of the patients. In addition, the medical data is increased day by day therefore, predicting the correct types of disease is become a challenging task. However, the process of the big data is very challenging and time consuming. Machine learning has been widely used in the field of disease prediction and has achieved great success in many areas. However, there are still some limitations to the current approaches. For example, they often require large amounts of labeled data, which can be expensive and time-consuming to collect. Additionally, they can be sensitive to noise and outliers in the data. To address these limitations, we propose a novel approach based on feature-based machine learning techniques. Our approach uses a small number of hand-crafted features to represent the data, which makes it more robust to noise and outliers. Additionally, our approach can be trained on a smaller dataset, which makes it more efficient.

The paper organised as follows: Sect. 2, explain the methodology, Sect. 3 presents the results and discussion and finally Sect. 4 concludes the paper.

2 Methodology

In this section, we explain the main framework to detect human diseases prediction. The Fig. 1 shows the overview of the framework to detect disease. The framework consists of four main steps:

- Data acquisition. The first step is to acquire data that can be used to train the disease detection model. This data can come from a variety of sources, such as electronic health records (EHRs), medical images, and clinical trials.
- Data preprocessing. The next step is to preprocess the data to make it suitable for training the model. This may involve cleaning the data, removing outliers, and normalizing the data.
- Model training. The third step is to train the disease detection model. This is done by feeding the preprocessed data to the model and allowing it to learn the patterns that are associated with different diseases.
- Model evaluation. The final step is to evaluate the performance of the disease detection model. This is done by testing the model on a held-out dataset that was not used for training.

The framework described has the potential to revolutionize the way that diseases are detected and predicted. By using machine learning techniques, it is possible to develop models that can identify diseases with a high performance. This could lead to earlier diagnosis and treatment of diseases, which could improve patient outcomes and save lives.

Fig. 1. Overview of data flow in our methodology

2.1 Dataset Description

The dataset consists of different symptoms along with predicted labels for each disease. Therefore, applying machine learning to the field of medical science can make the tasks of physicians easier. The dataset is made up of 132 different parameters, which can be used to predict 42 different types of diseases. Figure 2 visualizes the data to better understand disease prediction data. The dataset is a valuable resource for researchers and clinicians. It can be used to develop new machine learning models for disease prediction, and to improve the accuracy of existing models. Additionally, it can also be used to identify new patterns and associations between symptoms and diseases. This information can be used to develop new diagnostic and treatment strategies. Figure 2 shows a heatmap of the correlation between different symptoms and diseases. These correlations can be used to develop machine learning models for disease prediction. Machine learning models can be trained on a dataset of symptoms and diseases, and then used to predict the probability of a patient having a particular disease. This information can be used to help physicians diagnose diseases more accurately and quickly. The use of machine learning in the field of medical science has the potential to revolutionize the way that diseases are diagnosed and treated. By developing accurate and reliable machine learning models for disease prediction, it is possible to improve patient outcomes and save lives. [20]. In addition Fig. 3 displays the correlation matrix of the Fig. 2. Figure 3 provides features which have negative correlation with the target value while some have positive.

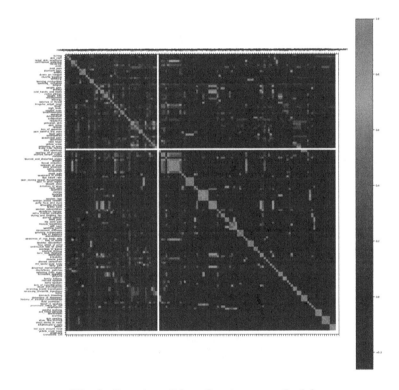

Fig. 2. Overview of data flow in our methodology

3 Experimental Results and Discussion

In order to evaluate the performance of the approach, we developed a novel framework based on machine learning and deep learning approaches. We used Keras, a high-level neural networks API, built on top of TensorFlow, to train the deep learning approaches, such as convolutional neural networks (CNNs), long short-term memory (LSTM) networks, and 2D-CNNs. We used scikit-learn to train the machine learning classifiers, such as support vector machines (SVMs), k-nearest neighbors (KNNs), random forests, and logistic regression. Table 1 shows the parameters used to train the machine learning and deep learning approaches.

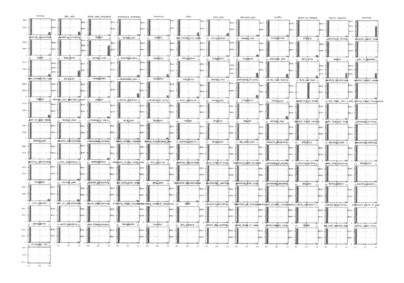

Fig. 3. Histogram for the Features

Table 2. Machine learning methods with their parameters

Algorithm	Parameters	Training time
KNN	Eculidean distance	1 m 20 s
SVM	RBF Kernal	2 m 4 s
Random Forest	Max depth = 2	1 m 5 s
Logistic regression	Penalty = 12	2 m 36 s
LSTM	10-layered	8 m 3 s
1D-CNN	dropout = 0.2	7 m 23 s
2D-CNN	dropout = 0.2	7 m 35 s

Table 2 present the summary of machine learning and deep learning approaches (Table 3).

It is difficult task to generate accurate diagnosis from patients based on the current symptoms because the patients required addition clinical examination and evaluation to provide accurate prediction of the type of disease. In addition, the dataset is small and it quite difficult to improve the performance for different types of diseases. In addition, there is only one subject has been used for most

Table 3. Summary of Results

ML/DL	Accuracy	Precision	Recall	F-measure
Logistic Regression	75.63	0.75	0.74	0.75
MLP	77.96	0.77	0.76	0.77
KNN	72	0.72	0.72	0.72
Random Forest	74.63	0.74	0.73	0.74
SVM	80.56	0.80	0.79	0.80
CNN	83.92	0.83	0.82	0.83
2D-CNN	86.73	0.86	0.85	0.86
LSTM	89.6	0.89	0.88	0.89

the diseases. Therefore, we are not certain that the current approach can widely detect different types of diseases on different subjects. Moreover, the current gender and age of the patient is not provided in the dataset.

4 Conclusion

In this paper, we propose a general disease prediction system based on machine learning and deep learning approaches. We utilized Support Vector Machine (SVM), Multilayer Perceptron (MLP), logistic regression, K-Nearest Neighbors (KNN), random forest, Convolutional Neural Network (CNN), 2D-CNN, and Long Short-Term Memory (LSTM) algorithms. Our results outperformed state-of-the-art approaches in terms of accuracy. However, we acknowledge that deep learning models have high time consumption during training. As future work, we aim to develop novel feature engineering techniques to extract the most valuable features for early detection of diseases.

References

1. Amin, R.U., et al.: Towards cloud-based and federated a-synchronous speech enhancement using deep neuro-fuzzy models: review, challenges & future directions. In: Proceedings of the AVSEC 2024, pp. 79–81 (2024)
2. Anwary, A.R., et al.: Target speaker direction estimation using eye gaze and head movement for hearing aids. In: Proceedings of the AVSEC 2024, pp. 73–74 (2024)
3. Chen, A.H., Huang, S.-Y., Hong, P.-S., Cheng, C.-H., Lin, E.-J.: Hdps: heart disease prediction system. In: 2011 Computing in Cardiology, pp. 557–560. IEEE (2011)
4. Dashtipour, K., et al.: Towards cross-lingual audio-visual speech enhancement. In: Proceedings of the AVSEC 2024, pp. 30–32 (2024)
5. Dashtipour, K., Gogate, M., Cambria,E., Hussain, A.: A novel context-aware multi-modal framework for persian sentiment analysis, arXiv preprint arXiv:2103.02636, 2021

6. Dashtipour, K., et al.: Evaluating the audio-visual speech enhancement challenge (AVSEC) baseline model using an out-of-domain free-flowing corpus. In: Proceedings of the AVSEC 2024, pp. 75–78 (2024)
7. Dashtipour, K., Gogate, M., Li, J., Jiang, F., Kong, B., Hussain, A.: A hybrid persian sentiment analysis framework: integrating dependency grammar based rules and deep neural networks. Neurocomputing **380**, 1–10 (2020)
8. Fu, Y., Du, Y., Cao, Z., Li, Q., Xiang, W.: A deep learning model for network intrusion detection with imbalanced data. Electronics **11**(6), 898 (2022)
9. Gavhane, A., Kokkula, G., Pandya, I., Devadkar, K.: Prediction of heart disease using machine learning. In: 2018 Second International Conference on Electronics, Communication and Aerospace Technology (ICECA), pp. 1275–1278. IEEE (2018)
10. Gogate, M., Dashtipour, K., Adeel, A., Hussain, A.: Cochleanet: a robust language-independent audio-visual model for real-time speech enhancement. Information Fusion **63**, 273–285 (2020)
11. Gogate, M., Dashtipour, K., Bell, P., Hussain, A.: Deep neural network driven binaural audio visual speech separation. In: 2020 International Joint Conference on Neural Networks (IJCNN), pp. 1–7. IEEE (2020)
12. Gogate, M., Dashtipour, K., Hussain, A.: Visual speech in real noisy environments (vision): a novel benchmark dataset and deep learning-based baseline system. In: Interspeech, 2020, pp. 4521–4525 (2020)
13. A lightweight real-time audio-visual speech enhancement framework. In: Proceedings of AVSEC 2024, pp. 19–23 (2024)
14. Gogate, M., Hussain, A., Dashtipour, K., Hussain, A.: Live demonstration: realtime multi-modal hearing assistive technology prototype. In: 2023 IEEE International Symposium on Circuits and Systems (ISCAS), p. 1. IEEE (2023)
15. Haq, A.U., Li, J.P., Memon, M.H., Nazir, S., Sun, R.: A hybrid intelligent system framework for the prediction of heart disease using machine learning algorithms. Mob. Inf. Syst. **2018** (2018)
16. He, K., Kim, D.D., Asghar, M.R.: Adversarial machine learning for network intrusion detection systems: a comprehensive survey. IEEE Commun. Surv. Tutor. (2023)
17. Hussain, A., et al.: Artificial intelligence-enabled analysis of public attitudes on facebook and twitter toward COVID-19 vaccines in the united kingdom and the United States: observational study. J. Med. Internet Res. **23**(4), e26627 (2021)
18. Hussain, T., et al.: A novel speech intelligibility enhancement model based on canonical correlation and deep learning. In: 2022 44th Annual International Conference of the IEEE Engineering in Medicine & Biology Society (EMBC), pp. 2581–2584. IEEE (2022)
19. Mohan, S., Thirumalai, C., Srivastava, G.: Effective heart disease prediction using hybrid machine learning techniques. IEEE Access **7**, 81 542–81 554 (2019)
20. Pal, A.K., Rawal, P., Ruwala, R., Patel, V.: Generic disease prediction using symptoms with supervised machine learning. Int. J. Sci. Res. Comput. Sci. Eng. Inf. Technol. **5**(2), 1082–1086 (2019)
21. Shen, W.X., Liang, S.R., Jiang, Y.Y., Chen, Y.Z.: Enhanced metagenomic deep learning for disease prediction and consistent signature recognition by restructured microbiome 2d representations. Patterns **4**(1), 100658 (2023)
22. Tiwari, U., et al.: Real-time audio visual speech enhancement: integrating visual cues for improved performance. In: Proceedings of the AVSEC 2024, pp. 38–42 (2024)

23. Uddin, S., Khan, A., Hossain, M.E., Moni, M.A.: Comparing different supervised machine learning algorithms for disease prediction. BMC Med. Inform. Decis. Mak. **19**(1), 1–16 (2019)
24. Zheng, X., Schweickert, R.: Differentiating dreaming and waking reports with automatic text analysis and support vector machines. Conscious. Cogn. **107**, 103439 (2023)

CAT-LCAN: A Multimodal Physiological Signal Fusion Framework for Emotion Recognition

Ao Li[1], Zhao Lv[1,2], and Xinhui Li[1(✉)]

[1] School of Computer Science and Technology, Anhui University, Hefei 230601, People's Republic of China
xinhuili@ahu.edu.cn
[2] The Key Laboratory of Flight Techniques and Flight Safety, CAAC, Hefei, China

Abstract. Emotion recognition is a complex task, especially in the fusion of multimodal physiological signals. Effectively capturing the dynamic characteristics and cross-modal information of signals is a primary challenge. To address this, we propose an emotion recognition method based on a Cross-Modal Attention Transformer and Learning-Classification Adversarial Network (CAT-LCAN), which effectively integrates multiple physiological signals. Cross-subject experiments conducted on two publicly available datasets, DEAP and WESAD, show that CAT-LCAN significantly outperforms several state-of-the-art baseline models. This innovative approach offers new insights into cross-subject multimodal emotion recognition and holds substantial research and practical significance.

Keywords: Emotion Recognition · Cross-modal Attention Transformer · Learning-Classify Adversarial Network · Physiological Signals

1 Introduction

Brain-computer interface (BCI) technology facilitates connecting the human brain and external devices [1]. Within BCIs, emotion recognition technology enhances the adaptability of these devices to human emotional states [2]. Unlike other methods based on non-physiological signals such as speech and facial expressions, emotion recognition using physiological signals offers advantages such as high real-time capability, rich data dimensions, and heightened sensitivity to individual differences, further strengthening its practical application potential [3]. However, limitations remain when using a single physiological signal to process complex emotions. Given the multidimensional nature of emotions, a solitary signal often fails to fully capture an individual's emotional state [4]. Additionally, the emotional information across different physiological signals varies significantly, making it challenging to achieve optimal results by merely concatenating features from multiple modalities, which may result in information redundancy or distortion [5]. Therefore, designing a multimodal fusion method that can effectively extract common modal features and complementary information is crucial.

Recently, advances in deep learning, particularly with the advent of Transformer architectures, have opened new avenues for multimodal emotion recognition. The Transformer, with its powerful feature extraction and fusion capabilities, is well-suited for processing high-dimensional data and capturing intricate inter-signal relationships [5]. This architecture significantly enhances emotion recognition accuracy, enabling researchers to explore richer information within multimodal signals [6]. In [7], a Cross-Modal Attention Transformer was introduced, allowing the extracted features of each modality to be independently represented while also obtaining complementary information through cross-modal interactions, thereby enhancing the expressive power of multimodal features. Despite the Cross-Modal Attention Transformer's ability to effectively extract complementary information from different modalities and its strong performance in emotion recognition tasks, it still faces a significant challenge: neglecting individual differences among subjects. Significant variations in physiological characteristics, emotional responses, and psychological states can lead to different physiological signal patterns for the same emotion across individuals [8]. Therefore, relying solely on cross-modal feature extraction while overlooking common feature extraction between subjects may limit the model's generalization ability and accuracy. To address this issue, it is crucial to introduce mechanisms that effectively align and fuse multimodal information from different subjects, allowing for a more comprehensive capture of individual-specific emotional expressions. For this purpose, we propose the Learning Classification Adversarial Network (LCAN). LCAN enhances cross-subject performance by leveraging adversarial training to optimize feature extraction across individuals, allowing the model to better capture shared features and improve its performance on unseen subjects. This approach not only increases the model's adaptability to individual variability but also significantly boosts emotion recognition accuracy.

Based on this, in this work, we introduce a novel multimodal emotion recognition framework, CAT-LCAN, which integrates the Cross-modal Attention Transformer with LCAN. Figure 1 illustrates the framework of CAT-LCAN. The main contributions of this work are as follows: (1) We introduce an innovative framework, named LCAN, designed to enhance the model's generalization through adversarial feature learning. This network can effectively align features across different modalities and subjects, ensuring that the model captures common features and enhances its generalization ability. (2) We design an end-to-end CAT-LCAN model that automates feature extraction, adversarial alignment, and emotion classification from multimodal inputs. This end-to-end framework simplifies the model design and training process, making better use of complementary information between modalities, and significantly improves cross-subject emotion recognition performance.

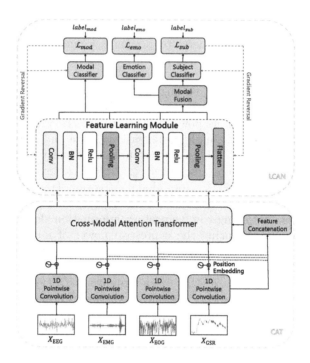

Fig. 1. The framework of CAT-LCAN.

2 Related Work

The deep learning-based multimodal representation learning framework can be categorized into multimodal joint representation and multimodal coordinated representation [9]. In multimodal joint representation, all modalities are treated as inputs, and signals from each modality are mapped into a shared representation space. For instance, Liu et al. [10] fused electroencephalogram (EEG) features with other attributes using a bimodal deep autoencoder (BDAE) to generate effective emotion-discriminative representations. Tang et al. [11] proposed a bimodal long short-term memory (LSTM) model that integrates time-domain and frequency-domain information from multimodal signals to achieve robust emotional discrimination. In contrast, multimodal coordination refers to mapping each modality into independent spaces that are coordinated with one another, allowing each modality to be used independently during the testing phase. Liu et al. [12] applied Deep Canonical Correlation Analysis (DCCA) to multimodal emotion recognition, and the results demonstrated that DCCA transformation preserves emotion-related information while eliminating irrelevant details. Tang et al. [13] proposed a multimodal ensemble neural network, RHPRNet, for emotion recognition based on central and peripheral nervous system signals. The model learns cross-modal and cross-domain emotion patterns by fusing EEG representations in the spatial frequency and complexity statistical domains with peripheral modalities, further enhancing the pattern fusion effect through a hierarchical global feature fusion module.

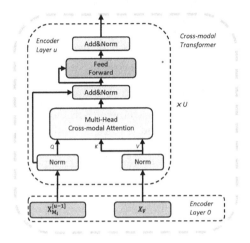

Fig. 2. The structure of Cross-modal Attention Transformer.

3 Methodology

3.1 CAT

Figure 2 shows the complete structure of CAT. Let $X_{m_i} \in \mathbb{R}^{L,D_i}$ represent the input signal of each modality, where $i = 1, 2, \ldots, u$ represent different physiological signals, L represents the length of the signal (sampling rate), and D_i represents the dimension of each signal (number of channels). Project them into the same dimension D using one-dimensional point convolution:

$$\bar{X}_{m_i} = Conv1D(X_{m_i}, 1) \quad (1)$$

where $\bar{X}_{m_i} \in \mathbb{R}^{L,D}$ represents an unimodal feature with the same dimension D. Then, in order to fully understand the relationship between adjacent elements, that is, the features of adjacent channels within a modal sequence, and thus obtain spatial information, traditional sine and cosine functions are used to encode $\bar{X}_{m_i}$ to obtain a primary unimodal feature $\bar{X}_{m_i}^p$ containing position information:

$$\bar{X}_{m_i}^p = \bar{X}_{m_i} + PE_{\bar{X}_{m_i}} \quad (2)$$

where $PE_{\bar{X}_{m_i}}$ represents the position information obtained by encoding the feature sequence $\bar{X}_{m_i}$, which can be defined as a matrix:

$$PE_{\bar{X}_{m_i}}[pos, 2k] = \sin\left(\frac{pos}{10000^{\frac{2k}{D}}}\right)$$
$$PE_{\bar{X}_{m_i}}[pos, 2k+1] = \cos\left(\frac{pos}{10000^{\frac{2k}{D}}}\right) \quad (3)$$

where *pos* represents the position of the token in the sequence, $pos \in [1, ..., L]$, $k \in [1, ..., \frac{D}{2}]$. In addition, concatenate the primary unimodal features of all modalities to obtain the primary multimodal feature $\bar{X}_F \in \mathbb{R}^{L_F, D}$:

$$\bar{X}_F = Concat(\bar{X}^p_{m_1}, \bar{X}^p_{m_2}, ..., \bar{X}^p_{m_u}) \quad (4)$$

To enhance the performance of LCAN's feature learning module in its adversarial interactions with the modality classifiers, we separately feed each modality's primary unimodal and multimodal features into a cross-modal attention module, which consists of several Cross-modal Transformer networks. The objective of this cross-modal attention mechanism is to compute the attention score between the target primary unimodal feature $\bar{X}^p_{m_i}$ and the primary multimodal feature $\bar{X}_F$. This attention score guides other primary unimodal features embedded within the multimodal feature to adapt and strengthen the target unimodal feature. The unimodal query (Q_{Uni}), fusion key (K_{Fus}), and fusion value (V_{Fus}) are defined as follows:

$$Q_{Uni} = \bar{X}^p_{m_i} \cdot W_{Q^{Uni}}$$
$$K_{Fus} = \bar{X}_F \cdot W_{K^{Fus}} \quad (5)$$
$$V_{Fus} = \bar{X}_F \cdot W_{V^{Fus}}$$

where $W_{Q^{Uni}} \in \mathbb{R}^{D, D_Q}, W_{K^{Fus}} \in \mathbb{R}^{D, D_K}$, and $W_{V^{Fus}} \in \mathbb{R}^{D, D_V}$ are learnable weights. Next, the potential adaptation and reinforcement of the primary multimodal features to the target unimodal features, i.e. the cross-modal attention $\bar{X}^{head_j}_{m_i} \in \mathbb{R}^{(L, D_V)}$ learned in the j-th head cross-modal attention, can be defined as:

$$\bar{X}^{head_j}_{m_i} = softmax\left(\frac{(Q^j_{Uni} \cdot K^j_{Fus})^\tau}{\sqrt{D_K}}\right) \cdot V^j_{Fus} \quad (6)$$

where $softmax(\cdot)$ represents the scaled cross-modal attention rating matrix between multimodal features and target unimodal features, and $\bar{X}^{head_j}_{m_i}$ is defined as single-head cross-modal attention. Therefore, the multi-head cross-modal attention between the i-th target modality and multimodal features can be expressed as:

$$\bar{X}^{Mul}_{m_i} = Concat\left(\bar{X}^{head_1}_{m_i}, ..., \bar{X}^{head_h}_{m_i}\right) \quad (7)$$

where h is the number of heads. Finally, the multi-head cross-modal attention representation is processed through a feedforward network that incorporates residual connections and layer normalization, enhancing both efficiency and stability. As a result, the intermediate unimodal feature $Y_{m_i} \in \mathbb{R}^{(L, D)}$ is obtained via the feedforward computation of a cross-modal Transformer network comprising U cross-modal attention encoder layers:

$$Y_{m_i} = LayerNorm\left(\bar{X}^{Mul}_{m_i} + FeedForward\left(\bar{X}^{Mul}_{m_i}\right)\right) \quad (8)$$

3.2 LCAN

The LCAN architecture is primarily composed of a modality classifier, a subject classifier, and a feature learning module. These three components collaborate through adversarial games. Initially, the intermediate unimodal feature Y_{M_i} is fed into the feature learning module, resulting in the unoptimized advanced unimodal feature $Z_{M_i} \in \mathbb{R}^{L,D}$:

$$Z_{m_i} = ConvBlock(Y_{m_i}) \tag{9}$$

where $ConvBlock(\cdot)$ is a continuous convolutional block, as shown in Fig. 1, consisting of a convolutional layer, normalization layer, activation function layer, and pooling layer. In addition, concatenate the advanced unimodal features Z_{M_i} of all modalities to obtain the unoptimized advanced multimodal features $Z_F \in \mathbb{R}^{L_F,D}$:

$$Z_F = Concat(Z_{m_1}, Z_{m_2}, ..., Z_{m_u}) \tag{10}$$

Next, the unoptimized advanced unimodal feature Z_{m_i} and the advanced multimodal feature Z_F are fed into a modality classifier and a subject classifier, both consisting of two fully connected (FC) layers, for adversarial classification learning. The modality classifier and subject classifier produce the modality classification output O^M and the subject classification output O^S, respectively.

$$\begin{aligned} O^M &= Softmax\left(f_{mod}(Z_{m_i})\right) \\ O^S &= Softmax\left(f_{sub}(Z_F)\right) \end{aligned} \tag{11}$$

where $f_{mod}(\cdot)$ and $f_{sub}(\cdot)$ are the mapping functions of the modal classifier and the subject classifier, respectively. The goal of a modal classifier is to distinguish which modality the current input unimodal feature comes from, and its loss function is defined as:

$$\mathcal{L}_{mod} = -\sum_{i=1}^{u} y_i log(O^M) \tag{12}$$

where y_i is the true label of the modality ID, and u is the number of modalities. Meanwhile, the goal of the subject classifier is to distinguish which subject the current multimodal feature comes from, and its loss function is defined as:

$$\mathcal{L}_{sub} = -\sum_{i=1}^{s} t_i log(O^S) \tag{13}$$

where t_i is the true label of the subject ID, and s is the number of subjects.

To optimize the generated advanced unimodal feature Z_{m_i} and advanced multimodal feature Z_F, we introduce an adversarial training mechanism. During adversarial loss training, the modality classifier aims to correctly classify modality IDs, optimizing its loss function as $\arg\min_{\theta_{mod}} \mathcal{L}_{mod}$, where θ_{mod} represents the parameters of the modality classifier. Simultaneously, the subject classifier focuses on correctly classifying subject IDs, optimizing its loss as $\arg\min_{\theta_{sub}} \mathcal{L}_{sub}$, where θ_{sub} represents the parameters of the subject classifier. Conversely, the feature learning module aims to hinder the classifiers from making accurate predictions by optimizing its parameters, expressed as $\arg\max_{\theta_{feat}}(\mathcal{L}_{mod} + \mathcal{L}_{sub})$, where θ_{feat} denotes the parameters of the feature learning module.

Finally, based on the concatenated advanced multimodal feature Z_F, the emotion classifier consisting of two FC layers can output the classification result O^E of the final emotion label:

$$O^E = Softmax\left(f_{emo}(Z_F)\right) \qquad (14)$$

where $f_{emo}(\cdot)$ is the mapping function of the modal sentiment classifier. At the same time, the calculated emotion category classification loss $\mathcal{L}_{emo}$ can be expressed as:

$$\mathcal{L}_{emo} = -\sum_{i=1}^{n} k_i log(O^S) \qquad (15)$$

where k_i represents the true label of the emotion category, and n is the total number of emotion categories. Additionally, because the optimization directions of the modality classifiers, subject classifiers, and feature learning module are inconsistent, applying standard gradient descent for deep neural network optimization becomes challenging. To address this, we employ a gradient reversal layer, which reverses the gradient flow from the adversarial loss for the feature learning module. This approach enables our model to be trained end-to-end by minimizing the total loss L as defined in Equation (16):

$$\mathcal{L} = \alpha \cdot \mathcal{L}_{mod} + \beta \cdot \mathcal{L}_{sub} + \gamma \cdot \mathcal{L}_{emo} \qquad (16)$$

where α, β, γ are the weights of the loss functions $\mathcal{L}_{mod}, \mathcal{L}_{sub}$, and $\mathcal{L}_{emo}$, respectively.

4 Experiments and Results

4.1 Dataset and Experimental Setup

This experiment selects two publicly available multimodal physiological signal datasets, DEAP [14] and WESAD [15]. Binary classification experiments are conducted on the DEAP dataset using Valence and Arousal dimensions, and ternary classification experiments on the WESAD dataset for neutral, stress, and amusement states.

We use the leave-one-subject-out (LOSO) strategy to divide the training and testing sets for both datasets. In each experimental round, we use the data of one subject as the testing set and the data of other subjects as the training set to ensure each subject is tested to evaluate model performance. Finally, we take the average of all subjects' results as the final model performance. We use Adam [16] as the optimizer, set the learning rate to 10–3, truncate gradients with absolute values exceeding 10 to stabilize the training, and set the training period to 50 epochs. All models are implemented using PyTorch and deployed on GeForce GTX 1080 Ti GPUs, with parallelism adjusted to accommodate resource constraints.

Table 1. LOSO accuracy results for DEAP and WESAD using different methods

Method	DEAP-Valence	DEAP-Arousal	WESAD
TARDGCN	58.24/ 5.87	59.71/9.14	–
DSSN	58.32/8.16	60.22/8.66	–
MCMT	61.96/4.28	61.80/8.45	65.78/8.10
DGR-ERPS	61.74/6.37	62.40/7.82	66.34/7.14
RDFKM	63.06/8.59	64.00/8.88	68.73/6.70
CAT-LCAN (ours)	65.11/8.53	66.53/12.03	69.93/9.11

4.2 Experimental Results

In the comparative experiment, we selected several state-of-the-art models, including EEG-based models for comparing the DEAP dataset: TARDGCN [17], DSSN [18], and multimodal-based models for comparing both the DEAP and WESAD datasets: MCMT [19], DGR-ERPS [20], RDFKM [21]. Table 1 shows the results of comparing our model with these baseline models. From the results presented in Table 1, it is evident that the proposed CAT-LCAN framework outperforms other baseline models across various evaluation metrics. For the DEAP dataset, our model achieved an average accuracy of 65.11% ± 8.53% in the valence dimension and 66.53% ± 12.03% in the arousal dimension. For the WESAD dataset, the average accuracy of the three classification tasks performed by CAT-LCAN is 69.93 ± 9.11%. Through comparative analysis, we have validated the effectiveness and robustness of the CAT-LCAN model in processing multimodal emotional data, providing strong support and a solid foundation for future research. Future research can focus on further optimizing model design and exploring additional potential application scenarios based on this foundation.

Furthermore, Fig. 3 illustrate the emotion recognition accuracy of each model for each subject on both the DEAP and WESAD datasets. It can be observed that although CAT-LCAN may not achieve the best performance on some subjects, potentially due to the varying adaptability of different models across subjects, the overall performance demonstrates that CAT-LCAN still maintains the highest recognition accuracy. These experimental results enable further analysis of the performance differences across different models and subjects, and facilitate the exploration of how individual characteristics impact the accuracy of emotion recognition.

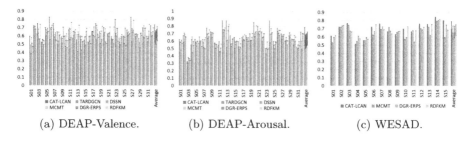

(a) DEAP-Valence.   (b) DEAP-Arousal.   (c) WESAD.

Fig. 3. LOSO accuracy results on the DEAP and WESAD datasets.

5 Conclusions

This paper introduces a novel multimodal emotion recognition framework, CAT-LCAN, which combines cross-modal attention transformer with learning classification adversarial networks. Experiments on the DEAP dataset confirmed CAT-LCAN's effectiveness, while comparisons with state-of-the-art models demonstrated its clear advantages. These results indicate that our method effectively captures the relationships among different physiological signals across various subjects, enabling accurate identification of emotional states. In future research, we aim to extend our method to additional datasets to validate its universality further.

Acknowledgments. This work is supported by the Natural Science Research Project of Anhui Educational Committee under Grant (No. 2024AH050054), Distinguished Youth Foundation of Anhui Scientific Committee (No. 2208085J05), National Natural Science Foundation of China (NSFC) (No.62476004), Cloud Ginger XR-1 platform.

References

1. Jaber, W., Jaber, H.A., Jaber, R., Saleh, Z.: The convergence of ai and bcis: a new era of brain-machine interfaces. Artificial Intelligence in the Age of Nanotechnology, pp. 98–113 (2024)
2. Samal, P., Hashmi, M.F.: Role of machine learning and deep learning techniques in EEG-based BCI emotion recognition system: a review. Artif. Intell. Rev. **57**(3), 50 (2024)
3. Houssein, E.H., Hammad, A., Ali, A.A.: Human emotion recognition from EEG-based brain-computer interface using machine learning: a comprehensive review. Neural Comput. Appl. **34**(15), 12527–12557 (2022)
4. Holm, K.: Classifying Emotions with Physiological Sensor Systems and Machine Learning. Master's thesis, University of Oslo (2024)
5. Kalateh, S., Estrada-Jimenez, L.A., Hojjati, S.N., Barata, J.: A systematic review on multimodal emotion recognition: building blocks, current state, applications, and challenges. IEEE Access (2024)
6. Rodriguez, J.F.V.: Multimodal transformers for emotion recognition. Ph.D. thesis, Université Grenoble Alpes [2020-....] (2023)

7. Wang, R., et al.: Husformer: a multi-modal transformer for multi-modal human state recognition. IEEE Trans. Cogn. Dev. Syst. (2024)
8. Van Doren, N., Dickens, C.N., Benson, L., Brick, T.R., Gatzke-Kopp, L., Oravecz, Z.: Capturing emotion coherence in daily life: using ambulatory physiology measures and ecological momentary assessments to examine within-person associations and individual differences. Biol. Psychol. **162**, 108074 (2021)
9. Baltrušaitis, T., Ahuja, C., Morency, L.P.: Multimodal machine learning: a survey and taxonomy. IEEE Trans. Pattern Anal. Mach. Intell. **41**(2), 423–443 (2018)
10. Liu, W., Zheng, W.L., Lu, B.L.: Emotion recognition using multimodal deep learning. In: Neural Information Processing: 23rd International Conference, ICONIP 2016, Kyoto, Japan, 16–21 October 2016, Proceedings, Part II 23, pp. 521–529. Springer (2016)
11. Tang, H., Liu, W., Zheng, W.L., Lu, B.L.: Multimodal emotion recognition using deep neural networks. In: Neural Information Processing: 24th International Conference, ICONIP 2017, Guangzhou, China, 14–18 November 2017, Proceedings, Part IV 24, pp. 811–819. Springer (2017)
12. Liu, W., Qiu, J.L., Zheng, W.L., Lu, B.L.: Multimodal emotion recognition using deep canonical correlation analysis. arXiv preprint arXiv:1908.05349 (2019)
13. Tang, J., Ma, Z., Gan, K., Zhang, J., Yin, Z.: Hierarchical multimodal-fusion of physiological signals for emotion recognition with scenario adaption and contrastive alignment. Information Fusion **103**, 102129 (2024)
14. Koelstra, S., et al.: Deap: a database for emotion analysis; using physiological signals. IEEE Trans. Affect. Comput. **3**(1), 18–31 (2011)
15. Schmidt, P., Reiss, A., Duerichen, R., Marberger, C., Van Laerhoven, K.: Introducing wesad, a multimodal dataset for wearable stress and affect detection. In: Proceedings of the 20th ACM International Conference on Multimodal Interaction, pp. 400–408 (2018)
16. Kingma, D.P.: Adam: a method for stochastic optimization. arXiv preprint arXiv:1412.6980 (2014)
17. Li, W., Wang, M., Zhu, J., Song, A.: EEG-based emotion recognition using trainable adjacency relation driven graph convolutional network. IEEE Trans. Cogn. Dev. Syst. **15**(4), 1656–1672 (2023)
18. Li, W., Dong, J., Liu, S., Fan, L., Wang, S.: Dynamic stream selection network for subject-independent eeg-based emotion recognition. IEEE Sens. J. (2024)
19. Li, J., Chen, N., Zhu, H., Li, G., Xu, Z., Chen, D.: Incongruity-aware multimodal physiology signals fusion for emotion recognition. Information Fusion **105**, 102220 (2024)
20. Li, J., Li, J., Wang, X., Zhan, X., Zeng, Z.: A domain generalization and residual network-based emotion recognition from physiological signals. Cyborg Bionic Syst. **5**, 0074 (2024)
21. Zhang, X., et al.: Emotion recognition from multimodal physiological signals using a regularized deep fusion of kernel machine. IEEE Trans. Cybern. **51**(9), 4386–4399 (2020)

A Novel Thermal Imaging and Machine Learning Based Privacy Preserving Framework for Efficient Space Allocation, Utilisation and Management

Maria Bruevich[1], Nilupulee A. Gunathilake[1], Mandar Gogate[1],
Adeel Hussain[1], Bin Luo[2], Jinchang Ren[3], Amir Hussain[1], Fengling Jiang[4],
and Kia Dashtipour[1(✉)]

[1] School of Computing, Edinburgh Napier University, Edinburgh EH10 5DT, UK
K.Dashtipour@napier.ac.uk
[2] Anhui University, Hefei, China
[3] National Subsea Centre, Robert Gordon University, Aberdeen AB21 0BH, UK
[4] Hefei Normal University, Hefei 230601, China
40546722@live.napier.ac.uk

Abstract. Most companies and organisations face notorious challenges in space management. The Internet-of-Things (IoT) networks attempt to pave the way to get space allocation information about the environment using sensors. However, the deployment of sensors and IoT tags is not cost-effective for large offices. Therefore, this work aims to develop a novel privacy-preserving framework based on thermal imaging cameras and machine learning (ML) techniques in order to monitor space allocation remotely. Hence, to utilise and manage dedicated and unused space appropriately. The work analyses the progress of the proposed methodology via several Deep Learning (DL) techniques based on convolutional neural networks (CNN) and vision transformers (ViT). The experimental results indicated that Visual Geometry Group (VGG) 16 model outperforms other models such as ViT, ResNet50, AlexNet and etc. In addition, a web application has been developed to select a model that is preferred to identify human occupancy in real-time and, therefore, to process space monitoring and utilisation remotely in large office settings.

Keywords: Machine Learning · Desk Allocation · Deep Learning

1 Introduction

The allocation of space in a workspace is crucial for establishing criteria to optimally allocate available space. It enhances employee productivity and the functionalities of the company by creating a comfortable environment that fosters efficient work. However, [24] indicates that 50% of assigned office spaces are underutilised, and 50% of large meeting rooms often remain empty. Additionally, it notes that 40% of scheduled meetings do not occur, resulting in allocated spaces being unused for those periods. These factors contribute to increased operational expenditure (OpEx). Consequently, effective space utilisation and management based on real-time insights into workspaces can significantly reduce OpEx while also increasing employee productivity. Therefore, the implementation of a cost-effective and privacy-preserving space monitoring system is essential for designing and optimising smart offices. In recent years, space allocation has garnered growing attention from researchers and innovators.

Many automated solutions leverage sensor technology based on Internet-of-Things (IoT) communication to optimise space monitoring, allocation, and utilisation and different tasks such as health monitoring [4,5,11,13–17,19,28]. The rationale for this approach lies in the capabilities of IoT methods to provide real-time data processing and remote operations through implemented sensor networks. However, deploying sensor networks with a large number of IoT tags is often not cost-effective in extensive office spaces [1,2,10]. Therefore, adapting machine learning (ML) algorithms, particularly through deep learning techniques, represents a more suitable approach for managing large spaces. Convolutional neural networks (CNNs), a type of deep learning model, have already demonstrated promising results in object detection using thermal cameras. The main contributions of this study are outlined as follows:

- This study introduces a novel privacy-preserved space monitoring framework, using thermal imaging and machine learning to detect and classify occupancy for the first time.
- A unique thermal imaging dataset is created, and the proposed model is benchmarked against seven advanced deep learning architectures, including CNN, LSTM, VGG16, ResNet50, AlexNet, and Vision Transformers (ViT).
- A user-centric smart web app has been developed and validated with real-world data from Edinburgh Napier University's library and hot desk office spaces, allowing users to remotely identify and book available spaces.

2 Related Work

In this section, we review the most recent studies in the field of desk allocation.

You Only Look Once (YOLO) is an intelligent version of a convolutional neural network (CNN) used for object detection in real-time. [26] presents the YOLO architecture in detail. Its unified architecture is identified to be extremely fast. The base version processes images at 45 frames per second, while fast YOLO, which is a reduced version of the network, processes 155 frames per second. On the other hand, it is identified to make significant localization errors, but is far less prone to predict false detection where nothing exists. The process of a YOLO implementation in Keras, a Python deep learning Application Programming Interface (API), can be referred to in [20]. Inception architecture, which is a deep neural network. It has repetitive components known as Inception modules that improve the utilization of the computing resources in the network. For optimizing the quality, the decisions are made based on the Hebbian theory and perception of multi-scale processing. It evaluates deep CNNs for large-scale image classification based on the VGG architecture. These networks exploit tiny 3 × 3 convolution filters, and the model improvement is delivered by increasing the depth of convolutional layers combined with those filters. The outcomes of it validate the necessity of depth in visual representation. MobileNetV2 [27] is a new mobile architecture based on an inverted residual structure. The input and the output of the residual block are thin bottleneck layers, and this model uses lightweight depthwise convolutions to filter. The work analyses the trade-offs between accuracy and several operations and parameters. Another CNN model with a Line-to-Line Ground (LG) - Radial Based Function (RBF) NN classifier, [3] demonstrates a CNN model based on LeNet with six inputs. It targets Human Action Recognition (HAR) for six action classes using thermal IR cameras, and classification accuracy of 87.44% has been gained for the test data.

[8] proposes the Cycle-consistent Generative Adversarial Network (Cycle-GAN) for halo effect removal. It detects 11 human actions using images generated from thermal camera videos. Furthermore, a combination of CNN and LSTM takes the skeleton frame as input, and it helps in extracting more spatial and temporal features from the frame. [22] discusses the effects of face/body regions considering clothing and AC on human skin temperature using images from Forward Looking IR (FLIR) A35 thermal camera. The system is based on descriptive statistics such as the mean, the standard deviation, the maximum and the minimum. Algorithms have been designed in [23] for indoor activity recognition including object identification as well as person recognition. It targets HAR from FLIR One Pro thermal images in residential spaces using TensorFlow's Object Detection API. The results confirm that the 3D model is better than the 2D model regarding temperature estimation.

In addition, there are some pre-trained models, alternatively known as transfer learning, available for image classification in deep learning. [25] is about VGGNet architecture, including VGG16 and VGG19 and functions based on Keras. Based on deep residual learning presented by a team of MS Research Asia (MSRA), Inception architecture for the 2nd, the 3rd, the 4th versions and InceptionResNetV2 provided by Google is in [7]. Furthermore, an extension of the Inception named Xception developed by Keras is in [9]. NasNetMobile model based on Nas architecture developed by Google based on [29] is identified to be most suitable for mobile and embedded applications.

3 Methodology

In our study, we utilised adopted pre-trained models and tested them on the desk allocation dataset. The pretrained models included basic CNN, CNN with regularization, CNN-LSTM, VGG16, ViT ResNet50, AlexNet.

VGG networks exploit tiny 3 × 3 convolution filters, and the model improvement is delivered by increasing the depth of convolutional layers combined with those filters. The base structure of VGG16 uses 13 convolutional layers, 5 max-pooling layers and 3 fully connected dense layers. ResNet50 is a substantially deeper network with 50 layers [18]. AlexNet has a more complex architecture than the basic CNN model featuring 5 convolutional layers and 3 fully connected dense layers [21]. The networks were trained for 80 epochs using the binary cross-entropy as a loss function for binary classification and accuracy as a metric, and an Adam optimiser with a learning rate of 0.001. Data augmentation is applied to the training set in all models to increase diversity samples through random transformations such as image rotation, cropping, changing contrast, etc.

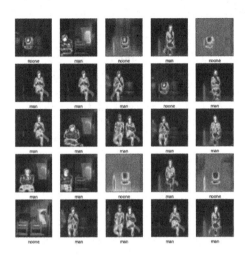

Fig. 1. Example of Thermal Imaging Corpus Collected for Desk Allocation

4 Experimental Results

In this section, we provide a detailed description of the corpus and the experimental setup employed in this study.

4.1 Desk Allocation Thermal Imaging Corpus

Two volunteer participants, one male and one female, both aged between 20 and 25, were recruited to participate in an experiment aimed at capturing and analysing thermal signatures under controlled conditions. The thermal data were recorded using two devices: a FLIR Thermal Imaging Camera, known for its high-resolution thermographic capabilities, and an Apple iPhone 12, which served as a complementary tool to compare consumer-grade imaging with specialized equipment.

The experiment was designed to collect thermal data under four distinct seating configurations. In the first condition, the male participant was seated in the left chair while the right chair remained unoccupied. In the second condition, the female participant was seated in the right chair, leaving the left chair empty. In the third scenario, both participants sat simultaneously in the left and right chairs. Finally, in the fourth condition, both chairs were left unoccupied, and the thermal signatures of the empty chairs were recorded to establish a baseline for comparison. Each scenario was carefully timed and repeated to ensure consistency and reliability in the data collection process. These experimental conditions were selected to examine the variations in thermal signatures depending on the seating arrangement and to assess the sensitivity of both devices to detect and distinguish human thermal patterns in a controlled indoor environment. The data obtained will provide insight into the applicability of thermal imaging technologies in settings such as human detection, occupancy monitoring, and potentially, more advanced applications in security and behavioural studies. The Fig. 1 shows example of thermal imaging corpus collected for desk allocation.

Images were adapted from short video clips obtained from the thermal camera. 1209 thermal images were used for the train set and 315 images were used for the test set. Therefore, the total size of the data set used is 1524. Types of models used to analyse the performances are basic CNN, CNN with regularisation, CNN-LTSM, VGG16, ResNet50, AlexNet and ViT. In the CNN with regularisation model, a dropout of 0.2 was applied after two dense layers to reduce overfitting.

4.2 Experimental Setup

In order to train the models, we used 80 epochs. The summary of the results obtained by each model is in Table 1. According to that, the highest accuracy has been gained by the CNN-LTSM model. The VGG16 model shows the highest score. The accuracy and loss variations for the models of basic CNN, CNN with regularisation, CNN-LTSM, VGG16 and ResNet50 are illustrated in Fig. 2, Fig. 3, Fig. 4, Fig. 5, Fig. 6 and Fig. 7 respectively.

Table 1. Summary of the results

Model	Accuracy	Precision	Recall	F1
Basic CNN	0.909	0.972	0.848	0.906
CNN with regularisation	0.959	1.000	0.921	0.959
CNN-LTSM	0.747	0.674	0.982	0.799
VGG16	0.962	0.932	1.000	0.965
ResNet50	0.684	0.619	1.000	0.765
AlexNet	0.953	0.981	0.927	0.953
ViT	0.650	0.64	0.64	0.64

Figure 8 shows the confusion matrices for all seven models. The top right square indicates the number of false positives, the bottom left square shows the number of false negatives and the bottom right square shows the number of true positives. False negatives are where it is unable to detect a human when a human is actually present, false positives are where empty space is taken as a human,

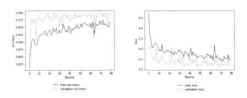

Fig. 2. Loss and accuracy changes in the basic CNN model

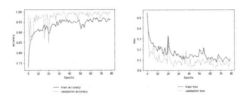

Fig. 3. Loss and accuracy changes in the CNN regularisation model

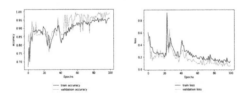

Fig. 4. Loss and accuracy changes in the CNN-LTSM model

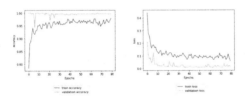

Fig. 5. Loss and accuracy changes in the VGG16 model

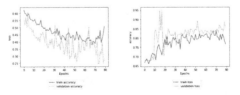

Fig. 6. Loss and accuracy changes in the ResNet50 model

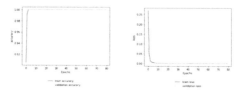

Fig. 7. Loss and accuracy changes in the ViT model

and true positives mean that a human is detected. Successful app development finely works with all seven models.

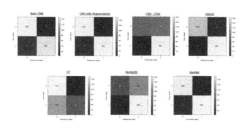

Fig. 8. Confusion matrices of the models

4.3 Discussion

The models of basic CNN, CNN with regularisation, VGG16 and AlextNet show successful performance because their test accuracy is greater than 0.9. The VGG16 and the CNN with regularisation architectures perform the best on the test data set, having a test accuracy of 0.962 and 0.959, respectively. The VGG16 model correctly identifies all the images with a human while misclassifying 12 images as false positives. Also, the VGG16 demonstrates smoother learning curves for accuracy and loss graphs as in Fig. 5. However, the CNN with regularisation model fails to detect a human in 13 images. The basic CNN model has a test accuracy of 0.909 and misclassified 33 images. Thus, compared to the basic CNN model, the CNN with regularisation architecture offers better accuracy and lower loss. The AlexNet model fails to detect a human in 12 images misclassifying the empty space in 3 cases.

The performance of the combined CNN-LTSM architecture is lower than the above four variants having the accuracy of 0.674 on the test data set. It is biased more towards predicting a human while there is no one because of 78 false positives and 3 false negatives. Although LTSM layers give the ability to process temporal features between sequential frames in a video, this may not be as important in this case due to the short duration of most video frames and a relatively small sample size. ResNet50 model misclassifies most of the images where no human is, having predictions of 101 false positives while its test accuracy is 0.684.

The self-attention layer of ViT lacks locality inductive bias which is the prior knowledge that the learning algorithm uses to predict outputs that it has not yet encountered. [12] notices that through the lack of some of the locality bias inherent in CNNs, such as translation equivariance and locality, transformers do not generalise well to smaller-scale data sets. [6] corroborates the intuition that ViT models which do not use convolutions require more data or stronger regularisation. On the other hand, CNNs look at images through spatial sliding windows which help them get better results with smaller data sets. This is consistent with the higher accuracy results on the test set in the CNN architectures such as the basic variant of the CNN network, VGG16, ResNet50 and AlexNet. Therefore, it can be assumed that ViT may perform better with a configuration that is trained with a large number of data.

5 Conclusions

Space allocation and management in large offices with the use of IoT tags are costly. Therefore, applying DL techniques using thermal images to identify human presence at a place is more cost-effective in such applications. Due the fact that, this work configures different ML configurations such as basic CNN, CNN with regularisation, CNN-LTSM, VGG16, Resnet50, AlexNet and ViT to analyse the performance of human detection using images captured from a thermal camera. Moreover, a web app has been developed in order to use these seven models to remotely monitor and manage space on the campus premises. According to the results and observations of this work, the VGG16 model has the best test accuracy as a metric (0.962), correctly identifying all images with a human present and the CNN with regularisation model performs better than the basic CNN model.

References

1. Adeel, A., Ahmad, J., Larijani, H., Hussain, A.: A novel real-time, lightweight chaotic-encryption scheme for next-generation audio-visual hearing aids. Cogn. Comput. **12**, 589–601 (2020)
2. Adeel, A., et al.: A survey on the role of wireless sensor networks and IoT in disaster management. In: Geological Disaster Monitoring Based on Sensor Networks, pp. 57–66 (2019)
3. Akula, A., Shah, A.K., Ghosh, R.: Deep learning approach for human action recognition in infrared images. Cogn. Syst. Res. **50**, 146–154 (2018)
4. Amin, R.U., et al.: Towards cloud-based and federated a-synchronous speech enhancement using deep neuro-fuzzy models: review, challenges & future directions. In: Proceedings of the AVSEC 2024, pp. 79–81 (2024)
5. Anwary, A.R., et al.: Target speaker direction estimation using eye gaze and head movement for hearing aids. In: Proceedings of the AVSEC 2024, pp. 73–74 (2024)

6. Arnab, A., Dehghani, M., Heigold, G., Sun, C., Lučić, M., Schmid, C.: ViViT: a video vision transformer. In: IEEE/CVF International Conference on Computer Vision (ICCV), pp. 6816–6826 (2021)
7. Baldassarre, F., Morín, D.G., Rodés-Guirao, L.: Deep koalarization: image colorization using CNNs and inception-ResNet-v2. arXiv preprint arXiv:1712.03400 (2017)
8. Batchuluun, G., Nguyen, D.T., Pham, T.D., Park, C., Park, K.R.: Action recognition from thermal videos. IEEE Access **7**, 103 893–103 917 (2019)
9. Chollet, F.: Xception: deep learning with depthwise separable convolutions. In: Proceedings of the IEEE Conference on Computer Vision and Pattern Recognition, pp. 1251–1258 (2017)
10. Dashtipour, K., et al.: Towards cross-lingual audio-visual speech enhancement. Proceedings of the AVSEC 2024, pp. 30–32 (2024)
11. Dashtipour, K., et al.: Evaluating the audio-visual speech enhancement challenge (AVSEC) baseline model using an out-of-domain free-flowing corpus. In: Proceedings of the AVSEC 2024, pp. 75–78 (2024)
12. Dosovitskiy, A., et al.: An image is worth 16×16 words: transformers for image recognition at scale. arXiv preprint arXiv:2010.11929, 2020
13. Gogate, M., Dashtipour, K., Adeel, A., Hussain, A.: CochleaNet: a robust language-independent audio-visual model for real-time speech enhancement. Inf. Fusion **63**, 273–285 (2020)
14. Gogate, M., Dashtipour, K., Bell, P., Hussain, A.: Deep neural network driven binaural audio visual speech separation. In: 2020 International Joint Conference on Neural Networks (IJCNN), pp. 1–7. IEEE (2020)
15. Gogate, M., Dashtipour, K., Hussain, A.: Visual speech in real noisy environments (vision): a novel benchmark dataset and deep learning-based baseline system. In: Interspeech, pp. 4521–4525 (2020)
16. Gogate, M., Dashtipour, K., Hussain, A.: A lightweight real-time audio-visual speech enhancement framework. In: Proceedings of AVSEC 2024, pp. 19–23 (2024)
17. Gogate, M., Hussain, A., Dashtipour, K., Hussain, A.: Live demonstration: realtime multi-modal hearing assistive technology prototype. In: 2023 IEEE International Symposium on Circuits and Systems (ISCAS), p. 1. IEEE (2023)
18. He, K., Zhang, X., Ren, S., Sun, J.: Deep residual learning for image recognition. In: IEEE Conference on Computer Vision and Pattern Recognition (CVPR), vol. abs/1512.03385, pp. 770–778 (2015)
19. Hussain, A., et al.: Artificial intelligence-enabled analysis of public attitudes on Facebook and twitter toward COVID-19 vaccines in the united kingdom and the united states: Observational study. J. Med. Internet Res. **23**(4), e26627 (2021)
20. Kang, K., et al.: T-CNN: tubelets with convolutional neural networks for object detection from videos. IEEE Trans. Circuits Syst. Video Technol. **28**(10), 2896–2907 (2017)
21. Krizhevsky, A., Sutskever, I., Hinton, G.E.: ImageNet classification with deep convolutional neural networks. Commun. ACM **60**(6), 84–90 (2017)
22. Metzmacher, H., Wölki, D., Schmidt, C., Frisch, J., van Treeck, C.: Real-time human skin temperature analysis using thermal image recognition for thermal comfort assessment. Energy Build. **158**, 1063–1078 (2018)
23. Naik, K., Pandit, T., Naik, N., Shah, P.: Activity recognition in residential spaces with internet of things devices and thermal imaging. Sensors **21**(3) (2021)
24. Pereira, R., Cummiskey, K., Kincaid, R.: Office space allocation optimization. In: 2010 IEEE Systems and Information Engineering Design Symposium, pp. 112–117. IEEE (2010)

25. Perez, L., Wang, J.: The effectiveness of data augmentation in image classification using deep learning. arXiv preprint arXiv:1712.04621 (2017)
26. Redmon, J., Divvala, S., Girshick, R., Farhadi, A.: You only look once: unified, real-time object detection. In: IEEE Conference on Computer Vision and Pattern Recognition (CVPR). arXiv, pp. 779–788 (2016)
27. Sandler, M., Howard, A., Zhu, M., Zhmoginov, A., Chen, L.-C.: MobileNetv2: inverted residuals and linear bottlenecks. In: Proceedings of the IEEE Conference on Computer Vision and Pattern Recognition, pp. 4510–4520 (2018)
28. Tiwari, U., et al.: Real-time audio visual speech enhancement: integrating visual cues for improved performance. In: Proceedings of the AVSEC 2024, pp. 38–42 (2024)
29. Zoph, B., Vasudevan, V., Shlens, J., Le, Q.V: Learning transferable architectures for scalable image recognition. In: Proceedings of the IEEE Conference on Computer Vision and Pattern Recognition, pp. 8697–8710 (2018)

Training Feature-Awared GPU-Memory Allocation and Management for Deep Neural Networks

Qintao Zhang[1], Xin Li[2], Chengchuang Huang[2], Ying Zhu[2], Jilin Zhang[2(✉)], and Meng Han[2]

[1] Huawei Technologies Co Ltd., Hangzhou 310056, Zhejiang, China
[2] School of Computer Science, Hangzhou Dianzi University, No. 1158, 2nd Avenue, Baiyang Street, Qiantang District, Hangzhou 310018, Zhejiang, China
jilin.zhang@hdu.edu.cn

Abstract. Memory limitation of a single GPU is an urgent problem for efficient training of deep learning models on multi-GPU clusters, as a large number of model parameters, intermediate states, and activation values need to be stored during training. Efficient management and allocation of GPU memory is the key to improving memory resource utilization and model training efficiency. The existing methods mainly optimize the layout of memory space when training data is allocated and released to reduce memory waste. These methods do not consider the structural differences between models and the memory access feature of different data (model parameters, intermediate data, etc.) during the model training, so there are still problems of memory fragmentation and low utilization. To address these problems, this paper proposes an efficient GPU-memory allocation and management method, TMManager. It proposes an access feature analyzer, sampling and analyzing the structure of the model and the access features of memory for different data in the model training. Then, we design a dual-level memory partition management method with block and chunk, and a time-sharing deque memory allocation method, to reduce memory fragmentation and improve memory utilization. The experiments demonstrate that TMManager can save up to 23.5%, 59.9% of memory space compared with the memory allocators of TensorFlow, and Pytorch. Compared to PagedAttention, TMManager also realizes a faster and more convenient way of memory allocation.

Keywords: Deep Neural Networks · Model Training · Memory Allocation · Memory Fragmentation

1 Introduction

The success of deep neural network (DNN) models in development and application in various fields, such as natural language processing (NLP), and computer vision (CV), has pushed the DNN model scale to billions of parameters. For example, Bert [1] with 110 million parameters, GPT-3.5 [2] with 175 billion parameters, LLama [3] with 650 billion parameters and etc. Training these large models requires a lot of computing and memory resources. The limited memory resource of a single GPU accelerator is one of the bottlenecks that lead to the insufficient utilization of single GPU-accelerator computing resources, which reduces the training performance of DNN models. In particular, scaling model parameters or batch sizes often leads to out-of-memory (OOM) errors, interrupting the training process. Therefore, efficiently managing GPU memory is a key challenge in training deep neural networks.

To address memory challenges, deep learning (DL) frameworks such as TensorFlow [4], PyTorch [5], and MindSpore [6] have implemented various memory management strategies. For instance, TensorFlow employs the BFC (Best Fit with Coalescing) allocation algorithm based on the buddy system [7], which reduces memory fragmentation by selecting appropriate free memory blocks and merging adjacent blocks. PyTorch adopts a cache allocator inspired by the slab allocation technique [8], which preserves freed memory for future requests to minimize frequent memory allocation and deallocation operations. Similarly, MindSpore combines dynamic memory allocation with memory pool management to optimize memory usage during model training. However, these frameworks primarily focus on the size of memory requests and often overlook the periodic nature of memory allocation and deallocation during the training process. This limitation causes inefficiencies in memory utilization, leading to memory fragmentation and underusing available resources.

Memory compression and swapping techniques offer additional ways to enhance memory efficiency. Compression methods reduce the memory footprint by compressing model parameters and activations while swapping offloads data to CPU memory or disk when GPU memory is insufficient. These approaches can significantly improve overall memory utilization and support larger models on resource-constrained devices.

Building on these advancements, recent research efforts have sought to overcome these limitations by proposing advanced memory optimization methods. Notable examples include ShortcutFusion [9], PagedAttention [10], SuperNeurons [11], and GMLAKE [12]. These methods employ various techniques, such as dynamic memory management and virtual memory stitching to enhance memory usage efficiency. For instance, ShortcutFusion optimizes data reuse in CNN accelerators by employing static memory allocation techniques, while PagedAttention uses virtual memory paging to store non-contiguous memory blocks, reducing memory demands. SuperNeurons enhances GPU memory usage by dynamically managing memory blocks during training, and GMLAKE reduces GPU memory consumption by merging non-contiguous memory blocks into virtual memory spaces. However, despite their advancements, these methods still face challenges when dealing with the structural disparities among different models and the specific memory access patterns of different data types, such as model parameters and intermediate data. This oversight results in persistent issues such as memory fragmentation and suboptimal memory utilization.

For the above problems, this paper proposes an efficient GPU memory allocation and management for deep neural network training with memory access features, TMManager. The contributions of this paper are as follows:

- An access feature analyzer employing sliding window and distance differential analysis is proposed to reveal periodic memory requirement variations in model training.
- We employ a multi-level memory partition management method that integrates the categorization of space into blocks and chunks, to efficiently handle small, regular, and temporary data requests.
- We design and implement a time-sharing deque memory allocation method, which utilizes time-sharing and dual-end management to reduce memory fragmentation, preserving contiguous space for increased reuse.

2 Methodology

In this paper, we propose TMManager, an efficient GPU-memory allocation and management method for deep neural network training. Figure 1 provides an overview of TMManager. It includes three key technologies, access feature analyzer, memory partition management, and time-sharing deque memory allocation.

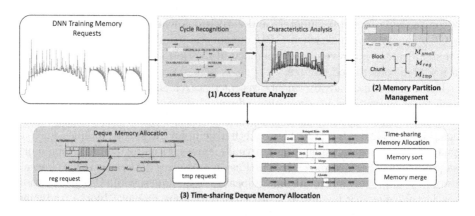

Fig. 1. Workflow of TMManager method

2.1 Access Feature Analyzer

The access feature analyzer is designed to sample and analyze DNN model training access features like training cycles, temporary requests, and so on. Then, it utilizes an access feature analysis method with the cooperation of sliding window and distance difference to efficiently analyze the memory access features of DNN training tasks.

2.1.1 Sampling

To understand memory access patterns for training the DNN model, TMManager samples and processes memory request data, such as request size $request\_size$, data type $type$, the total number of model parameters w_{model}, and the dynamic memory size $dynamic\_memory$. The memory cost $Memory\_cost$ for the model training can be described as follows:

$$Memory\_cost = request\_size * sizeof(type) * w_{model} + dynamic\_memory \quad (1)$$

$$\begin{aligned}dynamic\_memory = w_{intermediate\_activations} + w_{gradients} \\ w_{optimizer\_states} + w_{temporary\_variables}\end{aligned} \quad (2)$$

where $w_{intermediate\_activations}$, $w_{gradients}$, $w_{optimizer\_states}$ and $w_{temporary\_variables}$ are dynamic memory usage generated in model training, including activations, gradients, optimizer and other temporary results generated during the forward and backward passes. To capture these changes, we sampled memory access patterns over one hundred iterations and performed a detailed analysis. This allowed us to track the evolving memory demands and gain insights into the overall memory footprint during training. The details are shown in Fig. 2.

In the initial training phase, the sequence period is long, and the sequence of allocation requests is intensive and differs from subsequent periodic requests, due to data loading and parameter initialization. In the regular analysis phase, TMManager attempts to identify periodic patterns, which require at least two cycles due to the lack of periodicity during the initialization phase. In the periodic request phase, requests reflect various training stages.

2.1.2 Identifying Request Access Period

This section proposes a memory request cycle identification algorithm based on the ideas of sliding window, to dynamically identify and distinguish request cycles [13] and distance difference. It uses a sliding window approach to compare subsequences of the request sequence, adjusting the window size to find the period length and start subscript of the cycle.

Let the initial window size be max_period_length. It cuts two subsequences: subset1 moves forward from the end of the request sequence, and subset2 moves backward from the beginning of the request sequence. And then, it uses the

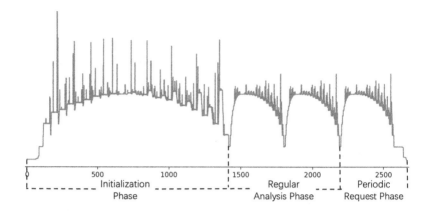

Fig. 2. Memory Request Partitioning Schematic

distance difference method to compare the differences between sequence subset1 and sequence subset2, as shown in Eq. 3.

$$distance = \sum_{i \in subset} |subset1_i - subset2_i| \qquad (3)$$

where $distance$ represents the total distance difference between the corresponding elements in subset1 and subset2, while $subset1_i$ and $subset2_i$ denote the $i - th$ elements of subset1 and subset2, respectively.

The memory request cycle identification method is shown in Algorithm 1. If the distance difference between subset1 and subset2 is not zero, subset2 slides back to continue the comparison. Otherwise, it means that the requested data in two sequences is equal. At this point, subset1 and subset2 simultaneously expand the comparison window to find the starting position of the cycle.

2.1.3 Analyzing Memory Access Characteristics Within a Period

This section analyzes data requests within the cycle and identifies data requests with temporary features. It compares each memory allocation request with a relative index labeled i with all release requests in the index range from i to ($i +$ TMP_SLOT]. Where TMP_SLOT is a temporary access interval, controlling the ratio of data requests handled by M_{tmp} and M_{reg}. Suppose there is a release request with the same address as the current allocation request. In that case, it indicates that the data corresponding to the allocation request is temporary data generated during the training process and will be released quickly. So, we label this temporary data to optimize resource allocation.

2.2 Memory Partition Management

This section proposes a memory data management approach based on multi-level partitioning to optimize the memory fragmentation problem in the model

Algorithm 1: Memory Request Cycle Identification Algorithm

Input : request list R, a new request r, max_period_length max,
　　　　　min_period_length min
Output: period_start_index S, period_length L
$R \leftarrow R + [r]$ // Add the new request to the request sequence for a new round of period identification.
$L \leftarrow max$ // Initialize period_length as max_period_length
while $L \geq min$ do
　　$subset1 \leftarrow R[-L:]$ // Extract the last L elements of R
　　$subset2 \leftarrow R[:L]$ // Extract the first L elements of R
　　$distance \leftarrow compare\_distance(subset1, subset2)$ // Calculate the distance difference
　　if $distance == 0$ then
　　　　// Expand $subset1$ and $subset2$ to find the start of the cycle
　　　　$large\_size \leftarrow expand(subset1, subset2)$ // Get the maximum enlargement size of the two subsets
　　　　if $large\_size > 0$ then
　　　　　　$L \leftarrow L + large\_size$
　　　　　　$S \leftarrow get\_index(R, L, subset1, subset2)$
　　　　　　return L, S
　　　　end
　　end
　　else
　　　　| $L \leftarrow L - 1$ // Reduce the period_length and continue searching
　　end
end
return $None, None$ // No period found

training. We use block with fixed size and chunk with dynamic size for a multilevel partitioned memory manager. Blocks, including multiple chunks, serve as the first level of memory management for preallocation, while chunks handle dynamic memory allocation based on task requirements as the second level.

The system requests a fixed-size memory space from the device, dividing it into two block regions: Block1, a fixed-size block reserved at the start of the memory area, and Block2, the remaining memory space. Additionally, the memory space is subdivided into areas for small request data M_{small}, regular request data M_{reg}, and temporary request data M_{tmp}, as in Fig. 3.

The small request data M_{small} manages small-sized memory requests through multiple chunks, to effectively prevent fragmentation caused by scattered small data blocks occupying contiguous memory space. The regular request data M_{reg}, consisting of multiple non-fixed-size chunks, manages non-periodic data requests and data from periodic requests that need to be stored in the cache for extended periods. The temporary request data M_{tmp}, including multiple dynamically sized chunks, manages data that is quickly released in periodic requests.

The dimensions of the M_{reg} and M_{tmp} regions are not fixed. The size of M_{reg} floats towards higher addresses. The size of M_{tmp} fluctuates towards lower

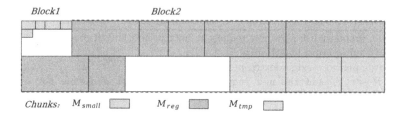

Fig. 3. Multilevel partitioned data structure

addresses. A continuous unused memory can be reserved between M_{reg} and M_{tmp}. This provides operating space for memory allocation in multitasking environments.

2.3 Time-Sharing Deque Memory Allocation

Frequent memory allocation and release in model training lead to memory fragmentation and wasted resources. We propose a time-sharing dual-end memory allocation method that manages memory space according to the cycle characteristics of memory requests and uses a double-ended data allocation approach for training data. This method manages temporary and non-temporary data generated during model training, enabling finer control over memory allocation and release, resulting in efficient memory resource usage.

2.3.1 Time-Sharing Memory Allocation

We divide the memory access requests into three phases: initialization, regularity analysis, and periodic request. With the memory partition, including small request data area M_{small}, regular request data area M_{reg}, and temporary request data area M_{tmp}, we set M as the fragmentation generated during memory management, the target of Time-sharing dequeue memory allocation is to minimize M as follows:

$$M_{min} = (M_{small} - M_{small_{used}}) + (M_{reg} - M_{reg_{used}}) + (M_{tmp} - M_{tmp_{used}}) \quad (4)$$

where, $M_{small_{used}}$, $M_{reg_{used}}$, and $M_{tmp_{used}}$ represent the used memory usage of M_{small}, M_{reg}, and M_{tmp} respectively.

It uses memory defragmentation to reorganize the spatial layout of data to optimize memory resource utilization, memory defragmentation technique to minimize memory footprint in the initialization and regularity analysis phase, and memory merging technique to reuse memory effectively. The details are described in Algorithm 1.

2.3.2 Deque Memory Allocation

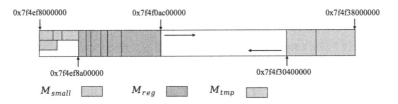

Fig. 4. Works of Deque Memory Allocation

The deque memory allocation strategy is proposed to reduce the memory fragments caused by a large number of temporary data requests and releases during model training, and maximize the retention of contiguous memory space.

Figure 4 shows the works of the deque memory allocation. In the periodic request phase, non-temporary data requests are allocated to M_{reg}, and memory is allocated by dividing chunks from front to back if no free chunks are available. Temporary data requests are allocated to M_{tmp}, with memory allocated by dividing chunks from back to front. When the low-address chunk in M_{tmp} is depleted, the memory is freed directly. The details of this method are described in Algorithm 2.

Algorithm 2: Time-sharing Deque Memory Allocation Algorithm

Input: request r, threshold $SMALL\_REQUEST\_SIZE$
if $not find\_request\_pattern(r)$ then
    if $size(r) < SMALL\_REQUEST\_SIZE$ then
    | Allocate(M_{small}, r) // Allocate memory for r in M_{small}
    end
    else
        Memory_defragmentation(M_{reg}) // Use memory defragmentation techniques to reorganize M_{reg}
        Allocate(M_{reg}, r)
    end
end
else
    if $size(r) < SMALL\_REQUEST\_SIZE$ then
    | Allocate(M_{small}, r)
    end
    else
        $type \leftarrow get\_request\_type(r)$ // Determine the type of request (non-temporary or temporary)
        Memory_Consolidation($type$) // Use memory consolidation techniques to reorganize M_{reg} or M_{tmp}
        Allocate(M_{type}, r) // Allocate r in M_{reg} or M_{tmp} using deque memory allocation based on request type
    end
end

3 Experiments

We developed TMManager, a CUDA allocation interface wrapper (cudaHostAlloc, cudaFreeHost), enhancing memory management without disrupting CUDA APIs. This paper uses a GPU server to take experiments. The server is equipped with an Intel(R) Xeon(R) CPU and 32 GB of RAM, as well as an NVIDIA Tesla P100 high-performance GPU with 12 GB of RAM. The server runs Ubuntu 18.04 as the operating system, providing a stable environment for our experiments. We makes experiments with models, including VGG16 [14], MobileNet [15], InceptionV3 [16], DenseNet-121 [17], ResNet-50 [18], ResNet-152 [18], NASNet [19] and EfficientNet [20], to validate our proposed methods. Compared with the BFCAllocator of TensorFlow2.10, CudaCachingAllocator of PyTorch2.2, and PagedAttention method, the results show that TMManager can greatly improve memory allocation performance, and minimize memory fragments.

3.1 Improvement of Memory Allocation

In this section, we validate the effectiveness of the time-sharing deque memory allocation strategy of the TMManager method. This primarily focuses on the impact of the time-sharing and deque memory allocation methods on memory management.

Effectiveness of Time-Sharing Allocation. We validate the effectiveness of the time-sharing memory allocation method by comparing the maximum memory usage between time-sharing data management and memory merging. The time-sharing data management approach integrates memory merging and memory defragmentation techniques to optimize memory usage. The time-sharing memory allocation method outperforms the memory merging method, reducing the average maximum memory requirement by 7.4% across various models, as shown in Fig. 5. In particular, the time-sharing memory allocation method excels on the InceptionV3 model, achieving a 17.4% reduction in memory requirement. Furthermore, it also demonstrates strong performance on the DenseNet-121 and ResNet-152 models, with reductions of 10.5% and 16.0% in memory requirement, respectively. This reduction is achieved by leveraging the periodic characteristics of the model training process and employing memory merging and defragmentation techniques for memory management. For complex models with vast parameter data, TMManager minimizes memory requirements during training by optimizing data layout through the time-sharing allocation method.

Effectiveness of Deque Data Allocation. We tested the free memory area size under different TMP_SLOT values to validate the effectiveness of the deque data allocation method. Figure 6 illustrates the specific experimental results, where TMP_SLOT was set to $\{0, 1, 2, 3, 4\}$. Compared to not using the deque data allocation method ($TMP\_SLOT = 0$), the improvements ranged from a minimum of approximately 1.06 times to a maximum of 1.65 times. For instance, InceptionV3 increases by 1.06 times with $TMP\_SLOT = 1$, while VGG16 shows a 1.65 times increase with $TMP\_SLOT = 4$. These results demonstrate that the

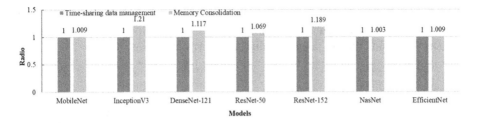

Fig. 5. Max memory usage Ratio Time-sharing data management and Memory Merging

deque data allocation method significantly enhances the free memory area size for each model, improving memory efficiency and reducing fragmentation by maintaining contiguous memory space through the allocation of temporary data to high addresses and non-temporary data to low addresses.

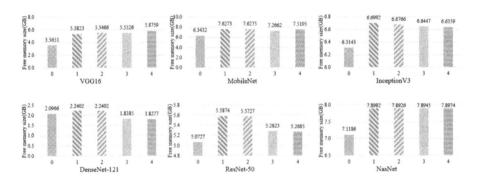

Fig. 6. Free memory area size for deque data allocation method (Unit: GB)

3.2 Analysis of Memory Requirement

Figure 7 illustrates the maximum memory consumption of different memory allocation methods, including PyTorch, TensorFlow, PagedAttention, and TMManager, under various models. For ease of comparison, the maximum memory consumption of each method is normalized to a ratio to the TMManager method, where the TMManager's memory consumption is set as 1. Compared to PyTorch, TensorFlow, and PagedAttention, the TMManager method reduces memory consumption by 59.9%, 23.5%, and 3.0%, respectively. As the model complexity increases, the reduction in memory consumption by the TMManager method may become less significant. For example, when comparing to PyTorch, the reduction in memory consumption decreases from 59.9% for VGG16 to 20.2% for EfficientNet; when comparing to TensorFlow, it decreases from 23.5% for

Nasnet to 3.2% for ResNet-152. Despite the trends, the TMManager method remains effective in optimizing memory management and significantly reducing memory demands during deep learning model training.

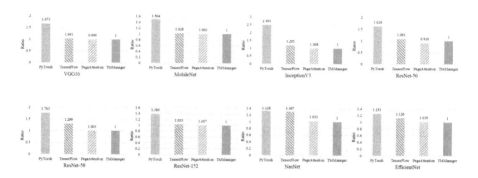

Fig. 7. Normalized maximum memory consumption ratios of various memory allocation methods.

3.3 Optimization in Memory Fragment

Table 1 shows the size of memory fragments produced by various memory management methods under different models. TMManager consistently produces the least amount of memory fragmentation across multiple models, compared to other methods. Specifically, the TMManager method reduces memory fragmentation by up to approximately 2250 times compared to PyTorch, 77105 times compared to TensorFlow, and 741 times compared to PagedAttention. We can find that these memory management methods exhibit significant differences. PyTorch has an average memory fragmentation of 45004.65 KB, TensorFlow exhibits a higher average memory fragmentation of 296335.3 KB, and PagedAttention shows an average memory fragmentation of 66456.62 KB. In contrast, TMManager achieves a remarkably low average memory fragmentation of only 140.59 KB, indicating its effectiveness in significantly reducing memory fragmentation compared to the other methods.

4 Conclusion

This paper addresses the issue of memory waste due to fragmentation in iterative model training. We propose TMManager, a memory allocation algorithm tailored to the memory access characteristics of AI training tasks. TMManager identifies request access periods and categorizes requests into provisional and non-provisional. It employs a multi-level partitioning approach using blocks and chunks to manage memory space efficiently. The time-sharing deque memory

Table 1. Memory fragmentation under different model memory request sequences (Unit: B)

Models\Plans	PyTorch	TensorFlow	PagedAttention	**TMManager**
VGG16	7864880	269566640	2401068	**3496**
MobileNet	8059691	276016755	16298139	**17063**
InceptionV3	34555771	251687623	40199263	**77987**
DenseNet-121	48176787	444874387	72742283	**134991**
ResNet-50	46869923	343224575	30490855	**55399**
ResNet-152	81665919	506723827	90492983	**139383**
NasNet	80318231	168420127	286025455	**385898**
EfficientNet	62905251	107985343	191487910	**336934**

allocation method optimizes data layout throughout the training process. Experiments show that TMManager reduces memory usage by up to 23.5% and 59.9% compared to TensorFlow and PyTorch, respectively, and outperforms PagedAttention in allocation speed and convenience.

Acknowledgements. This work is supported by the National Key Research and Development Program of China under Grant No. 2023YFB3001501; the "Pioneer" and "Leading Goose" RD Program of Zhejiang Province under Grant 2024C01104; the Natural Science Foundation of Zhejiang Province under Grant No. LQ23F020015.

References

1. Lee, J., Toutanova, K.: Pre-training of deep bidirectional transformers for language understanding. arXiv preprint arXiv:1810.04805, vol. 3, p. 8 (2018)
2. Brown, T., et al.: Language models are few-shot learners. In: Advances in Neural Information Processing Systems, vol. 33 (2020)
3. Touvron, H., et al.: LLaMA: open and efficient foundation language models. arXiv preprint arXiv:2302.13971 (2023)
4. Abadi, M., et al.: TensorFlow: large-scale machine learning on heterogeneous distributed systems. arXiv preprint arXiv:1603.04467 (2016)
5. Paszke, A., et al.: PyTorch: an imperative style, high-performance deep learning library. In: Advances in Neural Information Processing Systems, vol. 32 (2019)
6. Tong, Z., Du, N., Song, X., Wang, X.: Study on mindspore deep learning framework. In: 2021 17th International Conference on Computational Intelligence and Security (CIS), pp. 183–186. IEEE (2021)
7. Peterson, J.L., Norman, T.A.: Buddy systems. Commun. ACM **20**(6), 421–431 (1977)
8. Bonwick, J., et al.: The slab allocator: an object-caching kernel memory allocator. In: USENIX Summer, Boston, MA, USA, vol. 16 (1994)
9. Nguyen, D.T., Je, H., Nguyen, T.N., Ryu, S., Lee, K., Lee, H.-J.: ShortcutFusion: from tensorflow to FPGA-based accelerator with a reuse-aware memory allocation for shortcut data. IEEE Trans. Circuits Syst. I Regul. Pap. **69**(6), 2477–2489 (2022)

10. Kwon, W., et al.: Efficient memory management for large language model serving with pagedattention. In: Proceedings of the 29th Symposium on Operating Systems Principles, pp. 611–626 (2023)
11. Wang, L., et al.: Superneurons: dynamic GPU memory management for training deep neural networks. In: Proceedings of the 23rd ACM SIGPLAN Symposium on Principles and Practice of Parallel Programming, pp. 41–53 (2018)
12. Guo, C., et al.: GMLake: efficient and transparent GPU memory defragmentation for large-scale DNN training with virtual memory stitching. In: Proceedings of the 29th ACM International Conference on Architectural Support for Programming Languages and Operating Systems, vol. 2, pp. 450–466 (2024)
13. Tao, Y., Papadias, D.: Maintaining sliding window skylines on data streams. IEEE Trans. Knowl. Data Eng. **18**(3), 377–391 (2006)
14. Simonyan, K., Zisserman, A.: Very deep convolutional networks for large-scale image recognition. arXiv preprint arXiv:1409.1556 (2014)
15. Howard, A.G., et al.: MobileNets: efficient convolutional neural networks for mobile vision applications. arXiv preprint arXiv:1704.04861 (2017)
16. Szegedy, C., Vanhoucke, V., Ioffe, S., Shlens, J., Wojna, Z.: Rethinking the inception architecture for computer vision. In: Proceedings of the IEEE Conference on Computer Vision and Pattern Recognition, pp. 2818–2826 (2016)
17. Huang, G., Liu, Z., Van Der Maaten, L., Weinberger, K.Q.: Densely connected convolutional networks. In: Proceedings of the IEEE Conference on Computer Vision and Pattern Recognition, pp. 4700–4708 (2017)
18. Kaiming, H., Xiangyu, Z., Shaoqing, R., Jian, S., et al.: Deep residual learning for image recognition. In: Proceedings of the IEEE Conference on Computer Vision and Pattern Recognition, vol. 34, pp. 770–778 (2016)
19. Zoph, B., Vasudevan, V., Shlens, J., Le, Q.V.: Learning transferable architectures for scalable image recognition. In: Proceedings of the IEEE Conference on Computer Vision and Pattern Recognition, pp. 8697–8710 (2018)
20. Tan, M., Le, Q.: EfficientNet: rethinking model scaling for convolutional neural networks. In: International Conference on Machine Learning, pp. 6105–6114. PMLR (2019)

TR-LDA: An Improved Potential Topic Recognition Model

Anzhen Li[1], Shufan Qing[1], Weijie Qin[1], Liwen Qin[1], Jiawei Zhang[2], Meilin Shi[3], Jinchang Ren[4], and Mingchen Feng[1(✉)]

[1] College of Information Engineering, Northwest A&F University, Yangling 712100, Shaanxi, China
{3200305332,chenfan,2022013259,qlwww,mingchen}@nwafu.edu.cn
[2] School of Software, Shanxi Agricultural University, Taigu 030801, China
20231614345@stu.sxau.edu.cn
[3] Beijing Institute of Computer Technology and Applications, Beijing 100006, China
[4] National Subsea Centre, School of Computing and Engineering, Robert Gordon University, Aberdeen AB21 0BH, UK
jinchang.ren@ieee.org

Abstract. Latent topic recognition is a technique for extracting hidden themes or patterns from textual data. And it plays an important role in popular application fields such as text mining and information retrieval. It can improve search accuracy, achieve personalized recommendations and automate document classification. At present, traditional natural language processing methods suffer from problems such as long processing time on large-scale datasets and insufficient consideration of the correlation between words in text, resulting in inaccurate and unstable recognition of potential themes and internal keywords in document sets. To address the aforementioned issues, this paper proposes an improved latent topic recognition model called TR-LDA. It aims to combine the TF-IDF model with the LDA model. At the same time, by integrating the TextRank algorithm, a semantic analysis module is constructed to consider the co-occurrence relationships of words within the text. The experimental results on the THUCNews Dataset indicate that the average F1 score increased from 0.4263 with the original LDA to 0.5127 with TR-LDA. Semantic consistency improved by 54.20%, and perplexity was reduced by 98.74%. Furthermore, there was a time saving of 69.08%.

Keywords: Latent Topic Recognition · Co-occurrence Relationship of Vocabulary · TR-LDA Model · Natural Language Processing

1 Introduction

In recent years, with the rapid development of big data technology and artificial intelligence, latent topic recognition [4,8] has found widespread applications in the field of natural language processing. For example, in the field of social media analysis, companies can conduct brand monitoring and user behavior analysis by identifying popular topics, trends, and interests in user posts and comments. In the field of news classification, automatically categorizing news articles into predefined categories to improve the accuracy and user experience of news recommendation systems. However, in latent topic recognition techniques, particularly in capturing co-occurrence relationships and contextual information, there are challenges in accurately grasping these latent details, which affects the accuracy and utility of the extracted information.

Traditional text analysis methods such as the LDA model [1] and TF-IDF [9,11] model often overly rely on word frequency information while neglecting the relationships between words and contextual information. This can lead to an inability to accurately capture the deep semantics and latent topics within the text, thereby impacting the comprehensive understanding and accurate identification of these topics.

In this study, we propose the TR-LDA model based on the TextRank algorithm [6]. This allows for a more accurate, effective, and rapid identification of latent topics. The framework is illustrated in Fig. 1, which will be detailed in Sect. 3. The main contributions of this paper are as follows:

- Integrating a semantic analysis module into TF-IDF improves semantic consistency, reduces perplexity, and better identifies text features.
- Proposing the construction of a word co-occurrence network in text reduced the computation time for models on large-scale datasets.

In the rest of this paper, the related work on LDA, TF-IDF, and TextRank will be discussed in Sect. 2. Section 3 provides a detailed description of the model principles and main contributions. Section 4 presents the evaluation results on the THUCNews Dataset. Section 5 concludes the paper. Finally, Sect. 6 is our acknowledgments.

2 Related Work

In the past two decades, many scholars have conducted research and improvements on traditional latent topic recognition methods. Blei, Ng, and Jordan first introduced the Latent Dirichlet Allocation (LDA) model in 2003 [1]. Compared to the previous pLSI [5] model, LDA's topic selection is no longer constrained by the content of the training set, making it a significant tool in text modeling. Subsequently, Xingshu et al. [11] combined rank factor naive Bayes algorithms for intelligent language classification recommendation systems. Chen Xingshu et al. [2] proposed the ICE-LDA model, which improves the limitations of the traditional LDA model by incorporating cross-linguistic information, allowing for more accurate topic discovery in multilingual environments. Additionally, Chen Xingshu et al. [3] introduced the ccLDA model, which enhances hot topic detection by integrating information from multiple data sources. Zhu Zede et al. [14] improved the application of the LDA model in keyword extraction, enabling more effective extraction of keywords from texts. Jiang Haoda et al. [7] utilized the

BERT model to enhance the feature representation capability of TF-IDF, constructing a domain-specific sentiment lexicon to better capture text sentiment. These studies have provided new research directions and technical means for latent topic recognition methods. However, no scholars have yet made improvements or conducted research on optimizing the lack of word co-occurrence relationships and the time consumption of word frequency calculations.

3 Methodology

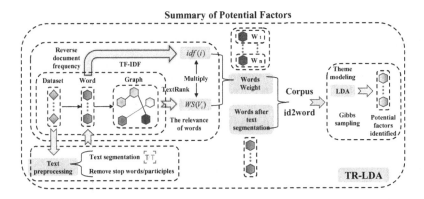

Fig. 1. Brief framework of TR-LDA

Based on the TextRank algorithm, this article improves the traditional method of calculating word contribution in topic modeling, and proposed the TR-LDA model. The Brief framework of TR-LDA is illustrated in Fig. 1. The TR-LDA model framework is mainly divided into three parts. The first part is the inverse document frequency calculation in the TF-IDF model. TF-IDF believes that the importance of words is not only related to their frequency of occurrence within the document, but also to their distribution throughout the entire corpus. After text preprocessing, text is segmented into phrases, which used to calculate the corresponding IDF value. The second part is the calculation of TextRank weight values. The TextRank algorithm is a graph sorting algorithm that divides text into several constituent unit words and uses a voting mechanism for sorting [14], thereby achieving word contribution calculation. Obtain WS values of each candidate keyword by constructing graph. Multiply to obtain the weights of all candidate keywords. The third part is LDA [1] for identifying potential themes and their internal keywords. The LDA model considers implicit topics as the probability distribution of vocabulary (Topic Word), and a single document is represented as the probability distribution of these implicit topics (Doc Topic) [10], with an internal structure such as Fig. 2. Convert to a dictionary and document collection through format conversion, input to the LDA model, and iteratively converge through Gibbs sampling. The next 10 steps explain the specific calculation method.

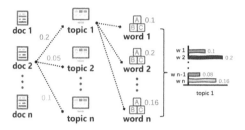

Fig. 2. Figure of LDA Latent Topic Topology

Step 1: Text preprocessing involves segmenting the text to obtain phrases. Common stop words are removed using a publicly available stop word list (stopword.txt), and remove participles, constructing the set of phrases.
Step 2: For the TextRank algorithm, the documents are concatenated, and using the set of phrases after text preprocessing, a vertex set of text units is constructed, representing the candidate keyword graph. Edges are created based on the relationships between the phrases.
Step 3: The document is treated as a network of words, where the links represent the semantic relationships between words. Weights are assigned on the constructed graph, and the weights of each node are iteratively propagated until convergence. The calculation formula is as follows in e2. The node weights are then sorted in descending order to filter the candidate keywords. $WS(V_i)$ represents the weight of a specific term i, W_{ji} indicates the co-occurrence relationship between term i and term j, $WS(V_j)$ denotes the weight of the previously selected term j, $Adj(V_i)$ is the set of adjacent vertices, and d is the damping factor, typically set to 0.85.

$$WS(V_i) = (1-d) + d \sum_{V_j \in Adj(V_i)} \frac{W_{ji}}{\sum_{V_k \in Adj(V_j)} W_{jk}} WS(V_j) \qquad (1)$$

Step 4: For the filtered candidate keywords, which serve as the vocabulary for different documents, the corresponding inverse document frequency $idf(w)$ for each term is calculated. The calculation formula is as follows in Eq. (2). N represents the total number of documents in the corpus, and $N(w)$ indicates the number of documents in which the term w appears.

$$idf(w) = log \frac{N}{N(w)+1} \qquad (2)$$

Step 5: The contribution of each term within the document is calculated, as shown in Eq. (3). $Word\_Re(i)$ represents the importance weight of each term.

$$Word\_Re(i) = WS(V_i) * idf(i) \qquad (3)$$

Step 6: Using the segmented corpus, a term-document matrix is constructed as shown in Eq. (4). In this matrix, each row represents a term, each column represents a document, and the elements of the matrix indicate the

$Word\_Re(i)$ value of term i in the corresponding document. 'Dictionary' is a dictionary composed of terms i and their corresponding weight values $Word\_Re(i)$.

$$Dictionary = \{i : Word\_Re(i)\} \quad (4)$$

Step 7: The term-document matrix is used as input for the LDA model. '*corpus*' refers to the collection of documents, and '*id2word*' is a dictionary that maps words to unique integer identifiers. Both of these components are part of the input data for the LDA model. The following parameters are set: the number of topics (*num_topics*), the number of keywords (*num_words*), the prior for document-topic distribution (*alpha*), the prior for topic-word distribution (*eta*), and the number of *passes* through the corpus during training (*passes*) are as follows. *alpha* = 'symmetric', *eta* = 'symmetric', *passes* = 60. The 'symmetric' parameter can explicitly provide an array of values.

Step 8: For each term in every document within the corpus, a topic number z is randomly assigned.

Step 9: The corpus is then rescanned, and for each term, Gibbs Sampling is used to sample its topic, as shown in Eq. (5), updating the topic assignment in the corpus accordingly. $p(z_i = k|\vec{z}_{\neg i}, \vec{w})$ represents the conditional distribution corresponding to any coordinate axis i. The parameters α_k and β_t denote the smoothing parameters related to category k and word t, respectively. The i-th word in the corpus $\vec{z}$ is denoted as z_i, where $i = (m, n)$ is a two-dimensional index corresponding to the m-th document and the n-th word. The notation $\neg i$ indicates the exclusion of the word indexed by i.

$$p(z_i = k|\vec{z}_{\neg i}, \vec{w}) \propto \frac{n_{m,\neg i}^{(k)} + \alpha_k}{\sum_{k=1}^{K}(n_{m,\neg i}^{(k)} + \alpha_k)} \cdot \frac{n_{k,\neg i}^{(t)} + \beta_t}{\sum_{t=1}^{V}(n_{k,\neg i}^{(t)} + \beta_t)} \quad (5)$$

Step 10: Repeat **Step 9** until Gibbs Sampling converges. This process identifies the final latent topics, keywords, and their weights.

4 Experimental Results

4.1 The THUCNews Dataset

The THUCNews news [13] text classification dataset provided by THUCTC. The THUCNews dataset comprises 10 categories: 'Sports', 'Finance', 'Real Estate', 'Home', 'Education', 'Technology', 'Fashion', 'Current Affairs', 'Games', and 'Entertainment'. Each document has a specific topic. The dataset consists of three files: cnews.train.txt (training set: 5,000 articles for each category), cnews.val.txt (validation set: 500 articles for each category), and cnews.test.txt (test set: 1,000 articles for each category). In the experiment, a subset of cnews.val.txt was selected for result calculation and evaluation. The subset is manually or randomly selected.

4.2 Experimental Evaluation Metrics

The experiment evaluates the effectiveness of latent topic identification using six metrics: Perplexity, Semantic Coherence (C_{UMass}), Precision, Recall, F1 score, and Similarity.

The formula for Perplexity [1] is defined as follows in Eq. (6). M represents the number of documents in the test set, N denotes the number of words in the d-th document, w_d represents the word in the d-th document, while $P(w_d)$ denotes the probability of the d-th document calculated.

$$Perplexity(D_{test}) = e^{-\sum_{d=1}^{M} \log P(w_d)/N} \tag{6}$$

The UMass coherence score [15] is a method developed by the University of Massachusetts to evaluate topic coherence. The formula for C_{UMass} is as follows in Eq. (7). C_{UMass} represents the UMass coherence score, N denotes the number of terms within a topic, $P(w_i, w_j)$ refers to the probability of terms w_i and w_j co-occurring within a particular topic, and $P(w_j)$ represents the probability of term w_j appearing in the entire corpus. The term $\log \frac{P(w_i, w_j)}{P(w_j)}$ is the log-likelihood ratio for the term pair (w_i, w_j) relative to the individual term w_j.

$$C_{UMass} = \frac{2}{N*(N-1)} \sum_{i=2}^{N} \sum_{j=1}^{i-1} \log \frac{P(w_i, w_j)}{P(w_j)} \tag{7}$$

The formulas for precision, recall, and F1 score are shown as Eq. (8), Eq. (9), and Eq. (10), respectively. Additionally, the experiment utilized the Chinese word vector version 0.2.0 released by Tencent AI Lab [12] on December 24, 2021, to calculate similarity. According to its rationale, a similarity of over 50% can be considered as indicating that the two words are similar.

$$Precision = \frac{TP}{TP + FP} \tag{8}$$

$$Recall = \frac{TP}{TP + FN} \tag{9}$$

$$F1 = 2 \times \frac{Precision \times Recall}{Precision + Recall} \tag{10}$$

4.3 Experiments on the THUCNews Dataset

First, Select 10–15 news articles from each label, resulting in a total of 130 articles. Each news was treated as a complete document, with parameters set to $num\_topics = 1$, $passes = 60$, and $num\_words = 10$. The overall similarity being derived from the highest similarity obtained by comparing the 10 keywords under each news topic with the actual label. Results presented in Table 1. It can be seen that the TR-LDA model has the highest similarity on different news, thus identifying keywords more accurately. "1.0000" represents that the identified internal keywords contain real labels, which are determined by Tencent AI Lab Embedding Corpora to be equal to the potential theme and the true theme of the text.

Next, on the dataset cnews.val.txt, The total precision, recall and F1 score were calculated for various randomly selected dataset sizes of [100, 150, 200, 250, 300, 350] news articles, as shown in Table 2. The average F1 score after predicting ten indicators increased from 0.4263 to 0.5127 in the original LDA model.

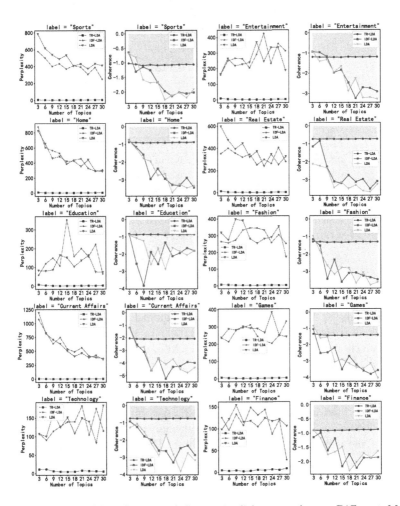

Fig. 3. Comparison of Perplexity and Semantic Coherence Across Different Models with Varying Topic Counts on the Same Dataset

Subsequently, Extract 300 articles from each label. With varying numbers of topics that $num\_topics = [3, 6, 9, 12, 15, 18, 21, 24, 27]$. The models were

Table 1. Comparison of similarity on the same dataset

News Number	Ours	IDF-LDA	LDA
0	**1.0000**	0.3987	0.4581
19	**1.0000**	0.4817	0.4547
49	**0.6454**	0.5211	0.6087
69	**0.5205**	0.4303	0.5081
89	**0.7964**	0.3990	0.7963
109	**1.0000**	0.6440	0.4837
119	**0.7490**	0.4964	0.7489
129	**0.5740**	0.5455	0.4554
Average	**0.7154**	0.4923	0.5231

Table 2. Precision, recall, and F1 score of different models on different sizes

Dataset size	Ours			IDF-LDA			LDA		
	Precision	Recall	F1 score	Precision	Recall	F1 score	Precision	Recall	F1 score
100	**0.5896**	**0.4590**	**0.5162**	0.2943	0.2901	0.2922	0.3823	0.3526	0.3669
150	**0.5451**	**0.5155**	**0.5299**	0.4192	0.3301	0.3694	0.4787	0.4070	0.4400
200	**0.5213**	**0.4475**	**0.4816**	0.2912	0.2612	0.2754	0.4921	0.4341	0.4613
250	**0.5545**	**0.4844**	**0.5171**	0.2797	0.2742	0.2769	0.4580	0.4252	0.4410
300	**0.5075**	**0.4731**	**0.4897**	0.2964	0.2885	0.2924	0.4008	0.4118	0.4062
350	0.4587	0.4259	0.4417	0.3224	0.2714	0.2947	**0.4674**	0.4181	0.4414

employed to predict the labels, and perplexity and semantic coherence were calculated for comparison, as shown in Fig. 3. The average semantic consistency increased from −2.4551 to −1.1245 in the original LDA model, and the average perplexity decreased from 310.6577 to 3.9293 in the original LDA model. In sum, semantic consistency improved by 54.20%, and perplexity was reduced by 98.74%. Additionally, this study compares the total time taken for news label prediction under different actual labels and calculates the label optimization percentage and overall optimization percentage for the TR-LDA model, with formulas provided in Eq. (11) and Eq. (12). A comparison of the total running times is shown in Table 3. From the time analysis, it can save approximately 69.08% of the time compared to the original LDA topic modeling.

$$Ci = \frac{1}{2}\left\{\frac{1}{T_{i(I-L)}}\left(T_{i(I-L)} - T_{i(T-L)}\right) + \frac{1}{T_{i(L)}}\left(T_{i(L)} - T_{i(T-L)}\right)\right\} \quad (11)$$

$$C = \frac{1}{n}\sum_{i=1}^{n} Ci \quad (12)$$

The optimization percentage of the label is C_i. $T_{i(I-L)}$ represents the total running time of IDF-LDA under the current label, $T_{i(T-L)}$ denotes the total running time of TR-LDA under the current label, and $T_{i(L)}$ indicates the total running time of LDA under the current label. The overall optimization percentage C is defined, where n represents the total number of labels.

Table 3. Comparison of Total Running Times (Unit: second)

label	Ours	IDF-LDA	LDA	Impl.TIME
Sports	**84.0947**	302.9236	311.6376	**72.63%**
Finance	**103.7714**	323.1738	317.9459	**67.63%**
Real Estate	**81.3591**	400.0100	399.7857	**79.66%**
Home	**77.4234**	380.1022	375.0133	**79.49%**
Education	**73.7014**	362.9297	380.6321	**80.16%**
Technology	**97.1222**	176.7362	174.2454	**44.65%**
Fashion	**68.7651**	155.5715	156.1097	**55.87%**
Current Affairs	**53.4526**	236.7639	231.7183	**77.18%**
Games	**77.1558**	213.5296	214.3896	**63.94%**
Entertainment	**71.2123**	237.4841	231.6396	**69.64%**

5 Conclusion

This paper presents an improved latent topic recognition model, referred to as the TR-LDA model. This approach not only accurately identifies latent topics and their associated keywords but also significantly optimizes the time consumption issues related to word frequency statistics in traditional models, demonstrating advantages on large-scale datasets. Experiments on the THUCNews dataset validate that the TR-LDA model outperforms conventional latent topic recognition methods across multiple metrics, exhibiting high accuracy and stability.

In the future work, we will conduct further research in areas such as multidimensional feature fusion, multimodal data latent topic recognition, cross domain topic recognition, and cross language topic recognition. In addition, we will strive to improve the interpretability of the model, enabling users to gain a deeper understanding and effectively apply the results of the model.

Acknowledgements. This work has been supported by National College Students' Science and Technology Innovation Project (Program No. X202410712560 and No. S202410712404) and the Natural Science Basic Research Program of Shaanxi (Program No. 2023-JC-QN-0684).

References

1. Blei, D., Ng, A., Jordan, M.: Latent Dirichlet allocation. J. Mach. Learn. Res. **3**, 993–1022 (2003). https://doi.org/10.1162/jmlr.2003.3.4-5.993
2. Chen, X., Luo, L., Wang, H., Wang, W., Gao, Y.: Analysis and research on cross language topic discovery in Chinese and English. Gongcheng Kexue Yu Jishu/Adv. Eng. Sci. **49**, 100–106 (2017). https://doi.org/10.15961/j.jsuese.201601032
3. Chen, X., Ma, C., Wang, W., Gao, Y., Wang, H.: Multi-source topic detection analysis based on improved cclda model. Gongcheng Kexue Yu Jishu/Adv. Eng. Sci. **50**, 141–147 (2018). https://doi.org/10.15961/j.jsuese.201700626
4. Chien, J.T., Huang, Y.H.: Latent semantic and disentangled attention. IEEE Trans. Pattern Anal. Mach. Intell. 1–12 (2024). https://doi.org/10.1109/TPAMI.2024.3432631
5. Chou, T.C., Chen, M.C.: Using incremental PLSI for threshold-resilient online event analysis. IEEE Trans. Knowl. Data Eng. **20**(3), 289–299 (2008). https://doi.org/10.1109/TKDE.2007.190702
6. Gao, X., Xu, W., Zhang, Z., Tang, Y., Chen, G.: Cross-platform event popularity analysis via dynamic time warping and neural prediction. IEEE Trans. Knowl. Data Eng. **35**(2), 1337–1350 (2023). https://doi.org/10.1109/TKDE.2021.3090663
7. Jlang, H., Zhao, C., Chen, H., Wang, C.: Construction method of domain sentiment lexicon based on improved TF-IDF and BERT. Comput. Sci. **51**(6A), 230800011 (2024). https://doi.org/10.11896/jsjkx.230800011
8. Kasaei, S.H., Lopes, L.S., Tomé, A.M.: Local-LDA: open-ended learning of latent topics for 3d object recognition. IEEE Trans. Pattern Anal. Mach. Intell. **42**(10), 2567–2580 (2020). https://doi.org/10.1109/TPAMI.2019.2926459
9. Kumar, R., Sinha, R., Saha, S., Jatowt, A.: Extracting the full story: a multimodal approach and dataset to crisis summarization in tweets. IEEE Trans. Comput. Soc. Syst. 1–11 (2024). https://doi.org/10.1109/TCSS.2024.3436690
10. Linshan, M.: Research on design of novelty retrieval aided analysis system based on LDA model. J. Mod. Inf. **38**(2), 111 (2018). https://doi.org/10.3969/j.issn.1008-0821.2018.02.018
11. Luo, Y., Lu, C.: TF-IDF combined rank factor naive Bayesian algorithm for intelligent language classification recommendation systems. Syst. Soft Comput. **6**, 200136 (2024). https://doi.org/10.1016/j.sasc.2024.200136
12. Song, Y., Shi, S., Li, J., Zhang, H.: Directional skip-gram: explicitly distinguishing left and right context for word embeddings, pp. 175–180 (2018). https://doi.org/10.18653/v1/N18-2028
13. Sun, M., et al.: THUCTC: an efficient Chinese text classifier (2016)
14. Zhu, Z., Li, M., Zhang, J., Zeng, W., Zeng, X.: A LDA-based approach to keyphrase extraction. Zhongnan Daxue Xuebao (Ziran Kexue Ban)/J. Cent. South Univ. (Sci. Technol.) **46**, 2142–2148 (2015). https://doi.org/10.11817/j.issn.1672-7207.2015.06.023
15. Zimmermann, J., Champagne, L.E., Dickens, J.M., Hazen, B.T.: Approaches to improve preprocessing for latent Dirichlet allocation topic modeling. Decis. Support Syst. **185**, 114310 (2024). https://doi.org/10.1016/j.dss.2024.114310

Brain-Inspired Object Domain Adaptive Segmentation

Mengyin Pang[1,2], Song Xu[3], Lina Wang[3(✉)], Zhenfei Liu[3], Meijun Sun[1,2], and Zheng Wang[1,2(✉)]

[1] College of Intelligence and Computing, Tianjin University, Tianjin, China
{pmy,sunmeijun,wzheng}@tju.edu.cn
[2] Tianjin Key Laboratory of Machine Learning, Tianjin University, Tianjin, China
[3] National Laboratory of Aerospace Intelligent Control Technology,
Beijing Aerospace Automatic Control Institute, Beijing, China
violina@126.com

Abstract. Addressing the domain shift challenge between datasets is critical to maintaining model performance. However, existing methods rarely consider the behavior of the human brain when dealing with the domain shift, resulting in poor model segmentation results. Brain-inspired strategies can enhance the migration ability and generalization performance of deep learning models across different data sets by simulating the mechanisms of the human brain in processing complex visual information. Inspired by this, we propose a Brain-inspired Domain Adaptation Network (BDANet) to solve the domain shift problem. Specifically, we adopt the teacher-student architecture for mutual learning and adversarial learning. Both the teacher and student models contain a dual-branch encoder and a brain-inspired decoder for object segmentation. Inspired by the cognitive process of the human brain, we propose a brain-inspired fusion module in the dual-branch encoder to effectively fuse the dual-branch features to obtain complete object information. In the brain-inspired decoder, we propose a brain-inspired refinement module to gradually refine the initial segmentation result and accurately segment the edges of objects. Extensive experiments demonstrate that the proposed method significantly outperforms the existing competitors on the challenging benchmark dataset under evaluation metrics.

Keywords: Brain-inspired · Domain adaptation · Object segmentation

1 Introduction

In the field of computer vision, object segmentation technology has always been a hot and difficult research topic. Object segmentation technology has broad application prospects in fields such as medicine [3], military [4], agriculture [17], and autonomous driving systems [18]. However, a long-standing problem that is difficult to ignore is that there are often significant differences between real-world data and training data. This difference is called "domain shift". This mismatch not only originates from external factors such as image acquisition equipment, lighting conditions, and background complexity, but may also involve internal characteristics such as the shape, color, and texture of the object itself.

To address this challenge, researchers have begun to explore object segmentation techniques based on domain adaptation (DA). This type of technology aims to enable the model to transfer knowledge between different domains, thereby achieving accurate segmentation of real data. Adaptive learning is an important machine learning method that aims to enable the model to transfer and migrate knowledge between domains with different data distributions to improve performance. However, although existing adaptive learning methods have achieved certain success, there are still many challenges in dealing with complex domain shift problems. The human brain is known for its excellent adaptability and pattern recognition capabilities [10], able to quickly adjust itself to diverse environments and conditions to identify and segment target-domain's objects accurately. Inspired by this powerful capability, we propose an innovative brain-inspired domain adaptation network (BDANet). This model achieves adaptive segmentation of objects in different domains by deeply simulating the working principle of the human brain, thereby demonstrating great potential in dealing with complex domain shift problems. The contributions are summarized:

- We introduce brain-inspired ideas into the DA task and propose the BDANet, which employs the teacher-student architecture with a dual-branch encoder (DE) and a brain-inspired decoder (BD) for object segmentation.
- A brain-inspired fusion module (BFM) in DE innovatively integrates "brain-inspired" concepts and designs an adaptive weight adjustment mechanism to obtain complement features between the two branches.
- The brain-inspired refinement module (BRM) in BD simulates the human brain to iteratively refine object edges and obtain accurate segmentation results, especially when faced with fuzzy boundaries.
- Extensive experimental evaluations show that our proposed BDANet outperforms state-of-the-art methods on the widely adopted benchmark dataset.

2 Related Work

2.1 Domain Adaptation

Domain-adaptive object segmentation methods address the domain shift problem. Many methods [2,8,16,20] have achieved good results. For example, domain adversarial neural network (DANN) [2] is the first to introduce adversarial-based methods into domain adaptation. Since then, many methods with various loss functions [16] or classifiers [8,20] have been proposed for image classification. Although domain adaptation methods have been successfully applied to image classification, pixel-level tasks such as object segmentation remain challenging.

2.2 Brain-Inspired Methods

Brain-inspired object segmentation methods [12] refer to object segmentation algorithms designed based on the working mechanism of the human brain's system. For example, SINet [1] emulates the progressive localization and exploration of objects, drawing inspiration from the foraging behavior of wild predators. ZoomNet [13] replicated the zoom-in and zoom-out behavior exhibited by humans in the perception of blurred images. PFNet [11] simulates the human brain mechanism by mining and eliminating distracting areas to eliminate interference from similar backgrounds and improve the accuracy of object segmentation.

3 Methodology

3.1 Overview

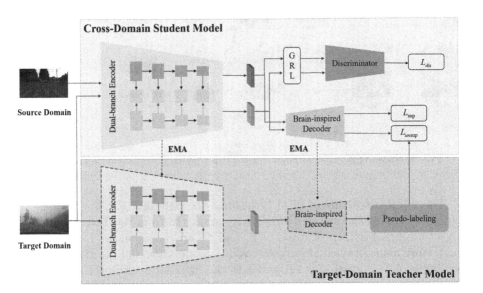

Fig. 1. The overall architecture of the proposed Brain-inspired Domain Adaptation Network (BDANet).

Figure 1 illustrates the overall architecture of our proposed Brain-inspired Domain Adaptation Network (BDANet). The BDANet adopts a teacher-student architecture, which mainly includes two models, the target-domain teacher model and the cross-domain student model. Inspired by the cognitive processes of the human brain, both models design a DE and a BD to obtain complete target information and refined segmentation results respectively.

3.2 Dual-Branch Encoder and Brain-Inspire Decoder

The architecture of the dual-branch encoder and brain-inspire decoder is shown in Fig. 2. First, the dual-branch encoder to extract features from both Transformer and CNN models. Then, the brain-inspired fusion module (BFM) dynamically fuses these features to obtain complete information. After that, the brain-inspired refine module (BRM) in the brain-inspired decoder to refine the segmentation and obtain more accurate prediction results.

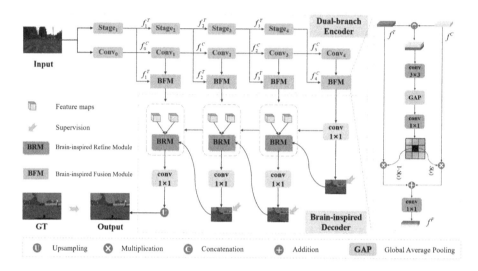

Fig. 2. The architecture of Dual-branch Encoder and Brain-inspire Decoder.

The specific structure of BFM is shown in the right of Fig. 2. Specifically, f^T and f^C represent the features respectively extracted from the Transformer and CNN branches. First, the features of f^T and f^C are concatenated. Subsequently, the feature map obtained after the convolution operation and the global average pooling (GAP) operation is subjected to the Sigmoid operation to obtain the brain-inspired weight (bw). This weight value is continuously adjusted during the model training process to integrate the features of different branches better. bw is defined as:

$$bw = S(Conv_{1\times1}(GAP(Conv_{1\times1}(Conv_{3\times3}(Cat(f^T, f^C))))))\quad(1)$$

where $Conv_{1\times1}(\cdot)$ and $Conv_{3\times3}(\cdot)$ represent the convolution operations with kernels of 1×1 and 3×3, respectively, $Cat(\cdot, \cdot)$ represents the concatenation operation, and $GAP(\cdot)$ represents the global average pooling operation. Finally, bw is multiplied by the original feature to enhance the original feature representation, and the two are added to obtain the final fused feature f^F:

$$f^F = Conv_{1\times1}(bw \times fT + (1 - bw) \times f^C)\quad(2)$$

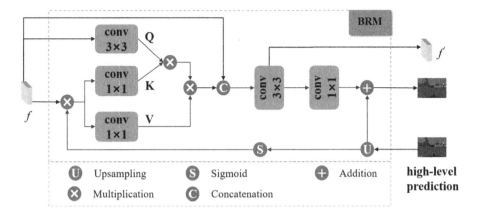

Fig. 3. The detailed structure of Brain-inspired Refine Module (BRM).

Brain-inspired Refine Module. The specific details of the BRM are depicted in Fig. 3. We upsample the higher-level prediction and normalize it with a sigmoid layer. Then we use this normalized map to calculate the weight. As commonly known, a pixel with a value of 1 represents the foreground, while a pixel with a value of 0 represents the background. Building upon this understanding, we hypothesize that pixels closer to a value of 0.5 exhibit higher uncertainty. Consequently, We denote the prediction input map as P and its weight calculation operation is as follows:

$$w = 0.5 - |0.5 - S(U(P))| \quad (3)$$

where $U(\cdot)$ and $S(\cdot)$ denote the upsampling layer and sigmoid layer, respectively. Given the input features f, We first employ three 3×3 convolution layers on the feature map f and reshape the convolution results to generate a new feature map query Q. Then the feature map f is multiplied by the calculated weight w and passed through a 1×1 convolution layer. The resulting output is reshaped to obtain two new feature maps, key K and value V, respectively. After that, we perform the following operations to generate the attention map X:

$$x_{ij} = \frac{exp(Q_{:i} \cdot K_{:j})}{\sum_j^N exp(Q_{:i} \cdot K_{:j})} \quad (4)$$

Meanwhile, we conduct the following operations to obtain the final output f':

$$f' = Conv_{3 \times 3} \left(Cat \left(\sum_j^N (V_{:j} \cdot x_{ji}), f_{:i} \right) \right) \quad (5)$$

where $Conv_{3 \times 3}(\cdot)$ and $Cat(\cdot, \cdot)$ denote the 3×3 convolution layer and concatenation operation, respectively. Finally, feature f' undergoes a 1×1 convolution layer and is added to the high-level prediction map after upsampling. This process yields an additional output, referred to as the refined prediction map.

The binary cross-entropy (L_{BCE}) and the IoU loss (L_{iou}^{w}) are commonly employed in segmentation tasks. Besides, we utilize multiple supervisions to guide the training process for the four side-output predictions ($P_i, i \in \{1, 2, 3, 4\}$) and the ground-truth mask (G). Finally, the segmentation loss is defined:

$$L_{seg}(P, G) = \sum_{i=1}^{4} 2^{-i}(L_{bce}(P_i, G) + L_{iou}(P_i, G)) \qquad (6)$$

3.3 Teacher-Student Network Framework

As shown in Fig. 1, the teacher-student architecture uses mutual learning and adversarial learning to train the entire network. The teacher model generates pseudo labels to train the student model, and the student updates the teacher model with an exponential moving average (EMA). In the student model, we use a discriminator and gradient reversal layer (GRL) for adaptive learning to align the distributions of the two domains. This makes the student model less domain-shifted and helps the teacher model generate more accurate pseudo labels.

Model Initialization. Initialization is very important for the self-training framework because the BDANet relies on the teacher model to generate reliable pseudo-labels for the target domain to optimize the student model. Therefore, we use the available supervised source data D_s and the segmentation loss in Formula 6 to optimize our model with a supervised loss L_{sup}:

$$L_{sup}(D_s, G) = L_{seg}(D_s, G) \qquad (7)$$

Optimize the Student Model. Since there are no labels in the target domain, we adopt the pseudo-labeling method to generate virtual labels on images in the target domain to train the student model. After obtaining the pseudo-labels (G_{pseudo}) of the teacher model on the target domain images (D_t), we can update the student model with the loss as:

$$L_{unsup}(D_t, G_{pseudo}) = L_{seg}(D_t, G_{pseudo}) \qquad (8)$$

Optimize the Teacher Model. To obtain strong pseudo labels, we apply EMA to update the weights of the teacher model (θ_t) by using the weights of the student model (θ_s). The update formula can be defined as:

$$\theta_t = \alpha \theta_t + (1 - \alpha)\theta_s \qquad (9)$$

Adversarial Learning. Since ground truth is only available on source data, both the teacher and student models can easily be biased towards the source domain during mutual learning. To avoid the teacher model generating noisy labels on target images and crashing the learning process, we need to bridge the domain bias between the source and target domains. Therefore, we introduce adversarial learning into the framework to align the distribution between the two domains. Specifically, given that the student model takes images from two domains at the same time, a domain discriminator (D) is placed after the dual-branch encoder (E) on the student model. For each input image (X), we can update the domain discriminator D using the binary cross entropy loss. Label the images from the source domain as d = 0 and the images from the target domain as d = 1. The discriminator loss L_{dis} can be expressed as:

$$L_{dis} = -d\log D(E(X)) - (1-d)\log(1 - D(E(X))) \tag{10}$$

In summary, the total loss L of our proposed BDANet training is as follows:

$$L = L_{sup} + \lambda_{unsup}L_{unsup} + \lambda_{dis}L_{dis} \tag{11}$$

where L_{sup} and L_{unsup} are used to train the dua-branch decoder and brain-inspire decoder, and L_{dis} is used to update the discriminator in the student model. While the teacher model is only updated through Formula 9.

4 Experiments

4.1 Experimental Setup

Datasets. We conduct experiments on the public datasets: Cityscapes (source domain), foggy Cityscapes (target domain), and Dark Zurich-N (target domain).

Implementation Details. We implement our network using the Pytorch toolbox [15]. An NVIDIA GeForce RTX 3090 GPU (24 GB) is used for both training and testing. The batch size is set to 8 and the optimizer is AdamW. The initial learning rate is set to 1e−4. In the initialization phase, only the source domain data is used for training for 10k iterations. Then, the weights are copied to the teacher and student models at the beginning of the mutual learning phase and trained for 50k iterations. For hyperparameters, all experiments set $\lambda_{unsup} = 1.0$, $\lambda_{dis} = 0.1$ in Formula 11 and $\alpha = 0.9996$ in Formula 9.

4.2 Comparison with State-of-the-Arts

To demonstrate the effectiveness of our BDANet, we compare it with 6 state-of-the-art methods, such as ProDA [19], DAFormer [5], HRDA [6], DiGA [14], MIC [7], and I2F [9]. Tables 1 and 2 show the adaptation results in different weather conditions and at different times. BDANet obtains consistent improvements over all categories and achieves +0.7 mIoU gain compared to I2F in Cityscapes → Foggy Cityscapes and achieves +1.1 mIoU gain compared to I2F in Cityscapes → Zurich-N. Furthermore, BDANet consistently surpasses all other models on evaluated datasets. In addition, Fig. 4 shows the segmentation results of BDANet compared to competitors, indicating that BDANet performs the best. This is mainly due to the BFM in BE to obtain comprehensive target information and the BRM in BD to refine the fuzzy areas such as the edge.

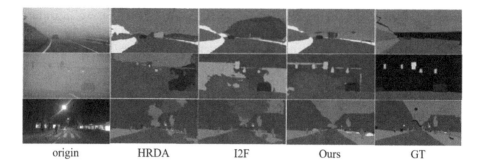

origin   HRDA   I2F   Ours   GT

Fig. 4. The Visual results on Foggy Cityscapes of BDANet and competitors.

Table 1. Foggy Cityscapes adaptation results.

Method	Bus	Bicycle	Car	Mcycle	Person	Rider	Train	Truck	mIou
ProDA	59.4	56.4	88.8	48.9	70.7	39.2	46.2	45.5	56.9
DAFormer	78.2	61.8	92.3	55.9	72.2	44.7	65.1	74.5	68.1
HRDA	77.9	62.1	91.9	59.9	74.4	51.3	73.9	77.8	71.1
DiGA	80.7	62.5	91.9	61.9	74.9	52.9	75.7	79.1	72.5
MIC	84.3	63.5	92.6	60.5	74.8	53.1	77.4	81.4	73.5
I2F	84.9	64.8	93.8	61.2	76.4	53.6	78.2	81.9	74.4
Ours	**85.7**	**65.3**	**94.2**	**62.5**	**77.8**	**53.8**	**78.9**	**82.6**	**75.1**

Table 2. Cityscapes to Zurich-N adaptation results.

Method	Road	Building	Person	Sky	Rider	Train	Truck	Bus	mIou
ProDA	64.2	59.9	38.8	69.5	32.8	65,7	16.2	65.2	49.5
DAFormer	78.9	65.1	40.6	74.8	38.2	68.2	19.7	71.4	57.1
HRDA	77.3	67.7	42.8	77.9	37.9	69.4	20.8	72.1	58.2
DiGA	83.4	67.8	42.7	77.3	39.2	70.9	21.1	72.7	59,4
MIC	86.7	70,2	44.2	78.5	41.7	71.4	22.4	73.2	59.7
I2F	89.2	72.1	45.9	80.9	40.2	71.8	22.9	74.7	62.2
Ours	**90.6**	**72.7**	**46.2**	**81.7**	**43.8**	**73.2**	**23.8**	**75.4**	**63.3**

4.3 Ablation Study

To verify the effectiveness of the proposed module, we conduct the following ablation studies on the Foggy Cityscapes and report the results in Table 3. The EMA Teacher means that EMA is used to update the parameters of the teacher model. Comparing (c) and (d) with (a) and (b) respectively, it can be seen that the BFM module improves segmentation performance by utilizing brain-inspired adaptive weight adjustment to integrate features from different branches effectively. Comparing (e) with (a) and (b), and comparing (f) with (c) and (d), it can

Table 3. Quantitative evaluation for ablation studies on Foggy Cityscapes.

Model	EMA Teacher	CNN	Transformer	BFM	BRM	mIou
(a)	✓	✓				58.4
(b)	✓		✓			67.5
(c)	✓	✓			✓	64.8
(d)	✓		✓		✓	71.4
(e)	✓	✓	✓	✓		71.2
(f)		✓	✓	✓	✓	72.3
(g)	✓	✓	✓	✓	✓	**75.1**

be seen that the use of the BRM module significantly improves the performance of the model, mainly due to the BRM module gradually refining the segmentation based on brain-inspired ideas to obtain more accurate segmentation results.

5 Conclusion

In this paper, we propose a BDANet for domain adaptation to improve segmentation performance in the target domain. Specifically, we design a dual-branch encoder to extract features from both Transformer and CNN models, and a brain-inspired fusion module dynamically combines these features for comprehensive information. After that, our brain-inspired decoder includes a BRM Module to refine the segmentation. In addition, we use mutual and adversarial learning to encourage the model to learn features from different domains. Extensive comparison experiments and ablation studies show that the proposed BDANet achieves superior performance over other state-of-the-art approaches. However, our model demands high computing resources and time. In the future, we plan to investigate model lightweight and explore the potential of our method.

Acknowledgements. The authors wish to acknowledge the support for the research work from the National Natural Science Foundation of China (CN) under grant Nos. [62076180] and [62376189].

References

1. Fan, D.P., Ji, G.P., Sun, G., Cheng, M.M., Shen, J., Shao, L.: Camouflaged object detection. In: Proceedings of the IEEE/CVF Conference on Computer Vision and Pattern Recognition, pp. 2777–2787 (2020)
2. Ganin, Y., et al.: Domain-adversarial training of neural networks. J. Mach. Learn. Res. **17**(59), 1–35 (2016)
3. Haithami, M., Ahmed, A., Liao, I.Y.: Enhancing generalizability of deep learning polyp segmentation using online spatial interpolation and hue transformation. In: International Conference on Brain Inspired Cognitive Systems, pp. 41–50 (2023)
4. Hall, J.R., et al.: A platform for initial testing of multiple camouflage patterns. Defence Technol. **17**(6), 1833–1839 (2021)
5. Hoyer, L., Dai, D., Van Gool, L.: DAFormer: improving network architectures and training strategies for domain-adaptive semantic segmentation. In: Proceedings of the IEEE/CVF Conference on Computer Vision and Pattern Recognition, pp. 9924–9935 (2022)
6. Hoyer, L., Dai, D., Van Gool, L.: HRDA: context-aware high-resolution domain-adaptive semantic segmentation. In: European Conference on Computer Vision, pp. 372–391. Springer (2022)
7. Hoyer, L., Dai, D., Wang, H., Van Gool, L.: MIC: masked image consistency for context-enhanced domain adaptation. In: Proceedings of the IEEE/CVF Conference on Computer Vision and Pattern Recognition, pp. 11721–11732 (2023)
8. Long, M., Zhu, H., Wang, J., Jordan, M.I.: Unsupervised domain adaptation with residual transfer networks. In: Advances in Neural Information Processing Systems, vol. 29 (2016)
9. Ma, H., Lin, X., Yu, Y.: I2F: a unified image-to-feature approach for domain adaptive semantic segmentation. IEEE Trans. Pattern Anal. Mach. Intell. **46**(3), 1695–1710 (2022)
10. Mehonic, A., Kenyon, A.J.: Brain-inspired computing needs a master plan. Nature **604**(7905), 255–260 (2022)
11. Mei, H., Ji, G.P., Wei, Z., Yang, X., Wei, X., Fan, D.P.: Camouflaged object segmentation with distraction mining. In: Proceedings of the IEEE/CVF Conference on Computer Vision and Pattern Recognition, pp. 8772–8781 (2021)
12. Pan, J., Jing, C., Zuo, Q., Nieuwoudt, M., Wang, S.: Cross-modal transformer GAN: a brain structure-function deep fusing framework for Alzheimer's disease. In: International Conference on Brain Inspired Cognitive Systems, pp. 82–92 (2023)
13. Pang, Y., Zhao, X., Xiang, T.Z., Zhang, L., Lu, H.: Zoom in and out: a mixed-scale triplet network for camouflaged object detection. In: Proceedings of the IEEE/CVF Conference on Computer Vision and Pattern Recognition, pp. 2160–2170 (2022)
14. Shen, F., Gurram, A., Liu, Z., Wang, H., Knoll, A.: DiGA: distil to generalize and then adapt for domain adaptive semantic segmentation. In: Proceedings of the IEEE/CVF Conference on Computer Vision and Pattern Recognition, pp. 15866–15877 (2023)
15. Su, J., Li, J., Zhang, Y., Xia, C., Tian, Y.: Selectivity or invariance: Boundary-aware salient object detection. In: Proceedings of the IEEE/CVF International Conference on Computer Vision, pp. 3799–3808 (2019)
16. Tzeng, E., Hoffman, J., Saenko, K., Darrell, T.: Adversarial discriminative domain adaptation. In: Proceedings of the IEEE Conference on Computer Vision and Pattern Recognition, pp. 7167–7176 (2017)

17. Wang, L., Yang, J., Zhang, Y., Wang, F., Zheng, F.: Depth-aware concealed crop detection in dense agricultural scenes. In: Proceedings of the IEEE/CVF Conference on Computer Vision and Pattern Recognition, pp. 17201–17211 (2024)
18. Wu, W., Deng, X., Jiang, P., Wan, S., Guo, Y.: CrossFuser: multi-modal feature fusion for end-to-end autonomous driving under unseen weather conditions. IEEE Trans. Intell. Transp. Syst. (2023)
19. Zhang, P., Zhang, B., Zhang, T., Chen, D., Wang, Y., Wen, F.: Prototypical pseudo label denoising and target structure learning for domain adaptive semantic segmentation. In: Proceedings of the IEEE/CVF Conference on Computer Vision and Pattern Recognition, pp. 12414–12424 (2021)
20. Zhou, L., Ye, M., Zhu, X., Li, S., Liu, Y.: Class discriminative adversarial learning for unsupervised domain adaptation. In: Proceedings of the 30th ACM International Conference on Multimedia, pp. 4318–4326 (2022)

Task Adaptive Feature Distribution Based Network for Few-Shot Fine-Grained Target Classification

Ping Li[1], Hongbo Wang[1], Jie Ren[2(✉)], Xin Mi[3], and Chao Shi[3]

[1] China People's Liberation Army National Defence University, Beijing, China
OBGNOHGNAW@Sohu.com
[2] College of Electronics and Information, Xi'an Polytechnic University, Xi'an, China
renjie@xpu.edu.cn
[3] North Automatic Control Technology Institute, Taiyuan, China

Abstract. Addressing the limitations of existing metric-based few-shot fine-grained classification methods, which often neglect task-specific nuances and suffer from inaccurate category descriptions and irrelevant information, we introduce TAFD-Net: a novel Task Adaptive Feature Distribution Network. Our approach innovatively integrates two key components: (1) a task-adaptive embedding module that captures fine-grained features tailored to each task, and (2) an asymmetric metric module that computes similarities between query samples and support categories based on their feature distributions. Through comprehensive experiments on three datasets, TAFD-Net demonstrates superior performance compared to recent incremental learning algorithms, highlighting its effectiveness in accurately describing categories.

Keywords: Few-shot learning · Fine-grained classification

1 Introduction

The task of Few-shot Fine-Grained Image Classification (FSFGIC) has witnessed remarkable advancements in recent years. This task aims to categorize images into fine-grained subordinate categories within a broader superclass, utilizing only a limited number of samples. The objective is to develop models that can effectively learn and generalize from minimal data, thereby enhancing their capability to distinguish between subtly different object classes.

This intricate challenge stems from two fundamental aspects: (1) The targeted objects in FSFGIC belong to sub-categories of a singular superordinate category, leading to their high visual similarity; (2) The fine-grained distinction between these sub-categories introduces nuanced inter-class differences, while simultaneously manifesting substantial intra-class diversity in aspects such as pose, scale, and rotation [11,13]. Furthermore, the annotation of fine-grained images is a process that is both labor-intensive and economically costly [14–16].

Owing to their simplicity and efficacy, metric-based methods have achieved state-of-the-art performance in FSFGIC tasks [10]. The essence of these metric-learning approaches lies in the concept of "Learning-to-Compare" which involves acquiring a similarity metric and assessing the relationships between query samples and various categories. The objective of an FSFGIC task is to accomplish classification among C categories (C-way) using only a limited number of samples (K-shot). However, most current metric-learning methods address each category independently, overlooking the correlations between different categories.

In this paper, we introduce a novel network framework for FSFGIC tasks, termed TAFD-Net. TAFD-Net comprises an embedding module and a metric module. Within the embedding module, each image is represented as a set of local descriptors, which offer more abundant feature representations when training samples are limited. We propose a task adaptive component (TAC) that traverses all local descriptors of support samples to capture task-level information. In the metric module, we utilize distribution-level information (i.e., mean vector and covariance matrix) of images to compute similarity scores.

The primary contributions of this paper are summarized as follows: 1) We introduce a Task-Adaptive Component (TAC), which incorporates a Common Feature Learner (CFL) and a Unique Feature Learner (UFL). CFL and UFL capture task-level information by examining the features of all support samples within the task. 2) We propose a distribution-level based asymmetric metric module that employs KL divergence to measure the asymmetric relationship between query and support samples. To further enhance the metric's performance, a Contrastive Measure Strategy (CMS) is integrated into the asymmetric metric module.

Comprehensive experiments are conducted on three benchmark datasets. In comparison to state-of-the-art methods, TAFD-Net demonstrates superior performance in most classification tasks.

2 Related Works

The current research methodologies in FSFGIC can be broadly categorized into three distinct approaches: data-augmentation-based, meta-learning-based, and metric-learning-based techniques.

Data-Augmentation-Based Methods: These methods focus on expanding the dataset by leveraging existing data. Zhu et al. [23] introduce a strategy that utilizes unlabeled samples for data augmentation, complemented by a self-training mechanism to carefully select these samples. In [5], a fully annotated auxiliary dataset, sharing a similar distribution with the target dataset, is employed to train a meta-learner capable of transferring knowledge across datasets. Wang et al. [21] propose Instance Credibility Inference (ICI), which harnesses unlabeled samples with high confidence levels to augment the training set.

Meta-learning-Based Methods: Meta-learning approaches aim to swiftly adapt to new tasks using prior knowledge. Wang et al. [20] introduce a Meta

Neural Architecture Search (M-NAS) method. Antoniou et al. [1] present Self-Critique and Adapt (SCA), an unsupervised loss function that can be integrated into a base model to learn from a critic loss network. Zhu et al. [24] propose Multi-Attention Meta-Learning (MattML), which applies attention mechanisms to both the base learner and task learner to capture local deep feature representations and fine-tune the initial classifier.

Metric-Learning-Based Methods: In the realm of metric learning, Li et al. [10] introduce the Deep Nearest Neighbor Neural Network (DN4), which learns an image-to-class metric through a k-nearest neighbor search over deep local descriptors of convolutional feature maps. Zhang et al. [22] employ a Deep Earth Mover's Distance (DeepEMD) to select local discriminative feature representations for optimal matching between query and support samples. Huang et al. [6] design a Low-Rank Pairwise Bilinear Pooling Operation Network (LRPABN) to obtain class-level deep feature representations between query and support samples.

The proposed method, TAFD-Net, falls under the metric-learning category. Unlike existing methods, TAFD-Net learns task-level features during the training phase and utilizes an asymmetric metric, KL divergence, to measure the asymmetric relationship between query and support samples.

3 Proposed Method

3.1 Problem Formulation

For the FSFGIC task, the dataset $\mathcal{D}$ comprises two components: a support set $\mathcal{S}$ and a query set $\mathcal{Q}$. The support set $\mathcal{S}$ contains C classes with K labeled samples per class. Meanwhile, the query set $\mathcal{Q}$ consists of J unlabeled samples. This can be mathematically represented as:

$$\mathcal{D} = \{\mathcal{S} = \{(x_i, y_i)_{i=1}^{C \times K}\} \cup \mathcal{Q} = \{(x_j)_{j=1}^{J}\}\}, \tag{1}$$

where $\mathcal{S} \cap \mathcal{Q} = \emptyset$, x_i and x_j denote fine-grained samples and $(x_i, x_j) \in C$, and $y_i \in C$ represents the ground truth label of x_i. The objective of FSFGIC is to accurately classify x_j into its corresponding class in C based on $\mathcal{S}$. Thus, this task is commonly referred to as a C-way K-shot problem.

Notably, the limited number of training samples per class in FSFGIC poses a challenge for effectively learning transferable knowledge. To address this problem, an episodic training paradigm is adopted, leveraging an auxiliary set $\mathcal{A}$ with a similar data distribution to $\mathcal{D}$. The auxiliary set is defined as:

$$\mathcal{A} = \{\mathcal{E} = \{(u_i, v_i)_{i=1}^{N}\} \cup \mathcal{F} = \{(u_j, v_j)_{j=1}^{M}\}\}, \tag{2}$$

where u_i and u_j are fine-grained images, v_i and v_j are their corresponding labels; $\mathcal{E} \cap \mathcal{F} = \emptyset$, $\mathcal{D} \cap \mathcal{A} = \emptyset$. The auxiliary set $\mathcal{A}$ encompasses numerous classes and labeled samples, significantly exceeding C and K in sample count.

In each training iteration, $\mathcal{A}$ is randomly partitioned into two parts: an auxiliary support set $\mathcal{G} = \{(u_i, v_i)_{i=1}^{C \times K}\}$ and an auxiliary query set $\mathcal{H} = \{(u_j, v_j)_{j=1}^{J}\}$. Given that $N \gg C \times K$, $\mathcal{E}$ can effectively simulate the composition of $\mathcal{S}$ in each iteration. Then $\mathcal{A}$ is utilized to acquire prior knowledge for training on $\mathcal{S}$.

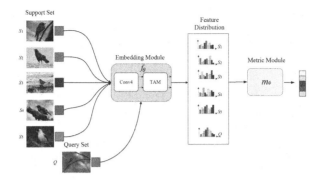

Fig. 1. The framework of TAFD-Net in 5-way 1-shot setting

3.2 Framework of TAFD-Net

The proposed TAFD-Net comprises two primary modules: an embedding module and an asymmetric metric module, as illustrated in Fig. 1. The embedding module incorporates a Conv4 [18] network (Ψ_θ), along with a task-adaptive component. In order to obtain local descriminitive descriptors, we remove both the fully connected layer and the global average pooling layer.

More specifically, for an input image X, $\Psi_\theta(X)$ is a $d \times h \times w$ three-dimensional tensor, which can be interpreted as a set of m ($m = h \times w$) d-dimensional local descriptors.

$$\Psi_\theta(X) = [x_1, x_2, ..., x_m] \quad (3)$$

where x_i is the i_{th} deep descriptor. These descriptors, extracted from the support samples, are subsequently fed into the task-adaptive component to capture task-specific information. Within the asymmetric metric module, we employ KL divergence and a contrastive measure strategy to assess the similarity between sample distributions. Finally, a non-parametric nearest neighbor classifier serves as the final classification mechanism.

3.3 Embedding Module

We utilize Conv4 as the backbone network for feature extraction. Conv4 comprises four convolution blocks, each integrating a convolutional layer, a batch normalization layer, and a Leaky ReLU activation function. Each convolutional layer is equipped with 64 filters of size 3×3. Additionally, the last two blocks incorporate a 2×2 max-pooling layer. When processing an image with size of 84×84, Conv4 yields a feature map of dimensions $64 \times 19 \times 19$, corresponding to 361 local descriptors, each with a size of 64. These descriptors, derived from the support samples, are subsequently input into the task-adaptive component, whose structure is depicted in Fig. 2.

In classification tasks, for samples belonging to the same category, despite variations in illumination, color, and angle, the objects inherently share common

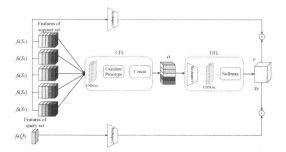

Fig. 2. Illustration of the task-adaptive component (TAC) in a 5-way 5-shot setting

features, which are crucial for determining the image category. Conversely, samples from different categories exhibit unique target features. These are referred to as intra-class common features and inter-class unique features, respectively. Within the TAC, we introduce a common feature learner (CFL) and a unique feature learner (UFL).

In a C-way K-shot task, for all the support samples, denote the output features from feature extractor as $\Psi_\theta(S)$: (CK, d_1, h_1, w_1), where d, h and w indicate the number of channels and the spatial size, respectively.

For the CFL, $\Psi_\theta(S)$ is reduced in dimension and then calculated the prototype for each category:

$$\Psi_\theta(S) : (CK, d_1, h_1, w_1) \to o : (C, d_2, h_2, w_2) \qquad (4)$$

where d_2, h_2 and w_2 indicate the output number of channels and the spatial size, respectively. The dimension reduction operation is implemented by convolutional neural network CNN_{CFL}, which aims to reduce the variability across categories. For the structure of CNN_{CFL}, we use two residual blocks [2]. The prototype computation process aligns with that of the prototype network [17], preserving common features for each class. Notably, no averaging operation is applied in 1-shot tasks. The final output o is obtained by concatenating all feature maps along the channel dimension.

For the UFL, to facilitate cross-category interaction, we reshape the tensor o by connecting its first dimension to the second:

$$o : (C, d_2, h_2, w_2) \to o^1 : (1, Cd_2, h_2, w_2) \qquad (5)$$

Same to CFL, we apply a convolutional neural network (CNN_{UFL}) for dimensional reduction. To identify the most discriminative dimensions in o, a softmax function is applied along the second dimension, iterating over all categories:

$$o^1 : (1, Cd_2, h_2, w_2) \to v : (1, d_3, h_3, w_3) \qquad (6)$$

The weight vector v contains task-level information, which we directly mask onto the original features. To align their dimensions, the original features are reshaped using a convolution layer (as denoted in Fig. 2 as Reshape2):

$$\Psi_\theta(\cdot) \to r_\theta(\cdot) : (Cd, d_3, h_3, w_3) \tag{7}$$

Dimensional transformations only carried out in channel dimension; thus, in the TAC, $h_1 = h_2 = h_3$, and $h = w$. Lastly, for the original feature maps of both query and support samples, we broadcast the value of the weight vector v along the first sample dimension. Subsequently, we concatenate r_θ and v to create the final representation.

3.4 Asymmetric Metric Module

In the metric module, similarity score is computed based on the distribution information of the descriptors. Specifically, we assume that the distributions of local descriptors extracted from either a query sample or a support class follow multivariate Gaussian distributions, which can be represented as $Q = \mathcal{N}(\mu_Q, \Sigma_Q)$, $S = \mathcal{N}(\mu_S, \Sigma_S)$, where μ and Σ denote the mean vector and covariance matrix of a specific distribution, respectively.

Then, the Kullback-Leibler (KL) divergence [9] between Q and S is adopted. Notably, the KL divergence is asymmetric, which helps to drive the distribution of the query sample closer to that of the support sample. In this manner, the feature distribution of the query image becomes more aligned with the true distribution of the corresponding class.

Furthermore, to validate the efficacy of the asymmetric metric, we employ the symmetric 2-Wasserstein distance [3] in the metric module, defined as:

$$D(Q,S)^2_{WA} = ||\mu_S - \mu_Q||^2_2 + ||\Sigma_S - \Sigma_Q||^2_F \tag{8}$$

where $||\cdot||_2$ represents the Euclidean distance and $||\cdot||_F$ represents the Frobenius norm. The 2-Wasserstein is symmetric, that is $D(Q,S)_{WA} = D(S,Q)_{WA}$.

To further boost the performance of the distribution metric, we incorporate a contrastive measure strategy (CMS) in the metric module. Specifically, in a classification task with a support $S = (S_1, ..., S_C)$. Where query sample Q belongs to S_i, S_i^1 is the set of other classes. CMS is denoted as follow:

$$(Q||S)_{KL} = (Q||S_i)_{KL} - (Q||S_i^1)_{KL} \tag{9}$$

Since the KL divergence decreases as the two distributions become closer, CMS serves to refine the metric function. Similar to DN4 [10], all descriptors of a support category are mapped into a feature space, and the nearest neighbor algorithm is utilized to identify the k descriptors that are most similar to the query image.

4 Experiments

4.1 Dataset

In this section, experiments are conducted on three benchmark fine-grained datasets: Stanford-Cars [8], Stanford-Dogs [7], and CUB-200-2011 [19]. The

Stanford-Cars dataset comprises 16,185 images depicting 196 car types, with 130, 17, and 49 classes designated for training (auxiliary), validation, and testing, respectively. The Stanford-Dogs dataset consists of 20,580 images representing 120 dog breeds, split into 70, 20, and 30 classes for training (auxiliary), validation, and testing, respectively. The CUB-200-2011 dataset includes 11,788 images of 200 bird species, with 130, 20, and 50 classes allocated for training (auxiliary), validation, and testing, respectively.

4.2 Experimental Setting

In each dataset, 5-way 5-shot and 5-way 1-shot tasks are performed. For each task, 15 query images per class are utilized in both training and testing stages, amounting to 75 query images in total. Specifically, the episodic training mechanism [18] is employed to train the models from scratch without pre-training. During the training stage, all models are trained for 40 epochs using the Adam algorithm. The evaluation criterion is based on the top-1 mean accuracy, and 95% confidence intervals are also reported.

4.3 Experimental Results

For comparison, five metric-learning based methods, namely P-Net [17], DN4 [10], LRPABN [6], QPN [12], and TOAN [4], were selected for comparison with TAFD-Net. The results obtained from our method and these other methods are presented in Table 1. As can be seen from Table 1, our method achieved the best performance in most tasks. Furthermore, the asymmetric metric (KL) demonstrated significantly better performance compared to the symmetric metric (WA). Specifically, in 1-shot and 5-shot tasks on the Stanford-Cars, Stanford-Dogs, and CUB-200-2011 datasets, TAFD-Net exhibited improvements of 8.97%, 0.05%, 7.47%, 5.48%, 7.64%, and 2.32% over DN4, and improvements of 4.58%, 5.41%, 3.90%, 6.47%, 6.75%, and 7.25% over TOAN. The performance enhancement of TAFD-Net was particularly pronounced in 1-shot tasks, thereby validating the efficacy of our approach when samples are limited.

Table 1. Comparison Results

Methods	Stanford-Cars		Stanford-Dogs		CUB-200-2011	
	1-shot	5-shot	1-shot	5-shot	1-shot	5-shot
P-Net	40.90 ± 1.01	52.93 ± 1.03	37.59 ± 1.00	48.19 ± 1.03	50.67 ± 0.88	75.06 ± 0.67
DN4	61.51 ± 0.85	89.60 ± 0.44	45.73 ± 0.76	66.33 ± 0.66	64.45 ± 0.89	85.36 ± 0.48
LRPABN	62.80 ± 0.76	73.29 ± 0.58	45.7 ± 0.75	60.94 ± 0.66	63.63 ± 0.77	76.06 ± 0.58
QPN	63.91 ± 0.58	89.27 ± 0.73	**53.69 ± 0.62**	70.98 ± 0.70	66.04 ± 0.82	82.85 ± 0.76
TOAN	65.90 ± 0.72	84.24 ± 0.48	49.30 ± 0.77	67.16 ± 0.49	65.34 ± 0.75	80.43 ± 0.60
TAFD-Net (WA)	68.13 ± 0.91	88.45 ± 0.75	51.67 ± 0.46	70.21 ± 0.66	69.18 ± 0.91	86.75 ± 0.54
TAFD-Net (KL)	**70.48 ± 0.68**	**89.65 ± 0.35**	53.20 ± 0.36	**71.81 ± 0.52**	**72.09 ± 0.66**	**87.68 ± 0.36**

4.4 Ablation Study

To further validate the efficacy of the proposed TAFD-Net, ablation experiments were conducted on the CUB-200-2011 dataset, with the KL divergence serving as the metric function throughout.

Impact of TAC: To demonstrate the effectiveness of TAC, experiments were performed under conditions where TAC was excluded (No TAC). The results are presented in Table 2. As shown in Table 2, TAC significantly enhances performance in 1-shot tasks. However, in 5-shot tasks, TAC may introduce extraneous noise, leading to only a marginal improvement.

Table 2. Ablation Experiment of TAC.

Methods	Accuracy	
	1-shot	5-shot
TAFD-Net (No TAC)	70.23 ± 0.67	87.49 ± 0.31
TAFD-Net	**72.09 ± 0.66**	**87.68 ± 0.36**

Impact of CMS. In metric module, CMS is employed to augment the performance of the metric function. To validate the utility of CMS, experiments were conducted under conditions where CMS was removed (No CMS). The results are presented in Table 3. As shown in Table 3, the incorporation of CMS enhances the network's performance, indicating the effectiveness of considering the context relationship within a task.

Table 3. Ablation Experiment of CMS

Methods	Accuracy	
	1-shot	5-shot
TAFD-Net (No CMS)	71.26 ± 0.77	87.43 ± 0.42
TAFD-Net	**72.09 ± 0.66**	**87.68 ± 0.36**

In TAC, two CNN were constructed to reduce the dimensions of feature maps. We construct the CNN_{CFL} and CNN_{UFL} with different depths (1, 2, 3 and 4 residual blocks). The results are presented in Table 4. It is evident from the table that as the depth of the network increases, the performance of the 5-shot task gradually declines.

Table 4. Ablation Experiment of CNN_{CFL} and CNN_{UFL}

Number of residual blocks	Accuracy	
	1-shot	5-shot
1	71.43 ± 0.71	**87.77 ± 0.42**
2	**72.09 ± 0.66**	87.68 ± 0.36
3	71.25 ± 0.74	85.36 ± 0.57
4	69.10 ± 0.31	84.21 ± 0.45

5 Conclusion

In this paper, a novel CNN architecture, named TAFD-Net, is introduced for the purpose of few-shot fine-grained image classification. This approach incorporates a task adaptive component that iterates through samples encompassing the entire task, thereby facilitating the acquisition of task-level information. Considering the asymmetric relationship that exists between query and support samples, an asymmetric divergence mechanism is employed to align the distribution of the query with that of the support class. Furthermore, a contrastive measure strategy is implemented to enhance the performance of the metric function. Experimental results conducted on three fine-grained image datasets corroborate the effectiveness of the proposed TAFD-Net.

References

1. Antoniou, A., Storkey, A.J.: Learning to learn by self-critique. In: Advances in Neural Information Processing Systems, vol. 32 (2019)
2. He, K., Zhang, X., Ren, S., Sun, J.: Deep residual learning for image recognition. In: Proceedings of the IEEE Conference on Computer Vision and Pattern Recognition, pp. 770–778 (2016)
3. He, R., Wu, X., Sun, Z., Tan, T.: Wasserstein CNN: learning invariant features for NIR-VIS face recognition. IEEE Trans. Pattern Anal. Mach. Intell. **41**(7), 1761–1773 (2018)
4. Huang, H., Zhang, J., Yu, L., Zhang, J., Wu, Q., Xu, C.: TOAN: target-oriented alignment network for fine-grained image categorization with few labeled samples. IEEE Trans. Circuits Syst. Video Technol. **32**(2), 853–866 (2021)
5. Huang, H., Zhang, J., Zhang, J., Wu, Q., Xu, J.: Compare more nuanced: pairwise alignment bilinear network for few-shot fine-grained learning. In: 2019 IEEE International Conference on Multimedia and Expo (ICME), pp. 91–96. IEEE (2019)
6. Huang, H., Zhang, J., Zhang, J., Xu, J., Wu, Q.: Low-rank pairwise alignment bilinear network for few-shot fine-grained image classification. IEEE Trans. Multimed. **23**, 1666–1680 (2020)
7. Khosla, A., Jayadevaprakash, N., Yao, B., Li, F.F.: Novel dataset for fine-grained image categorization: stanford dogs. In: Proceedings of the CVPR Workshop on Fine-Grained Visual Categorization (FGVC), vol. 2 (2011)

8. Krause, J., Stark, M., Deng, J., Fei-Fei, L.: 3D object representations for fine-grained categorization. In: Proceedings of the IEEE International Conference on Computer Vision Workshops, pp. 554–561 (2013)
9. Li, W., Wang, L., Huo, J., Shi, Y., Gao, Y., Luo, J.: Asymmetric distribution measure for few-shot learning. In: Proceedings of the Twenty-Ninth International Conference on International Joint Conferences on Artificial Intelligence, pp. 2957–2963 (2021)
10. Li, W., Wang, L., Xu, J., Huo, J., Gao, Y., Luo, J.: Revisiting local descriptor based image-to-class measure for few-shot learning. In: Proceedings of the IEEE/CVF Conference on Computer Vision and Pattern Recognition, pp. 7260–7268 (2019)
11. Li, X., Wu, J., Sun, Z., Ma, Z., Cao, J., Xue, J.H.: BSNet: bi-similarity network for few-shot fine-grained image classification. IEEE Trans. Image Process. **30**, 1318–1331 (2020)
12. Li, Y., Li, H., Chen, H., Chen, C.: Hierarchical representation based query-specific prototypical network for few-shot image classification. arXiv preprint arXiv:2103.11384 (2021)
13. Lu, L., Cai, Y., Huang, H., Wang, P.: An efficient fine-grained vehicle recognition method based on part-level feature optimization. Neurocomputing **536**, 40–49 (2023)
14. Lu, L., Huang, H.: Component-based feature extraction and representation schemes for vehicle make and model recognition. Neurocomputing **372**, 92–99 (2020)
15. Lu, L., Wang, P., Cao, Y.: A novel part-level feature extraction method for fine-grained vehicle recognition. Pattern Recogn. **131**, 108869 (2022)
16. Lu, L., Wang, P., Huang, H.: A large-scale frontal vehicle image dataset for fine-grained vehicle categorization. IEEE Trans. Intell. Transp. Syst. **23**(3), 1818–1828 (2020)
17. Snell, J., Swersky, K., Zemel, R.: Prototypical networks for few-shot learning. In: Advances in Neural Information Processing Systems, vol. 30 (2017)
18. Vinyals, O., Blundell, C., Lillicrap, T., Wierstra, D.: Matching networks for one shot learning. In: Advances in Neural Information Processing Systems, vol. 29, pp. 3630–3638 (2016)
19. Wah, C., Branson, S., Welinder, P., Perona, P., Belongie, S.: The caltech-UCSD birds-200-2011 dataset (2011)
20. Wang, J., Wu, J., Bai, H., Cheng, J.: M-NAS: meta neural architecture search. In: Proceedings of the AAAI Conference on Artificial Intelligence, vol. 34, pp. 6186–6193 (2020)
21. Wang, Y., Xu, C., Liu, C., Zhang, L., Fu, Y.: Instance credibility inference for few-shot learning. In: Proceedings of the IEEE/CVF Conference on Computer Vision and Pattern Recognition, pp. 12836–12845 (2020)
22. Zhang, C., Cai, Y., Lin, G., Shen, C.: DeepemD: few-shot image classification with differentiable earth mover's distance and structured classifiers. In: Proceedings of the IEEE/CVF Conference on Computer Vision and Pattern Recognition, pp. 12203–12213 (2020)

23. Zhu, P., Gu, M., Li, W., Zhang, C., Hu, Q.: Progressive point to set metric learning for semi-supervised few-shot classification. In: 2020 IEEE International Conference on Image Processing (ICIP), pp. 196–200. IEEE (2020)
24. Zhu, Y., Liu, C., Jiang, S., et al.: Multi-attention meta learning for few-shot fine-grained image recognition. In: IJCAI, Beijing, pp. 1090–1096 (2020)

ST_TransNeXt: A Novel Pig Behavior Recognition Model

Wangli Hao[1(✉)], Hao Shu[1], Xinyuan Hu[1], Meng Han[1,2], and Fuzhong Li[1]

[1] Shanxi Agricultural University, Taigu, Jinzhong, Shanxi, China
haowangli@sxau.edu.cn
[2] Hangzhou Dianzi University, Hangzhou, Zhejiang, China

Abstract. Pig behavior recognition serves as a crucial indicator for monitoring health and environmental conditions. However, traditional pig behavior recognition methods are limited in their processing capabilities, struggling to accurately extract image features and dynamically analyze sequence data. This paper introduces a novel ST_TransNeXt model, which ingeniously integrates the TransNeXt module with the sLSTM, enabling a profound understanding of the dynamic characteristics of pig group behaviors. Specifically, the Bio-inspired Aggregated Attention within the TransNeXt module is inspired by biological vision system to efficiently fuses local and global image features, sLSTM processes multi-frame data to capture temporal dependencies, the combination of these two techniques constructs a powerful temporal feature vector, enhancing model performance. Experimental results demonstrate that the proposed ST_TransNeXt model achieves an accuracy rate of 93.98%, outperforming existing models by more than 1.1% and achieving a maximum reduction in loss value of 0.4737.

Keywords: TransNeXt · sLSTM · Behavior recognition · Temporal sequence

1 Introduction

Pig behavior recognition is crucial for feeding management, aiding in the understanding pigs' needs and optimizing feeding conditions [1]. Traditional methods, including manual observation and sensor technology, have limitations, necessitating the development of improved approaches [2].

Technological advancements have led researchers to explore deep learning for pig behavior recognition, boosting accuracy, efficiency, real-time monitoring, and optimization of feeding management [3].

Ji et al. [4] pioneered the integration of the Temporal Shift Module (TSM) into various mainstream 2D convolutional neural network architectures, including ResNet50, ResNeXt50, DenseNet201, and ConvNeXt-t. This innovation significantly enhanced the model's capability to recognize pig aggression behaviors, achieving an impressive accuracy rate of 95.69% in experiments.

Gao et al. [5] focused on behavior analysis within video sequences by introducing a hybrid model that integrates Convolutional Neural Networks (CNNs) with Gated Recurrent Units (GRUs). This model effectively leverages the spatial feature extraction capabilities of CNNs and the temporal sequence processing strengths of GRUs, achieving an accuracy rate of 94.8% in experiments.

Despite some progress in deep learning for pig behavior recognition [6], challenges remain in accurately extracting image features and simultaneously analyzing long temporal sequence data. To overcome these difficulties, we propose the following strategies.

To further improve the performance of pig behavior recognition, we develop a novel ST_TransNeXt model based on TransNeXt and sLSTM. TransNeXt simulates biological vision through a Bio-inspired Aggregated Attention Mechanism to enhance global token perception and reduce information decay [7]. Additionally, TransNeXt incorporates a Convolutional GLU for adaptive local feature channel weight adjustment. Furthermore, sLSTM improves the stability and accuracy of the model in processing long sequences through a novel gating mechanism and normalization techniques [8], thereby complementing TransNeXt's limitations in capturing long-term dependencies.

In summary, the contributions of this paper are as follows:

(1) This paper proposes a novel ST_TransNeXt model for pig behavior recognition. Specifically, TransNeXt extracts key features from videos, sLSTM captures temporal information. Together, they enable ST_TransNeXt to detect behavioral changes and temporal patterns, thereby improving recognition accuracy.
(2) Several ablation studies confirm the effectiveness of the ST_TransNeXt model in pig behavior recognition. Specifically, ST_TransNeXt achieves a state-of-the-art accuracy of 93.98%, which is at most 10.77% higher than that of existing models. Additionally, it reduces loss by 47.37%.

2 Materials and Methods

2.1 Dataset

The pig behavior recognition data was collected from the pig breeding base of Xiangfen County Nongluyuan Agriculture Co., Ltd., Linfen City, Shanxi Province. Specifically, the data collection period ranged from August 12, 2022, to September 25, 2022. The farm comprises six pig houses, each housing 6-month-old ternary breed pigs, with an average of 10 pigs per house. Cameras have been installed in all six pig houses.

After data collection, 1.5 TB of pig video data (over 5,000 files) was obtained. Video clips ranging from 5 to 10 s were selected and preprocessed, resulting in six behavior categories: Fighting, Drinking, Investigating, Walking, Eating, and Lying, with about 450 videos per category. The final dataset comprised a total of 2,775 videos, as illustrated in Fig. 1.

Fig. 1. Examples of several pig behaviors.

2.2 ST_TransNeXt

In order to improve the performance of pig behavior recognition, this paper introduces a novel model, ST_TransNeXt, which is based on TransNeXt [7] and sLSTM [8], as shown in Fig. 2.

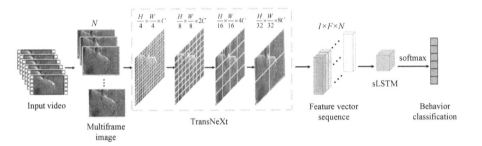

Fig. 2. The diagram of the ST_TransNeXt model.

The core advantage of TransNeXt is its ability to efficiently extract global context information. Additionally, the sLSTM network can capture the long-term temporal information underlying the corresponding videos. Based on these advantages, ST_TransNeXt is able to capture subtle changes in pig behavior, thereby improving the performance of pig behavior recognition [9].

The next two subsections will introduce the two modules in ST_TransNeXt, TransNeXt and sLSTM.

2.3 TransNeXt Module

TransNeXt [7] is a type of visual backbone network model that simulates the biological visual system by integrating aggregated pixel-focused attention and Convolutional GLU (Convolutional Gate-controlled Linear Unit) [10] (as shown in Fig. 3 (b)). This model particularly imitates biological macular vision and continuous eye movement, allowing each pixel to be perceived.

In terms of architecture, as shown in Fig. 3(a), the TransNeXt model adopts a staged network structure that processes input images in multiple stages. Each processing stage consists of a Convolutional GLU layer and an aggregation attention mechanism layer. This layout is designed to optimize model performance and ensure stability in processing image data.

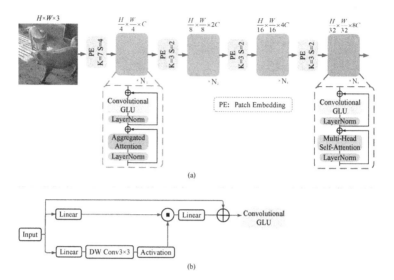

Fig. 3. The overview of the TransNeXt model.

2.4 sLSTM Module

As an extended variant of LSTM [11], the core improvement of sLSTM [8] lies in the introduction of an exponential gated activation function and a normalized state, which enables the sLSTM model to excel in capturing long temporal sequence dependencies. In addition, sLSTM proposes a cyclic connection mechanism to realize a paradigm of memory mixing. The architecture of sLSTM is depicted in Fig. 4. The specific formula is as follows:

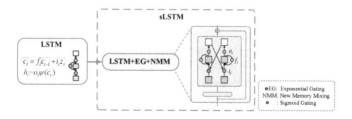

Fig. 4. The diagram of the sLSTM architecture.

$$c_t = f_t c_{t-1} + i_t z_t \tag{1}$$

where c_t represents the state held by the cell at time step t, f_t and i_t correspond to the forget gate and the input gate, respectively.

$$z_t = \varphi(\tilde{z}_t) \,, \quad \tilde{z}_t = w_z^T x_t + r_z h_{t-1} + b_z \tag{2}$$

where z_t denotes the unit input. Here, φ represents the activation function of the unit input gate, $\tilde{z}_t$ is an intermediate variable, w^T denotes the weight matrix, r indicates a learnable parameter, and b represents the bias.

$$n_t = f_t n_{t-1} + i_t \tag{3}$$

where n_t denotes the normalized state at time step t, summarizing the product of the input gate with all future forgetting gates.

$$h_t = o_t \tilde{h}_t \ , \ \tilde{h}_t = c_t/n_t \tag{4}$$

where h_t represents the hidden state and o_t denotes the output gate, while $\tilde{h}_t$ indicates an intermediate variable. The n_t is employed to prevent overflow issues.

$$i_t = exp(\tilde{i}_t) \ , \ \tilde{i}_t = w_i^T x_t + r_i h_{t-1} + b_i \tag{5}$$

where exp indicates the exponential function. and $\tilde{i}_t$ represents an intermediate variable.

$$f_t = \sigma(\tilde{f}_t) \ , \ \tilde{f}_t = w_f^T x_t + r_f h_{t-1} + b_f \tag{6}$$

where σ represents the gated activation function *sigmoid*, and $\tilde{f}_t$ denotes an intermediate variable.

$$o_t = \sigma(\tilde{o}_t) \ , \ \tilde{o}_t = w_o^T x_t + r_o h_{t-1} + b_o \tag{7}$$

where $\tilde{o}_t$ indicates an intermediate variable.

$$m_t = \max(\log(f_t) + m_{t-1}, \log(i_t)) \tag{8}$$

where m_t denotes the stabilizer state. *log* indicates the exponential inverse operation, which is equivalent to using *log* to degrade i_t and f_t to avoid overflow.

$$i'_t = exp(\log(i_t) - m_t) = exp(\tilde{i}_t - m_t) \tag{9}$$

$$f'_t = exp(\log(f_t) + m_{t-1} - m_t) \tag{10}$$

where i'_t denotes the stabilized input gate, and f'_t indicates the stabilized forget gate, both are adjusted according to the value of m_t.

3 Experiments and Analysis

To evaluate the performance of the proposed model effectively, we conducted a series of experiments from various perspectives. The detailed design and documentation of these experiments are presented below.

3.1 Experimental Setup

All experiments conducted in this paper were carried out on the Ubuntu 18.04 hardware system. The experimental parameters were set as follows: the batch size is 4, the initial learning rate is 0.001, and the total number of training rounds is 200 epochs. Additionally, the Adam optimizer was employed during the training process.

3.2 Evaluation of the Effectiveness of Different Models

To verify the effectiveness of our proposed model, ST_TransNeXt, several popular models, including ViT (Vision Transformer) [12], Swin Transformer [13], ConvNeXt [14], and TransNeXt [7], were employed for comparison. The comparison results are shown in Table 1.

Table 1. Comparison of different models.

Model	batch size	Accuracy (%)	Loss
ViT	4	83.21	0.8597
Swin Transformer	4	90.51	0.4902
ConvNeXt	4	92.52	0.3991
TransNeXt	4	92.88	0.4777
ST_TransNeXt	**4**	**93.98**	**0.3860**

Table 1 demonstrates that our proposed ST_TransNeXt model achieves the best performance across all evaluation criteria. Specifically, the ST_TransNeXt model achieves 93.98%, which is 10.77%, 3.47%, 1.46%, and 1.1% higher than the performances of the ViT, Swin Transformer, ConvNeXt, and TransNeXt models. Furthermore, the loss value of ST_TransNeXt is 0.3860, which is 0.4737, 0.1042, 0.0131, and 0.0917 lower than the loss values of the other four models, respectively. Similarly, this result also supports the validity of the TransNeXt module in the ST_TransNeXt model.

To further compare the capabilities of these models, Fig. 5 shows the accuracy and loss curves of the comparison models under different epochs.

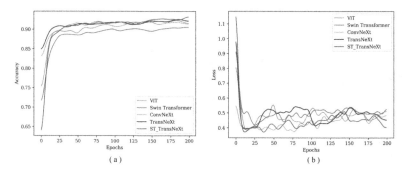

Fig. 5. The accuracy and loss curves under different epochs for different models. (a) is for accuracy and (b) is loss.

In Fig. 5(a), the accuracy of ST_TransNeXt uniformly surpasses that of the ViT, Swin Transformer, ConvNeXt, and TransNeXt models. Furthermore, in

Fig. 5(b), the ST_TransNeXt model achieves the lowest loss. These results further validate the effectiveness of ST_TransNeXt.

The superiority of the ST_TransNeXt model for pig behavior recognition can be attributed to the following reasons. The biologically inspired attention mechanism in ST_TransNeXt allows the model to capture global features effectively. Furthermore, the exponentially gated in ST_TransNeXt enables the model to establish long temporal dependencies underlying the corresponding video. Consequently, the promising global feature extraction and long temporal dependencies significantly enhance pig behavior recognition performance.

3.3 Evaluation of the Effectiveness of sLSTM

To evaluate the efficacy of the sLSTM [8] module within ST_TransNeXt, several models incorporating diverse temporal networks (LSTM [11] and GRU [10]) are utilized for comparative analysis. The results are shown in Table 2.

Table 2. Comparison of different models.

Model	batch size	Accuracy (%)	Loss
TransNeXt_GRU	4	91.42	0.6981
TransNeXt_LSTM	4	90.88	0.7579
ST_TransNeXt	4	**93.98**	**0.3860**

Table 2 illustrates that ST_TransNeXt outperforms both TransNeXt_LSTM and TransNeXt_GRU. Specifically, the ST_TransNeXt model achieves an accuracy of 93.98% with a loss value of 0.3860.

In order to evaluate the performance of different models, Fig. 6 reports the accuracy and loss values of these models under different training epochs.

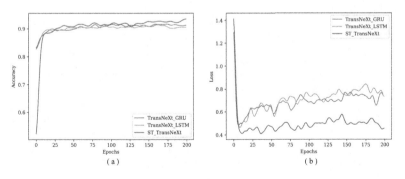

Fig. 6. The accuracy and loss curves under different epochs for TransNeXt_GRU, TransNeXt_LSTM and ST_TransNeXt. (a) is for accuracy and (b) is loss.

In Fig. 6(a), ST_TransNeXt demonstrates higher accuracy than both TransNeXt_GRU and TransNeXt_LSTM. Figure 6(b) shows that ST_TransNeXt's loss during training is significantly lower. In summary, ST_TransNeXt performs the best in pig behavior recognition.

The exceptional performance of the sLSTM within the ST_TransNeXt framework for analyzing pig behavior is due to its integration of an exponential gated activation function, which enables the ST_TransNeXt model to capture long temporal dependencies in the video data. Therefore, compared to the other two models, ST_TransNeXt achieves the best performance in the pig behavior recognition task.

3.4 Evaluation of the Effectiveness of the Number of Hidden Layers

To validate the influence of varying hidden layers on the performance of ST_TransNeXt, models with 1, 2, and 3 hidden layers were selected to compare, and the results are reported in Table 3.

Table 3. Comparison of different models with various number of hidden layers.

Model	batch size	hidden layers	Accuracy (%)	Loss
ST_TransNeXt	4	1	92.34	0.5925
ST_TransNeXt	**4**	**2**	**93.98**	**0.3860**
ST_TransNeXt	4	3	90.69	0.6054

Based on Table 3, when the number of hidden layers is 2, the ST_TransNeXt model achieves optimal performance with an accuracy of 93.98%. Furthermore, its loss value is lower, decreasing by 0.2065 and 0.2194 compared to models with 1 and 3 layers, respectively.

To evaluate the influence of varying hidden layers on the model's performance, Fig. 7 presents the accuracy and loss curves for different models, each with a distinct number of hidden layers.

In Fig. 7(a), it can be observed that when the number of hidden layers in the ST_TransNeXt model is 2, it shows the highest accuracy. Meanwhile, in Fig. 7(b), the model with 2 hidden layers achieves the lowest loss. These results further validate the effectiveness of ST_TransNeXt when the number of hidden layers is set to 2.

The impact of hidden layers on the model's performance can be attributed to the following reasons: If the number of layers is too small, the model may struggle to distinguish between different categories of data. Conversely, having too many layers can enhance the model's expressive ability, but it may lead to excessive feature abstraction. After analysis, we propose that setting the number of hidden layers to 2 allows the ST_TransNeXt model to achieve the best performance in pig behavior recognition.

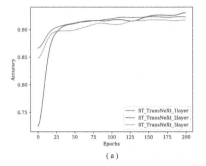 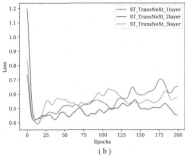

Fig. 7. The accuracy and loss curves under different epochs for different models with different number of hidden layers. (a) is for accuracy and (b) is loss.

4 Conclusions

In this paper, we propose a novel model named ST_TransNeXt for pig behavior recognition, which is based on TransNeXt and sLSTM. Specifically, the bio-inspired aggregated attention within the TransNeXt module efficiently fuses local and global image features. The sLSTM captures long temporal dependencies in videos, enhancing feature discriminability and improving recognition accuracy. Several ablation studies demonstrate that ST_TransNeXt achieves state-of-the-art performance, with accuracy improvements ranging from 1.1% to 10.77% and loss reductions between 1.31% and 47.37%. These results validate the effectiveness of ST_TransNeXt for pig behavior recognition.

Funding. This work was supported by the Shanxi Province Basic Research Program [20220- 3021212444]; Shanxi Agricultural University Science and Technology Innovation Enhancement Project [CXGC2023045]; Shanxi Province Higher Education Teaching Reform and Innovation Project [J20220274]; Shanxi Postgraduate Education and Teaching Reform Project Fund [2022YJJG094]; Shanxi Agricultural University doctoral research startup project [2021BQ88]; Shanxi Agricultural University Academic Restoration Research Project [2020xshf38]; Young and Middle-aged Top-notch Innovative Talent Cultivation Program of the Software College, Shanxi Agricultural University [SXAUKY2024005].

References

1. Hao, W.L., et al.: TSML: a new pig behavior recognition method based on two-stream mutual learning network. IEEE Access (2023)
2. Melfsen, A., Lepsien, A., Bosselmann, J., Koschmider, A., Hartung, E.: Describing behavior sequences of fattening pigs using process mining on video data and automated pig behavior recognition. Agriculture **8**, 1639 (2023)
3. Wei, J.C., Tang, X., Liu, J.X., Zhang, Z.Y.: Detection of pig movement and aggression using deep learning approaches. Animals **19**, 3074 (2023)

4. Ji, H.Y., Teng, G.H., Yu, J.H., Wen, Y.B., Deng, H.X., Zhuang, Y.R.: Efficient aggressive behavior recognition of pigs based on temporal shift module. Animals **13**, 2078 (2023)
5. Gao Y., et sl.: Recognition of aggressive behavior of group-housed pigs based on CNN-GRU hybrid model with spatio-temporal attention mechanism. Comput. Electron. Agric. **205** (2023)
6. Qi, L.T., et al.: An improved YOLOv5 model based on visual attention mechanism: application to recognition of tomato virus disease. Comput. Electron. Agric. **194**, 106780 (2022)
7. Shi, D.: TransNeXt: robust foveal visual perception for vision transformers. IEEE (2024)
8. Beck, M., et al.: xLSTM: extended long short-term memory. arXiv preprint arXiv:2405.04517 (2024)
9. Hakansson, F., Jensen, D.B.: Automatic monitoring and detection of tail-biting behavior in groups of pigs using video-based deep learning methods. Front. Vet. Sci. Front. Media SA **9**, 1099347 (2023)
10. Liu, C., Zhen, J.T., Shan, W.: Time series classification based on convolutional network with a gated linear units kernel. Eng. Appl. Arti. Intell. **123**, 106296 (2023)
11. Shiri, F.M., Perumal, T., Mustapha, N., Mohamed, R.: A comprehensive overview and comparative analysis on deep learning models: CNN, RNN, LSTM, GRU. arXiv preprint arXiv:2305.17473 (2023)
12. Han, K., et al.: A survey on vision transformer. IEEE Trans. Pattern Anal. Mach. Intell. (2022)
13. Liu, Z., et al.: Swin transformer: hierarchical vision transformer using shifted windows. IEEE (2021)
14. An, Y.Y., et al.: A hybrid attention-guided ConvNeXt-GRU network for action recognition. Elsevier (2024)

A Method for Predicting the RUL of HDDs Based on Bidirectional LSTM and Transformer

Zehong Wu, Jinghui Qin[✉], Zhijing Yang, and Yongyi Lu

School of Information Engineering, Guangdong University of Technology, West Waihuan Street No.100, Guangzhou City 510006, Guangdong Province, China
2112303025@mail2.gdut.edu.cn, {qinjinghui,yzhj,yylu}@gdut.edu.cn

Abstract. As the core equipment of the storage system, a hard disk may cause data loss, system crash, and even business interruption if it fails. In order to improve the reliability and security of data centers, more and more deep learning methods have emerged to predict the remaining useful life (RUL) of hard disk drives (HDDs), but these methods have the problem of poor long-term failure prediction due to ignoring the time information in the data. In order to solve these problems, we propose a method for predicting the remaining useful life of HDD based on bidirectional LSTM and Transformer. The data is first preliminarily extracted by bidirectional LSTM, and then the features are input to the Transformer for further optimization, and finally the remaining service life of the HDD is predicted according to the features. Experiments have shown that the proposed method is very effective and achieves better performance compared to many state-of-the-art methods.

Keywords: failure prediction · remaining useful life · Bidirectional LSTM · Transformer

1 Introduction

As the core equipment of the storage system, a hard disk may cause data loss, system crash, and even business interruption if it fails [1]. By predicting the failure of the hard drive in advance, users can back up important data in time and avoid irreversible data corruption. At the same time, forecasting can provide a valuable maintenance window to prevent downtime caused by sudden hardware failures, thereby ensuring the continuous operation of the system and reducing maintenance costs. With the expansion of data centers and the increasing reliance on hard disk equipment, hard disk failure prediction is not only a key measure to ensure data security, but also an important means to improve system reliability and optimize operation and maintenance management. Therefore, it is an important research topic to explore the prediction methods of HDD failure and take measures to prolong the service life of HDD.

SMART (Self-Monitoring, Analysis, and Reporting Technology) is a monitoring system used in hard disk drives (HDDs) and solid-state drives (SSDs) [2]. SMART technology helps users or system administrators prevent potential hard drive failures by collecting and reporting a series of parameters so that they can be backed up or replaced in a timely manner to avoid data loss. Researchers also use the SMART index to study the technology related to the prediction of the remaining useful life of HDDs. At present, most of the methods can be divided into two categories, one of which is traditional machine learning methods. I. C. Chaves et al. [3] proposed a Bayesian network-based hard drive failure prediction method that uses the deterioration of HDDs over time to predict the eventual failure through SMART (Self-Monitoring Analysis and Reporting Technology) attribute calculations. Shen J et al. [4] proposed a random forest-based prediction method that uses different combinations of decision trees to vote on classification results for different relevant SMART attributes of various types of HDD failures, so as to achieve good prediction results. Zhu B et al. [5] proposed an improved Support Vector Machine (SVM) model that uses SMART metrics to predict hard drive failures, with up to 95% prediction accuracy and low FAR. The other is the method of deep learning. Xu, C et al. [6] introduced a new method based on recurrent neural networks (RNNs) to assess the health of hard drives based on sequential SMART attributes that change gradually. A state-of-health assessment is more valuable in practice than a simple failure prediction method because it enables technicians to schedule the recovery of different hard drives depending on the level of urgency. J. Wu et al. [7] optimizes the Self-Monitoring, Analysis, and Reporting Technology (SMART) attributes by using information entropy to select the most relevant prediction attributes, and proposes a multichannel convolutional neural network-based LSTM (MCCNN-LSTM) model to predict whether a given disk will fail in the next few days. T. Jiang et al. [8] proposed an anomaly detection method based on a generative adversarial network (GAN) that models normal data and avoids data imbalance. Although these methods can achieve good results in short-term fault prediction, they are easy to perform poorly because they cannot obtain optimal feature information when extracting data feature information in long-term fault prediction tasks.

In addition, in recent years, Transformer models have been widely used in many fields due to their powerful sequence processing and feature extraction capabilities. Originally, Transformer was mainly used for natural language processing (NLP), but with the deepening of research, it has expanded to many fields such as computer vision, speech processing, and recommendation systems. However, there are few related studies on the application of Transformer in the field of hard disk failure prediction, so this paper tries to use the Transformer model to study the remaining service life of HDD. In order to solve the problems of the existing methods and apply Transformer to the task of hard disk failure prediction, we propose A HDD Failure Predict Approach based on Bidirectional LSTM and transformer to predict the remaining service life of hard disk.

2 Related Work

The reliability of hard drives is one of the key challenges faced by data centers, which often contain a large number of storage devices, such as HDDs. However, because there is often not enough time to migrate and back up important data from a hard drive in advance, it is especially important to accurately predict the remaining useful life of a hard drive before it fails. Through this prediction, data corruption can be effectively avoided and the reliability and stability of the system can be improved. In this section, several methods for predicting the remaining useful life are discussed.

In the early years, the methods used to detect the remaining useful life were mainly machine learning-related techniques. J. F. Murray et al. [9] used a naïve Bayesian model to predict the failure of hard drives, and proposed an NB model as well as an SVM model with prediction performance much higher than that of the threshold method while maintaining a low false positive rate, where the SVM model exhibited higher performance than the proposed NB model, but with a large number of parameters and the time-consuming problem of non-optimal grid search, which was computationally expensive. Jing Li et al. [10] proposed various prediction models for hard disk failures using classification and regression trees, which have better prediction performance than traditional backpropagation (BP) and artificial neural networks (ANNs). Subsequently, a hardware fault prediction technology based on classification trees (CT) and gradient boosted regression trees (GBRD) was proposed. As a result, more than 93% of failures at a false alarm rate (FAR) can be predicted on the CT model with a false alarm rate (FDR) of less than 0.01%, and 90% of failures at a false alarm rate (FAR) can be achieved with no false positives on the GBRT model.

In recent years, with the development of the field of deep learning, the focus of related technologies for predicting the remaining useful life has gradually shifted from the field of machine learning to the field of deep learning, and many more effective prediction methods have been developed. Shuangwang Zhang et al. [11] were able to learn sequence features from raw data by using LSTM networks. At the same time, the proposed attention mechanism can automatically identify the importance of different time steps and estimate the health state of the hard disk based on the failure time. Qinda Hai et al. [12] proposed a disk failure prediction method based on SMART attribute gated recursive unit (GRU) neural network and TimeGAN adversarial network, which first solved the data imbalance problem through TimeGAN, and then completed the hard disk failure prediction through GRU neural network. Anhui Bai et al. [13] proposed a method based on attention mechanism for bidirectional long short-term memory (LSTM) networks combined with differential features.In this method, the extracted general feature information is applied to the attention-based bidirectional LSTM network, and the features containing useful degradation information are captured for prediction by assigning higher weights to the key features. Experimental results show that the proposed method performs better than the traditional LSTM method in predicting the remaining useful life (RUL) of the hard disk 60 days before failure, and achieves a fault detection rate (FDR) of 97.83%.

3 Method

3.1 Proposed Framework

The proposed network framework is mainly composed of SMART feature extraction, bidirectional LSTM, Transformer, and classifier. Firstly, the data with SMART IDs of 1, 3, 5, 7, 9, 187, 189, 194, and 197 were extracted from the original data as the model input sequence. Then, the sequence was input into the bidirectional LSTM model for preliminary sequence feature extraction. Secondly, the extracted sequence features are positionally encoded and then input to the Transformer model for further sequence feature information extraction, so as to obtain better sequence feature information. Finally, the remaining service life of the HDD is predicted by using the obtained feature sequence, and the final prediction result is obtained (Fig. 1).

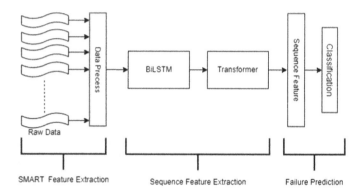

Fig. 1. The framework of the proposed network

3.2 Feature Extraction Mode

Since the LSTM network unit has good information extraction ability, we consider using this network as the network unit for the preliminary extraction of data features in our model. An LSTM neural network consists of four main components: a neural state, an amnesia gate, an input gate, and an output gate. Since the LSTM network uses a single memory cell state to transmit information for a long time, even if the four gates are interconnected, there will be no gradient vanishing or gradient explosion in the traditional cyclic convolutional network. Assuming that the forgetting gate unit is f_t, the memory gate unit is i_t and $\tilde{C}_t$, the update gate unit is C_t, and the output gate unit is h_t in the LSTM, the corresponding formula for each gate element is shown below.

$$f_t = \sigma\left(W_f \cdot [h_{t-1}, x_t] + b_f\right) \quad (1)$$
$$i_t = \sigma\left(W_i \cdot [h_{t-1}, x_t] + b_i\right) \quad (2)$$
$$\tilde{C} = \tanh\left(W_C \cdot [h_{t-1}, x_t] + b_C\right) \quad (3)$$
$$C_t = f_t * C_{t-1} + i_t * \tilde{C}_t \quad (4)$$
$$h_t = \sigma\left(W_o\left[h_{t-1}, x_t\right] + b_o\right) * \tanh(C_t) \quad (5)$$

Standard recurrent convolutional network RNNs, as well as traditional LSTM, GRU process sequences on time series, often ignoring future contextual information. However, when the HDD fails, the value of the smart attribute drops sharply, and when the current state of the HDD is in a critical state between intact and faulty, it can only be judged more accurately by combining the state of historical time and the state of future time. Therefore, in the task of this paper, we choose a bidirectional LSTM as the preliminary data feature extractor.

3.3 Feature Optimization

The feature optimization network we use is mainly the encoder part of the Transformer. The feature processing network here is composed of the same network layers as the N layer, and each network layer is roughly composed of four parts, namely the multi-head attention mechanism layer, the feedforward neural network layer, and two normalization layers, as shown in the Fig. 2.

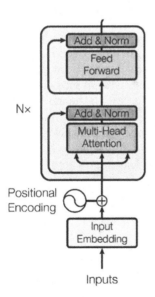

Fig. 2. This is the network structure of Transformmer [15]

Among them, the main function of the multi-head attention mechanism layer is to integrate the representations of subsequences at different positions into one information and capture the long-distance dependencies in the sequence. Then, the results of the multi-head attention mechanism after residual and layer normalization are input into the position-based feedforward neural network, and the feature information of different positions is further extracted and fused to enhance the nonlinear ability and expression ability of the model. The formula for the self-attention mechanism of the bull is as follows.

$$MultiHead(Q, K, V) = Concat(head_1, \cdots, head_h)W^o \qquad (6)$$

$$head_i = Attention(QW_i^Q, KW_i^K, VW_i^V) \qquad (7)$$

$$Attention(Q, K, V) = softmax(\frac{QK^t}{\sqrt{d_k}})V \qquad (8)$$

In addition, before the transformer is inputted, the input needs to be positionally encoded, as shown in the following equation, where pos is the position and i is the dimension.

$$PE_{(pos, 2i)} = \sin(pos/10000^{2i/d_{model}}) \qquad (9)$$

$$PE_{(pos, 2i+1)} = \cos(pos/10000^{2i/d_{model}}) \qquad (10)$$

4 Experiments and Performance

4.1 Datasets and Parameter Settings

The dataset [14] used in the experiment was provided by Backblaze, an online backup and cloud storage provider. Backblaze provides a daily snapshot of HDD's SMART attributes and health from 2013 to the present, and to date, the dataset contains more than 2TB of managed hard drives. In this article, we mainly select the hard drive data with the model number ST4000DM1000 as the dataset for our model training and testing. Since the prediction of the remaining useful life of HDD in this paper is a multi-classification task, CrossEntropyLoss is used as the loss function. In addition, we choose the ADAM optimizer as the model optimizer, and set the learning rate to 0.0001, the data batch size to 256, and the epoch size to 200.

4.2 Data Preprocessing

Since the results of the experiment are mainly related to the degree of health, they can be divided into four aspects: Good, Very Fair, Warning and Alert. Our methodology is designed to predict the remaining useful life of an HDD in the 60 days prior to failure. Thus, the degree of fitness was classified between 60 days before the failure. Details of the health classification are provided in the table (Table 1).

Table 1. SMART Feature

Health Degree	DayToFailure
GOOD	0–9
Very Fair	10–21
Warning	22–30
Alert	31–60

4.3 Evaluation Criteria

To evaluate the performance of our methodology, we have introduced several metrics as evaluation criteria. Specifically, ACC is the accuracy of the model after the validation set, the FAR false positive rate (FPR) is used to measure the accuracy of the model in predicting positive results, which specifically refers to the proportion of samples that are actually negative cases that are incorrectly predicted as positive examples, and the FDR fault detection rate refers to the proportion of the total fault data in the test set that are accurately detected as faults by the algorithm to the total number of fault data.

4.4 Ablation Experiments

As can be seen from the Table 2, when we do not use bidirectional LSTM and only use traditional transformer, the model decreases by 3.14%, 2.09% and 2.1% in the accuracy, FDR and FAR, respectively, while when we remove the transformer model and only use the bidirectional LSTM to predict the remaining service life in HDD, the model decreases by 6.66% and 1.36% in the correctness, FDR and FAR, respectively. The increase is 0.66%, which indicates that Transformer can further optimize the data features extracted by the bidirectional LSTM, which greatly improves the performance of the model. In addition, it can be seen from the table that the ability of Transformer to extract data features is better than that of bidirectional LSTM, so placing Transformer in bidirectional LSTM can further extract more information.

Table 2. Results of Ablation Experiments

METHOD	ACC	FDR	FAR
BiLSTM	91.07%	97.43%	1.55%
Transformer	93.59%	96.70%	2.99%
Our method	97.73%	98.79%	0.89%

4.5 Comparison of Methods

As can be seen from the table, we compared our proposed method with other methods in the literature based on the same data set. The comparison of the GRU-based prediction method [12], the attention mechanism-based LSTM [11], the attention-based bidirectional LSTM [13] and our method is shown in Table 3. This shows that our method outperforms other methods in terms of accuracy, FDA, and FAR. It can be found that the bidirectional LSTM can extract the features in the data well, and the Transformer model can optimize the feature information again in the data features extracted by the bidirectional LSTM. It can also be seen from the results that the Transformer is more effective in extracting and fusing feature information than the self-attention mechanism.

Table 3. Algorithm Comparison

METHOD	ACC	FDR	FAR
Timegan-GRU	92.87%	92.33%	2.88%
LSTM-Attention	89.76%	89.08%	3.99%
BiLSTM-Attention	96.24%	97.85%	1.83%
Our method	97.73%	98.79%	0.89%

5 Conclusion

In this paper, we propose a network framework for A HDD Failure Predict Approach based on Bidirectional LSTM and transformer to predict the remaining useful life of hard drives. In this method, the bidirectional LSTM is used to preliminarily extract the features in the sequence, and then the extracted sequence features are input into the Transformer model to capture the better sequence feature information in the sequence, and finally these features are predicted by the classifier to the remaining service life of the HDD. A large number of experimental results show that the performance of our proposed model is better than that of other models.

References

1. Wang, G., Zhang, L., Xu, W.: What can we learn from four years of data center hardware failures? In: 2017 47th Annual IEEE/IFIP International Conference on Dependable Systems and Networks (DSN), Denver, CO, USA, pp. 25–36 (2017)
2. Sankar, S., Shaw, M., Vaid, K.: Impact of temperature on hard disk drive reliability in large datacenters. In: 2011 IEEE/IFIP 41st International Conference on Dependable Systems & Networks (DSN), Hong Kong, China, pp. 530–537 (2011)

3. Chaves, I.C., de Paula, M.R.P., Leite, L.G., Queiroz, L.P., Gomes, J.P.P., Machado, J.C.: BaNHFaP: a bayesian network based failure prediction approach for hard disk drives. In: 2016 5th Brazilian Conference on Intelligent Systems (BRACIS), Recife, Brazil, pp. 427–32 (2016)
4. Shen, J., Wan, J., Lim, S. J., Yu, L.: Random-forest-based failure prediction for hard disk drives. Int. J. Distrib. Sens. Netw. **14**(11) (2018)
5. Zhu, B., Wang, G., Liu, X., Hu, D., Lin, S., Ma, J.: Proactive drive failure prediction for large scale storage systems. In: IEEE 29th Symposium on Mass Storage Systems and Technologies (MSST), Long Beach, CA, USA, pp. 1–5 (2013)
6. Xu, C., Wang, G., Liu, X., Guo, D., Liu, T.Y.: Health status assessment and failure prediction for hard drives with recurrent neural networks. IEEE Trans. Comput. **65**(11), 3502–3508 (2016)
7. Wu, J., Yu, H., Yang, Z., Yin, R.: Disk failure prediction with multiple channel convolutional neural network. In: 2021 International Joint Conference on Neural Networks (IJCNN), Shenzhen, China, pp. 1–8 (2021)
8. Jiang, T., Zeng, J., Zhou, K., Huang, P., Yang, T.: Lifelong disk failure prediction via gan-based anomaly detection. In: 2019 IEEE 37th International Conference on Computer Design (ICCD), Abu Dhabi, United Arab Emirates, pp. 199–207 (2019)
9. Murray, J.F., Hughes, G.F., Kreutz-Delgado, K.: Machine learning methods for predicting failures in hard drives: a multiple-instance application. J. Mach. Learn. Res. **6**, 783–816 (2005)
10. Li, J., et al.: Hard drive failure prediction using classification and regression trees. In: 2014 44th Annual IEEE/IFIP International Conference on Dependable Systems and Networks, Atlanta, GA, USA, pp. 383–394 (2014)
11. Zhang, S., Hai, Q., Wu, W., Han, G.: Hard disk drives failure prediction using the deep learning method based on attention mechanism. In: 2023 5th International Conference on Electronics and Communication, Network and Computer Technology (ECNCT), Guangzhou, China, pp. 133–137 (2023)
12. Hai, Q., Zhang, S., Liu, C., Han, G.: Hard disk drive failure prediction based on GRU neural network. In: 2022 IEEE/CIC International Conference on Communications in China (ICCC), Sanshui, Foshan, China, pp. 696–701 (2022)
13. Bai, A., Chen, M., Peng, S., Han, G., Yang, Z.: Attention-based bidirectional LSTM with differential features for disk RUL prediction. In: 2022 IEEE 5th International Conference on Electronic Information and Communication Technology (ICEICT), Hefei, China, pp. 684–689 (2022)
14. https://www.backblaze.com/cloud-storage/resources/hard-drive-test-data
15. Vaswani, A., et al.: Attention is all you need (2023). arXiv:1706.03762 [cs.CL]

Spatio-temporal Graph Learning on Adaptive Mined Key Frames for High-Performance Multi-Object Tracking

Futian Wang[1,2], Fengxiang Liu[1], and Xiao Wang[1,2(✉)]

[1] School of Computer Science and Technology, Anhui University, Hefei, China
xiaowang@ahu.edu.cn
[2] Anhui Provincial Key Laboratory of Multimodal Cognitive Computation, Hefei, China

Abstract. In this work, we propose a novel adaptive key frame mining strategy to guide multi-object tracking, which combines the advantages of short-term and long-term correlation to capture both the structured spatial relationships between different objects as well as the temporal relationships of objects in different frames. We designed the KFE (Key Frame Extraction) module to adaptively perform video segmentation using the reinforcement learning KFE module, which guides the tracker to mine the intrinsic logic of the video to make the correlation effect more significant. In video, mutual occlusion between objects frequently occurs, so the interactions between objects within frames cannot be ignored. Most of the current graph-based work utilizes the features of inter-frame objects for mutual fusion. Based on this, we design an IFF (intra-frame feature fusion) module that passes information between the target and the surrounding objects through GCN to make the target recognizable, thus solving the problem of tracking loss and similar appearance due to object occlusion. Our proposed tracker utilizes both long and short trajectories and takes into account the spatial relationship between objects while leveraging the intrinsic logic of the video clips, thus further improving the performance. It achieves impressive results: 68.6 HOTA, 81.0 IDF1, 66.6 AssA, and 893 IDS on the MOT17 dataset, proving its effectiveness and accuracy.

Keywords: Reinforcement Learning · Intra-frame Feature Fusion · Multi-Object Tracking

1 Introduction

Multi-Object Tracking (MOT) aims at tracking multiple objects simultaneously in a video, maintaining the identity of the objects, and generating their motion trajectories. It has wide applications in different fields such as video surveillance, autonomous driving, and video analytics. Despite many approaches proposed for multi-object tracking, fragmented trajectories or ID Switching (IDS) problems caused by frequent occlusions in crowded scenes and similar appearances are still significant challenges (Fig. 1).

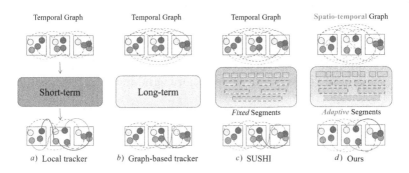

Fig. 1. Comparison between (a, b, c) existing algorithms and (d) our newly proposed MOT tracking framework

To address the ID Switching (IDS) issue, most state-of-the-art trackers use a combination of short and long trajectories for tracking [2,6]. Due to the variability of scenarios, different specific schemes are required, for example, motion-based local trackers and appearance-based trackers [10,11,14,15]. When there is heavy occlusion and uniformity of clothing, these methods often become highly specific to particular scenarios, making them not easily scalable to broader applications. Gao et al. [6] propose a hierarchical approach to processing video, with lower levels focusing on short-term associations and higher levels focusing on increasingly long-term scenes. [2] uses the same learnable model for all time scales, thus improving scalability. Many state-of-the-art MOT trackers use graph neural networks [7] to handle similar appearance problems [3,12]. However, there are still some issues that exist in current works, e.g., many of these methods are highly specific. Cetintas et al. [2] use the same learnable model for all time scales, obviously, the scalability and flexibility can be improved. In addition, we are also inspired by the scenarios as shown in Fig. 2. More in detail, scenario 1 demonstrates that the time interval between the appearance of a tagged object and its occlusion to reappearance is not consistent. There is a tendency in Scenario 2 to follow the wrong phenomenon due to similar clothing and similar location. However, seldom of existing MOT trackers consider the spatial relationship between different objects in a single frame.

To address the issues mentioned above, in this paper, we propose a novel multi-object tracking by spatio-temporal graph learning on adaptive mined key frames. As shown in Fig. 3, given the input video frames, we first adopt a KFE (Key Frame Extraction) module to divide them into frame segments adaptively. It formulates the key frame extraction as a decision-making problem using the Q-learning algorithm which makes full use of the advantage of high performance of the short trajectory. In addition, we also exploit spatio-temporal graph learning with the Intra-frame Feature Fusion (IFF) module to boost the interactions between targets in the same frame and nearby frames. By integrating the two modules into the baseline tracker SUSHI [2], the final performance can be further enhanced.

To sum up, we draw the main contributions of this paper as the following three aspects:

- We propose a novel adaptive key frames mining strategy guided multi-object tracking algorithm based on reinforcement learning. It is a unified, scalable, and hierarchical tracker that models short-term and long-term associations simultaneously.
- We propose a new spatio-temporal graph learning module that captures the structured spatial relations between different pedestrians in a single frame and temporal relations between different frames.
- Extensive experiments on the public MOT17 benchmark dataset fully validated the effectiveness of our proposed strategies for the MOT task.

Fig. 2. In the first row, the individuals depicted in the image are derived from the objects in the MOT17 video sequence MOT17-09. Notably, the reappearance of the girl with a high ponytail is subject to variable occlusion intervals in the video. In the second row, the two remarked objects in the image show similar appearances and are positioned close to each other. We can consider separating the two by using information about the surrounding objects

2 Our Proposed Approach

2.1 Reinforcement Learning Based Key Frame Mining

The KFE module is based on Q-learning, and we need to design the corresponding action selection, reward mechanism, active exploration strength, and table

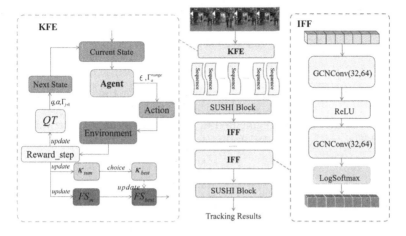

Fig. 3. An overview of our proposed multi-object tracking framework

update strategy. The optimal segmentation strategy FS_{best} and the corresponding optimal reward score κ_{best} are iterated during the learning process. The action selection is divided into the first frame of the video segment Γ_a, and the last frame action selection Γ_b, both of which have the same selection strategy, in order to avoid redundancy, we only illustrate the action selection strategy for the first frame. Γ^{i+1} is the action selection used to record the choice that will be made next, QT is the Q-value used to record the total return expected to be obtained by taking a specific action in a certain state, F denotes the state, $QT[F_a^i, \Gamma_a^j]$ represents the expected return of the next choice Γ_a^{j+1} in state F_a^i, and Γ_a^{range} represents the ranges of values of Γ_a. We choose the action with the highest reported value in the Q-value as the next action, in addition to randomly choosing the next action with a certain exploration rate ϵ.

$$\Gamma_a^{j+1} = \begin{cases} \max(QT[F_a^i, \Gamma_a^j]), \eta < \epsilon \\ random(\Gamma_a^{range}), \text{ otherwise} \end{cases} \quad (1)$$

where $\eta \in (0, 1), j \in (0, N)$ and $\epsilon = 0.1$. We need to develop a reasonable reward mechanism to determine the direction of model optimization, we use the feature difference between the first and last frames between the video segments as the reward value, κ_{i+1} is the reward for the $i+1$th segmentation result of the current round.

$$\kappa_{i+1} = ((1 - \varphi(f_{F_a^i}, f_{F_a^{i+1}}) + (1 - \varphi(f_{F_b^i}, f_{F_b^{i+1}})) \times \delta + \xi \quad (2)$$

where $f_{F_a^i}$ represents the feature of the first frame of the ith video segment, $f_{F_b^i}$ represents the feature of the last frame of the ith segment, $f_{F_a^j}$ represents the feature of the first frame of the jth video segment, and $f_{F_b^j}$ represents the feature of the last frame of the jth segment. Additionally, δ, ξ are constants, and φ denotes the cosine similarity. Regarding the update of the Q-value, we follow the approach of classical algorithms, where the Q-value is dependent on F^i,

F^{i+1}, the learning rate $q = 0.1$, the reward Γ, and the discount factor $\alpha = 0.99$.

$$\Gamma_{best}^{i+1} = max(QT(F^{i+1}, \Gamma^i) \quad (3)$$

$$QT[F^i, \Gamma] = \lambda \times q + QT[F^i, \Gamma] \quad (4)$$

where Γ_{best}^{i+1} represents the optimal action selection for the next step, $\lambda = \kappa_i + \alpha \times QT[F^{i+1}, \Gamma_{best}^{i+1}] - QT[F^i, \Gamma]$.

When the video is fully segmented after the mth iteration, the final reward score for this round κ_{sum} is denoted as

$$\kappa_{sum} = \frac{\sum_{i=1}^{n} \kappa_a^i}{len(FS_m)} \quad (5)$$

where κ_a^i is the reward for the ith segmentation result of the first frame of the current round and FS_m is the segmentation strategy in the mth iteration. We evaluate whether the current segmentation result is attributed to the advanced optimal reward κ_{best}. If it is superior, we proceed to update the parameters FS_{best} and κ_{best} accordingly

$$\begin{cases} \kappa_{best} = max(\kappa_{best}, \kappa_{sum}) \\ FS_{best} = max(FS_{best}, FS_m) \end{cases} \quad (6)$$

Through the aforementioned steps, we are able to achieve efficient segmentation results that fully exploit the inherent logic of the video, effectively combining the advantages of both short and long trajectories. Compared to SUSHI, our approach achieves up to +2.5 AssA and up to +1.7 IDF1 improvements, with a decrease in IDS. This indicates that the KFE module can effectively mitigate the issue of identity loss when objects reappear after occlusion. Thereby validating the effectiveness of our method. The complete inference procedure is outlined in Algorithm 1, where χ_a^j represents the state selection of the first frame in the jth video, χ_b^j represents the state selection of the last frame in the jth video.

2.2 Spatial-temporal Relation Mining

When occlusion of an object occurs, we can combine the surrounding objects to perform object recognition where the objects are too similar to each other and their positions are close. Inspired by the above ideas, we add IFF to solve the problem of occlusion and similar appearance. In performing the fusion of different levels, we first construct a graph $G = (V, E)$ for the level, where the object is represented by node $v_i \in V$ and the edge e_i represent hypotheses for judging the relationship between nodes. We perform in-frame feature fusion before hierarchy merging. The initial scheme is to complement the features of the object with a certain percentage after averaging the surrounding objects.

$$f_{v_i} = a \times f_{v_i} + b \times avg(f_{V_i^m}) \quad (7)$$

where f_{v_i} represents the feature of node v_i, V_i^m represents the set of nodes closest to v_i, $a, b \in (0, 1), b = 1 - a$. However, GCN is able to capture more complex

Algorithm 1. KFE Algorithm

Require:
env: The environment with two state spaces and action spaces
agent: A QAgent instance
episodes: The number of episodes to train
ϵ: The exploration rate
q: The learning rate for Q-table updates
α: The discount factor for future rewards
Γ_a^{range}: The range for selecting future state
Ensure:
FS_{best}: The states recorded during the best episode
1: Initialize κ_{best},best_episode, FS_{best}
2: **for** episode = 1 to episodes **do**
3:    Initialize χ_a^j,χ_b^j,done, κ_{sum},FS_m
4:    **while** not done **do**
5:       $\Gamma_a^{j+1} \leftarrow$ agent. select_action(χ_a^j,ϵ)
6:       $\Gamma_b^{j+1} \leftarrow$ agent. select_endaction(χ_b^j,ϵ)
7:       ($\chi_a^{j+1},\chi_b^{j+1},\kappa_a^{j+1},\kappa_b^{j+1}$,done) $\leftarrow$ env. step($\Gamma_a^{j+1},\Gamma_b^{j+1}$)
8:       agent. update_q_table($\chi_a^j,\Gamma_a^{j+1},\kappa_a^{j+1},\chi_a^{j+1},q,\alpha$)
9:       agent. update_qend_table($\chi_b^j,\Gamma_b^{j+1},\kappa_b^{j+1},\chi_b^{j+1},q,\alpha$)
10:      $\chi_a^j \leftarrow \chi_a^{j+1}$
11:      $\chi_b^j \leftarrow \chi_b^{j+1}$
12:      $\kappa_{sum} \leftarrow \kappa_{sum} + \kappa_a^{j+1} + \kappa_b^{j+1}$
13:      **if** not done **then**
14:          FS_m.append($(\chi_a^{j+1},\chi_b^{j+1})$)
15:      **end if**
16:    **end while**
17:    **if** $\kappa_{sum} > \kappa_{best}$ **then**
18:       $\kappa_{best} \leftarrow \kappa_{sum}$
19:       best_episode $\leftarrow$ episode
20:       $FS_{best} \leftarrow FS_m$
21:    **end if**
22: **end for**
23: **return** FS_{best}

contextual information than the simple average through the information transfer and aggregation between nodes, adaptively learns the relationship weights between different objects, and can automatically adjust the degree of influence of different neighboring nodes on the features of the central node according to the actual situation, so as to more accurately reflect the real relationship between the objects. In the case where the object is occluded, GCN can supplement the feature representation of the occluded object by fusing the information of the surrounding unoccluded objects, which helps to alleviate the impact of occlusion on recognition performance. When the objects are too similar to each other and close to each other in terms of location, GCN can utilize the structured spatial relations between objects. Therefore, we chose the method GCN for feature fusion with graphs.

$$f_{v_i} = a \times f_{v_i} + b \times GCN(f_{v_i}, f_{V_i^m}) \tag{8}$$

Utilizing GCN, we integrate features from neighboring objects, enabling each object to encapsulate a more comprehensive global context. This approach leads to improvements of up to +2.0 in AssA and +1.1 in IDF1, empirically validating the efficacy of our method within the realm of computer vision research.

2.3 Loss Function

Our proposed KFE method stands independent of the SUSHI Block and IFF frameworks. To determine the number of learning epochs, denoted as $M = (LN - u) \times (LN - n) \times 100$, where u represents the shortest length of the video segment, n represents the maximum length of the video segment, and LN represents the total length of the video. We adopt the unfreezing strategy from the SUSHI Block, where subsequent levels are unfrozen after 500 iterations. We apply the focal loss on the generated edge classification scores and sum these losses across all levels to obtain the final loss [8].

3 Experiments

3.1 Datasets and Evaluation Metrics

We conducted our experiments on the public dataset MOT17. We used MOT17-Private for experimental effect validation. We follow the HOTA protocol [9] for quantitative evaluation, where HOTA focuses on overall tracking quality, AssA is used to measure association accuracy, and IDS focuses on measuring the number of identity switches during tracking. In addition, we also adopt the metrics MOTA and IDF1, which reflect the overall performance of detection and tracking as well as association accuracy, respectively, providing us with a multidimensional evaluation reference.

3.2 Ablation Study

To demonstrate the effectiveness of the proposed modules, in Table 1 we conducted experiments on MOT17, where we describe the added features as: KFE (Key Frame Extraction), and IFF (Intra-frame Feature Fusion). We found that KFE and IFE improved the performance on the MOT17 set, where the IFF module improved by 1.1 HOTA and KFN improved by 1.2 HOTA. Adding both modules together improves 1.6 HOTA, 2.3 IDF1, and 3.3 AssA.

In Table 1 we started to use dynamic programming (DP) for segmentation, which has poor learnability and limited scalability, so we used the reinforcement learning Q-learning method, which has good applicability and scalability, and can be obtained from the experiment, Q-learning is more effective and improves the performance of the algorithm by 1.2 HOTA, 1.7 IDF1, and IDS decreased, so we finally chose the Q-learning algorithm. In the IFF module, there is a fusion between the Intra-frame features, we used a fixed ratio, and after the experiment in Table 2, we finally set a = 0.4.

Table 1. Validating Module Effectiveness

	HOTA↑	IDF1↑	AssA↑	MOTA↑	IDS↓
SUSHI	67.0	78.7	62.9	**85.1**	930
SUSHI+IFF	68.1	79.8	64.9	84.8	915
SUSHI+DP	67.8	79.7	64.8	84.8	924
SUSHI+KFE	68.2	80.4	65.4	85.0	855
Ours	**68.6**	**81.0**	**66.6**	84.0	**893**

Table 2. Ablation experiments on feature fusion with varying proportions

(a, b)	HOTA↑	IDF1↑	AssA↑	MOTA↑	IDS↓
(0, 1.0)	67.0	78.7	62.9	85.1	930
(0.1, 0.9)	67.2	78.0	63.7	84.9	963
(0.2, 0.8)	67.6	79.2	63.9	84.7	921
(0.3, 0.7)	68.0	**80.3**	64.2	85.0	936
(0.4, 0.6)	**68.1**	79.8	**64.9**	84.8	915
(0.5, 0.5)	67.8	78.9	63.9	84.7	987
(0.6, 0.4)	67.9	79.7	64.3	85.0	939
(0.7, 0.3)	67.8	79.2	63.6	**85.2**	939
(0.8, 0.2)	67.4	78.9	63.4	**85.2**	912
(0.9, 0.1)	67.2	78.7	63.2	**85.2**	**909**
(1.0, 0)	67.0	78.6	63.5	84.4	**909**

3.3 Comparison on Public Benchmarks

We conducted experiments on the MOT17 [4] public dataset. As shown in Table 3 our tracker achieves 68.6 HOTA, and our method outperforms all methods under HOTA ordering in the MOT- challenge. On the dataset, using the same detection, our method outperforms existing comparisons including SUSHI [2], CoNo-Link [6], ByteTrack [13], OC-SORT [1] and StrongSORT++ [5], where StrongSORT++ is an offline version of StrongSORT that enhances trajectories by post-processing tracking offline. The benchmark results strongly demonstrate the advanced performance of our tracker.

Table 3. Test set results on MOT17 benchmark. Our tracker achieves State-Of-The-Art (SOTA) performance in the HOTA metric

	HOTA↑	IDF1↑	AssA↑	MOTA↑	IDS↓
ByteTrack [13]	63.1	77.3	62.0	80.3	2196
FairMOT [14]	59.3	72.3	58.0	73.7	3303
OC-SORT [1]	63.2	77.5	63.2	78.0	1950
*StrongSORT++ [5]	64.4	79.5	64.4	79.6	1194
CoNo-Link [6]	67.6	79.3	65.6	**86.5**	909
SUSHI [2]	67.0	78.7	62.9	85.1	930
Ours	**68.6**	**81.0**	**66.6**	84.0	**893**

4 Conclusion

In this study, we start with the offline multi-object tracking algorithm SUSHI and propose a new approach of adaptive video segmentation combined with Intra-frame surrounding object feature complementation. The proposed KFE takes full advantage of short trajectories and solves the IDS problem to some extent, and the feature complementation of the surrounding environment within the frame of the IFF module makes the object more informative, thus solving the problem of following the wrong object due to occlusion or similarity in appearance. Both modules improve the performance of multi-object tracking in terms of correlation in a good way.

Acknowledgment. This work is supported by the University Synergy Innovation Program of Anhui Province under Grant(No. GXXT-2022-042), Anhui Province Higher Education Scientific Research Project (No. 2023AH052574), National Natural Science Foundation of China under Grant 62102205, the Anhui Provincial Natural Science Foundation under Grant 2408085Y032. The authors acknowledge the High-performance Computing Platform of Anhui University for providing computing resources.

References

1. Cao, J., Pang, J., Weng, X., Khirodkar, R., Kitani, K.: Observation-centric sort: rethinking sort for robust multi-object tracking. In: Proceedings of the IEEE/CVF Conference on Computer Vision and Pattern Recognition, pp. 9686–9696 (2023)
2. Cetintas, O., Brasó, G., Leal-Taixé, L.: Unifying short and long-term tracking with graph hierarchies. In: Proceedings of the IEEE/CVF Conference on Computer Vision and Pattern Recognition, pp. 22877–22887 (2023)
3. Cheng, C.C., Qiu, M.X., Chiang, C.K., Lai, S.H.: Rest: a reconfigurable spatial-temporal graph model for multi-camera multi-object tracking. In: Proceedings of the IEEE/CVF International Conference on Computer Vision, pp. 10051–10060 (2023)

4. Dendorfer, P., et al.: MOTChallenge: a benchmark for single-camera multiple target tracking. Int. J. Comput. Vis. **129**, 845–881 (2021)
5. Du, Y., et al.: StrongSORT: make DeepSORT great again. IEEE Trans. Multimedia **25**, 8725–8737 (2023)
6. Gao, Y., Xu, H., Li, J., Wang, N., Gao, X.: Multi-scene generalized trajectory global graph solver with composite nodes for multiple object tracking. In: Proceedings of the AAAI Conference on Artificial Intelligence, vol. 38, pp. 1842–1850 (2024)
7. Kipf, T.N., Welling, M.: Semi-supervised classification with graph convolutional networks. arXiv preprint arXiv:1609.02907 (2016)
8. Lin, T.: Focal loss for dense object detection. arXiv preprint arXiv:1708.02002 (2017)
9. Luiten, J., et al.: HOTA: a higher order metric for evaluating multi-object tracking. Int. J. Comput. Visi. **129**, 548–578 (2021)
10. Xiao, C., et al.: MotionTrack: learning motion predictor for multiple object tracking. Neural Netw. **179**, 106539 (2024)
11. Yi, K., et al.: Ucmctrack: Multi-object tracking with uniform camera motion compensation. In: Proceedings of the AAAI Conference on Artificial Intelligence, vol. 38, pp. 6702–6710 (2024)
12. You, S., Yao, H., Bao, B.K., Xu, C.: UTM: a unified multiple object tracking model with identity-aware feature enhancement. In: Proceedings of the IEEE/CVF Conference on Computer Vision and Pattern Recognition, pp. 21876–21886 (2023)
13. Zhang, Y., et al.: Bytetrack: multi-object tracking by associating every detection box. In: European Conference on Computer Vision, pp. 1–21. Springer (2022)
14. Zhang, Y., Wang, C., Wang, X., Zeng, W., Liu, W.: FairMOT: on the fairness of detection and re-identification in multiple object tracking. Int. J. Comput. Vis. **129**, 3069–3087 (2021)
15. Zhao, H., Wang, D., Lu, H.: Representation learning for visual object tracking by masked appearance transfer. In: Proceedings of the IEEE/CVF Conference on Computer Vision and Pattern Recognition, pp. 18696–18705 (2023)

From Image to the Ground: Recover the Ground Location of Vehicles from Traffic Cameras Using Neural Networks

Xuzhen Wang[1,2], Wenzhong Wang[1,2(✉)], and Jin Tang[1,2]

[1] School of Computer Science and Technology, Anhui University, Hefei, China
wenzhong@ahu.edu.cn
[2] Anhui Provincial Key Laboratory of Multimodal Cognitive Computation, Hefei, China

Abstract. Robust estimation of on-road vehicle positions is important for traffic monitoring. We propose a neural network based approach to recover the ground locations of vehicles in traffic surveillance videos. We frame the location estimation problem as an object detection task. In contrast to the current state-of-the-art image object detection methods which only roughly estimate the object bounding boxes on the image plane, we aim to estimate the vehicle locations on the ground (the location in the frame of the road ground plane). In order to achieve this goal, we first transform the image into the bird's eye view (BEV) image, and then detect the bounding box of vehicles in the BEV images. However, the BEV images exhibit severe distortions which hinders the detection accuracy significantly. Furthermore, the detections from consecutive images are inconsistent, i.e. the bounding boxes of the same vehicle in different images are of different sizes. To address these two challenges, we present a multi-task detection model which significantly boost the detection rate, and we design a novel consistency loss which constrains the size of detection boxes of identical vehicles to be the same. In order to effectively train and evaluate the model, we collect a dataset of traffic videos, which contains synchronous videos captured from both frontal view and bird's eye view. Our experiments show that our multi-task detection model and consistency loss significantly improve the detection accuracy and stabilize detection results.

Keywords: Object Detection · Vehicle Position Determination · Consistency Loss · Multi-task Model

1 Introduction

Traffic cameras are widely used in modern traffic environments. In order to monitor traffic flow using cameras, vehicles must be detected and tracked in the images, and their behaviors are then inferred from the tracks. There are lots of research on object detection and tracking in computer vision field. However, almost all the object detectors work on the image plane, yielding object positions and bounding boxes in the image coordinate frame. Though it is useful to have image position of vehicles, the positions and tracks are distorted representation of the vehicle behaviors. It is hard to infer traffic situations and driving behaviors such as lane crossing, lane deviation, overtaking, acceleration/deceleration, congestion, etc., from such tracks. A more appropriate representation of vehicle positions would be its locations on the physical ground plane of the road, rather than its pixel locations on the image plane. In Fig. 1, the red boxes in the left image are obtained by current state-of-the-art object detector. These boxes typically occupy multiple lanes, and it is hard to infer the real location of vehicles on the road. The filled boxes in the right image shows the bounding box of vehicles on the ground plane of the road. These boxes tell the true location of vehicles, and we can easily estimate many traffic parameters from these boxes. We project the ground boxes onto the image plane, shown with yellow dashed lines in the left image, for comparison. The difference between these two types of bounding box is very obvious. The projected ground box is no longer rectangles, and at least one of its corners are invisible in the image. It is hard to manually label these boxes, and also hard to detect those boxes from the images.

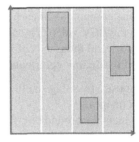

Fig. 1. Vehicle bounding boxes on the image plane (left) v.s. on the ground plane (right). (Color figure online)

This paper aims to determine the ground location of vehicles from monocular traffic surveillance cameras. By ground location, we mean location in a planar coordinate frame of the ground plane of the road. We propose a two-stage approach to estimate the ground location (Fig. 2). In the first stage, the frontal view image (image from the traffic camera) is transformed to the Bird's Eye View (BEV). In the BEV image, the ground plane is parallel to the image plane, so the bounding boxes and positions in the BEV image is a scaled representation of its real position in the ground plane. In the second stage, we detect the bounding box in the BEV image using a convolutional neural network, modified from the widely used YOLO [9] model.

Detecting vehicles in the BEV images using the state-of-the-art object detectors confronts two problems. Firstly, the region of vehicles are highly distorted due to perspective transforms, and yet the bounding box on the ground (where the vehicle contacts the ground) only occupy a small portion of this region (See Fig. 2), and even more, there are not enough discriminative features in this small region. Secondly, the detections in consecutive images are inconsistent, i.e. there are considerable jitters in the detections of the same vehicle in consecutive images, since YOLO is independently applied on each image.

In order to deal with the first problem, we incorporate an auxiliary detection task into the YOLO model, the auxiliary task is to detect the outer bounding box (shown in yellow in Fig. 2.) of the occupied region by a vehicle in the BEV image. We represent the region where the vehicle contacts the ground plane of the road as an oriented bounding box, which is called ground box (red boxes in Fig. 2.). We've found that simultaneously detect the outer and ground box significantly boost the detection rate.

In order to deal with the second problem, we propose a novel consistency loss, which we call WHCLoss (Width-Height-Consistency Loss). The proposed loss constraints the width and height of the ground boxes of the same vehicle in consecutive images to be the same. The WHCLoss effectively reduces the jitters in the size of bounding boxes, resulting in more stable and consistent detections.

Training the proposed model requires a large annotated dataset. However, there are no public dataset available, and manually labelling the ground box of each vehicle is a tedious and tough endeavor. To mitigate this problem, we used two drones to collect video data from two different perspectives, one shot from a normal traffic monitoring perspective and the other from a bird's-eye view. The two video streams are synchronized. We label the ground box of each vehicle in the BEV images as the ground-truth. Label the box in the BEV video is more approachable. By this way, we collect a large dataset consists of several thousands of images.

To summarize, our contributions are following:

1. We propose a two-stage method for obtaining the ground box of vehicles from traffic camera images.
2. We augment the ground box detection task with an auxiliary task, which significantly boost the detection accuracy.
3. We introduce a width-height consistency loss (WHCLoss). The WHCLoss ensures the detected bounding boxes of the same vehicle in consecutive images are consistent in size.
4. We propose a method for constructing a labelled dataset of traffic videos. The dataset contains 18 videos shot in different places, in which we labelled the ground box of each vehicle.

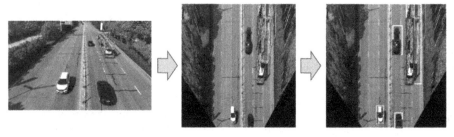

Fig. 2. Overview of our two-stage detection approach. The frontal view image is first warped into a BEV image using perspective transform, and then the ground boxes of each vehicle are detected in the BEV image using modified YOLO model. (Color figure online)

2 Related Works

2.1 Object Detection Algorithm

The development of Convolutional Neural Networks has propelled advancements in object detection algorithms, significantly enhancing both detection speed and accuracy. Common detection networks include: DETR [2], CenterNet [3], YOLO [4], RCNN [5], RetinaNet [8], etc. The YOLO series represents a one-stage object detection algorithm that treats the detection task as a regression problem, directly predicting bounding boxes and class probabilities from images, making it suitable for real-time applications. The initial YOLO model exhibited poor performance in small object detection and accuracy, but with continuous iterations, YOLOv8 has further optimized the network structure and training methods, leading to significant improvements in both detection accuracy and speed.

2.2 Vehicle Position Inspection

Many vehicle detection studies in traffic scenarios focus on the perspective of autonomous driving, completing vehicle detection within the vehicle coordinate system. For instance, some approaches utilize Multi-Layer Perceptrons (MLPs) as general mapping functions to model the transformation from front view to bird's-eye view, processing view transformations in a forward manner while implicitly learning the geometric mapping matrix. The paper PYVA [13] combines MLPs with CycleGAN, proposing a cyclical self-supervised scheme to enhance view projection, further introducing an attention-based feature selection process to relate the two views. However, methods using MLPs neglect geometric priors and thus lose spatial information between pixel. In contrast, Transformer-based methods adopt a top-down strategy by constructing queries and utilizing attention mechanisms to search for corresponding image features. BEVFormer [6] employs deformable attention to identify relevant image features and additionally integrates BEV representations from previous frames. However, such methods require multi-camera support and entail significant detection time overhead.

There are some studies on vehicle position detection in traffic surveillance scenarios. VPOE [12] uses a detection network to predict vehicle positions based on homography transformations. M. Zhu et al. [10] predict 3D bounding boxes of vehicles using rotational boxes with tails in distorted BEV image. While these works achieve vehicle detection, they do not focus sufficiently on precise localization that adheres to physical constraints.

3 Method

Our goal is to predict the ground box of each vehicle from the frontal view image I. We use oriented bounding box to encode the ground box, so the i th ground box is represented as $G_i = (x_i, y_i, w_i, h_i, \theta_i)$. We formulate the task as the following mapping:

$$O = f(I) \tag{1}$$

where $O = \{G_i\}$ is the set of ground boxes in image I.

We decompose f into two separate processes, g, h:

$$\begin{cases} I' = g(I) \\ O = h(I') \end{cases} \tag{2}$$

where g transform the frontal view image into its BEV I', and h is a convolutional neural network which predict ground boxes O from the BEV image I'.

3.1 Transform Frontal View Images into BEV Images

It is convenient to work on BEV images [1, 10, 11]. First of all, the ground box in BEV can be efficiently represented as an oriented rectangle (5 d.o.fs), while in the frontal view, it is a convex quadrilateral, and it needs 8 d.o.fs to uniquely specify a ground box. In addition, the scale of bounding boxes in the BEV images at different locations range in a small interval, making it easier to detect distant vehicles.

In most cases, the road can be approximately viewed as a flat plane. According to perspective geometry, there exist an invertible transformation, H, between the image plane and the road plane. H is the 3×3 homography matrix, which can be estimated in many ways [7, 10]. Given H, we can warp the frontal view image into its BEV counterpart (c.f. the leftmost two images in Fig. 2.).

3.2 Ground Box Detection in BEV Images.

There are two challenges in BEV vehicle detection. First, there are severe distortion in the BEV images, as shown in Fig. 3. Due to the perspective transformation, the vehicles are stretched. The image region of a vehicle extends far beyond its ground box. Furthermore, in crowded roads, the ground boxes are heavily occluded by other vehicles. This makes the detection of ground box very challenging.

Second, due to perspective transform, the stretching effects varies at different distance, and an identical vehicle exhibits different appearance at different positions, as shown in Fig. 4. When independently detect on these consecutive images, the size of detected boxes varies a lot. Such results are unrealistic, since the ground box of the same vehicle would not change in consecutive video frames.

To address these two problems, we propose a novel detection method which is modified from the widely used YOLOv8 model. We made two modifications to the original model which are detailed in following subsections.

3.2.1 Multi-task Detection

Humans can infer the presence of a vehicle from the visible deformed features, even if the position is not precisely accurate. The recognition of deformed vehicle features prompts us to extract more information to ascertain the vehicle's location. From a physical standpoint, a vehicle's features represent a natural whole, encompassing both the deformed parts and the bottom. The presence of one implies the existence of the other, creating mutual detection constraints. In order to deal with the first problem, we design a new auxiliary task based on the above heuristics to facilitate ground box detection. The auxiliary task is to detect the bounding box of the whole projected region of a vehicle (the outer box shown in green line in Fig. 3).

Each outer box is also represented as an oriented rectangle, and encoded using 5 parameters. Let $G'_i = (x'_i, y'_i, w'_i, h'_i, \theta'_i)$ denote the i th outer box. We augment the detection process in Eq. (2): $O = \{(G_i, G'_i)\}_{i=1}^{n}$.

In realization of the multi-task detection, we treat the ground box and the outer box as two different object category, and use two-category YOLO as our detection model. However, our design is different from multi-category object detection in that our model will predict two distinct and overlapping boxes for each single object.

Fig. 3. Distortion in the BEV images. Left: the vehicle is stretched, and its image region (green box) is much larger than its ground box (red box). Right: heavy occlusion in crowded traffic. (Color figure online)

Fig. 4. Jitters in detected boxes due to appearance variation. The same car shows different appearance, which makes the scale of detection results vary a lot.

3.2.2 Temporal Consistent Detection

In order to cope with the second problem, we design a consistency loss for the ground box G_i.

Let $G_i^{(t)} = x_i^{(t)}, y_i^{(t)}, w_i^{(t)}, h_i^{(t)}, \theta_i^{(t)}$ be the detection of the i th ground box in image frame t. Since the vehicle's physical width and length would not change in short period, the width and height of its ground box should be constant, i.e. they should meet the following temporal consistency condition:

$$w_j^{(t+1)} = w_i^{(t)}, h_j^{(t+1)} = h_i^{(t)}, if \delta_{ij} = 1 \qquad (3)$$

where δ_{ij} is an indicator variable. If $G_i^{(t)}, G_j^{(t+1)}$ are detections of the same vehicle, then $\delta_{ij} = 1$, and otherwise $\delta_{ij} = 0$.

We encode the above condition using the following Width-Height-Consistency Loss (WHCLoss):

$$L_{WH} = \frac{1}{C} \sum_{i=1}^{n_t} \sum_{j=1}^{n_{t+1}} \delta_{ij} \left[\left(h_j^{(t+1)} - h_i^{(t)}\right)^2 + \left(w_j^{(t+1)} - w_i^{(t)}\right)^2 \right] \quad (4)$$

where $C = \sum_i \sum_j \delta_{ij}$, $n_t = |O_t|$, $n_{t+1} = |O_{t+1}|$. The WHCLoss aims to guide the network h to learn the consistency constraint of vehicle width and height between adjacent frames. Hopefully, this loss function will resolve the significant size jitter in box detections. We expect that by incorporating this loss function, the detection results between consecutive frames will adhere to the physical constraint of constant vehicle width and length, thereby ensuring the width-height consistency of the detection bounding boxes. When combined with other detection loss functions in YOLOv8, this improves the accuracy of the vehicle localization.

When train the modified YOLO with the WHCLoss, we feed m pairs of images into the network, with each pair consists two consecutive images $I^{(t)}, I^{(t+1)}$ from the same video. The loss for each pair is:

$$L = w_{prev} \times L_{yolo}^{(t)} + w_{curr} \times L_{yolo}^{(t+1)} + w_{consis} \times L_{WH} \quad (5)$$

where $L_{yolo}^{(t)}$ is the original loss used by YOLO for image $I^{(t)}$, w_{prev}, w_{curr} and w_{consis} are the corresponding weights for these losses.

3.3 Data Collection and Annotation

Since there are no publicly available dataset with ground box annotation, and manually label the ground boxes is very tedious and error prone, we propose to collect traffic videos using two drones shooting from two different viewpoints.

The configuration of the two drones is illustrated in Fig. 5. We use the top drone to capture top view videos, and use the lateral view drone to captures the frontal view videos. There is sufficient overlap between these two views.

Fig. 5. Video capture configuration. Yellow: Top-view drone (Color figure online)

We estimate the homography matrix between road image in the frontal view and the top-view, and then warp the frontal view images into BEV images, giving image samples $\{I'^{(t)}\}$. This process is illustrated in Fig. 6.

Fig. 6. Left: frontal view video, Middle: top-view video, Right: BEV of the frontal view video.

We then manually labeled the vehicle bounding boxes in the top-view images, and transferred the annotation to the BEV images, as shown in Fig. 7. We also manually labeled the outer box for each ground box. In this way, we obtain the complete annotation for each BEV image $I'^{(t)}$.

Using the above approach, we collected and annotated a dataset consists of 5k images, sampled from 18 pair of videos. The dataset is randomly split into training set and test set with ratio 8:2.

Fig. 7. Labelling the ground box (red) and outer box(yellow) in the BEV images. (Color figure online)

4 Experiments

4.1 Settings

We use the Adam optimizer with an initial learning rate set to 0.001, applying a decay strategy to reduce it to a final value of 0.00001. The batch size is set to 16, with a total training epoch of 120, and the input image size is 640. The hardware environment includes an Intel i9 12900KF CPU and an NVIDIA GeForce RTX 3090 GPU.

4.2 Results

We conducted experiments on our dataset, evaluating the following situations: (1) using only the YOLOv8 detection network; (2) the YOLOv8 network combined with the multi-task detection strategy; (3) the YOLOv8 network with the WHCLoss; (4) the WHCLoss combined with the multi-task detection strategy. All the above 4 models are trained using our training set and evaluated using the test set.

We measure the detection performance using the widely adopted metric, mAP50-95. The results of these four situations are listed in Table 1. Our approach shows significant improvement over the vanilla YOLOv8. Both the multi-task detection strategy and the consistency loss contribute a lot to the improvement. The best performance is achieved by combining the two.

Table 1. Results of ablation study.

Network settings	$mAP_{50\text{-}95}$
YOLOv8	38
YOLOv8 + multi-task detection	55
WHCLoss	89
WHCLoss + multi-task detection	**94**

We show some results in Fig. 8. The left panel shows the detection results of one image using YOLOv8 with our WHCLoss and multi-task strategy, as well as those using the plain YOLOv8. It is obvious from the comparison that our approach significantly boosts the detection rate. The right panel shows more detection results using our approach, we transform the detections in BEV into the frontal view, and show the results using green boxes in the top row.

Fig. 8. Left panel: Detection results of YOLOv8 with (Left) and without (Right) WHCLoss and multi-task strategy. Right panel: detection results shown in the BEV images (Bottom) and the frontal view images (Top). (Coloe figure online)

5 Conclusion

We've proposed a new method to estimate the ground position of vehicles from frontal view images. We framed the detection task as a two-stage process, and propose to detect the ground box in the BEV images. Our detection model is built upon the YOLOv8 model. We augmented YOLOv8 with two techniques, the first one is a multi-task detection strategy which detect two boxes for each object, the second one is the width-height consistency loss which ensures the temporal consistency of the detection results. In addition, we designed a data collection and annotation strategy which can efficiently annotate the ground boxes.

We performed experiments using our dataset, the results showed that our approach significantly improve the performance over the baseline YOLOv8, and hence verified the effectiveness of our approach.

References

1. Can, Y.B., Liniger, A., Unal, O., Paudel, D., Van Gool, L.: Understanding bird's-eye view of road semantics using an onboard camera. IEEE Robot. Autom. Lett. **7**(2), 3302–3309 (2022)
2. Carion, N., Massa, F., Synnaeve, G., Usunier, N., Kirillov, A., Zagoruyko, S.: End-to-end object detection with transformers. In: European Conference on Computer Vision, pp. 213–229. Springer (2020)
3. Duan, K., Bai, S., Xie, L., Qi, H., Huang, Q., Tian, Q.: Centernet: keypoint triplets for object detection. In: Proceedings of the IEEE/CVF International Conference on Computer Vision, pp. 6569–6578 (2019)
4. Farhadi, A., Redmon, J.: Yolov3: an incremental improvement. In: Computer Vision and Pattern Recognition, vol. 1804, pp. 1–6. Springer, Heidelberg (2018)
5. Girshick, R., Donahue, J., Darrell, T., Malik, J.: Rich feature hierarchies for accurate object detection and semantic segmentation. In: Proceedings of the IEEE Conference on Computer Vision and Pattern Recognition, pp. 580–587 (2014)

6. Li, Z., et al.: Bevformer: learning bird's-eye-view representation from multi-camera images via spatiotemporal transformers. In: European Conference on Computer Vision, pp. 1–18. Springer (2022)
7. Mallot, H.A., Bülthoff, H.H., Little, J.J., Bohrer, S.: Inverse perspective mapping simplifies optical flow computation and obstacle detection. Biol. Cybern. **64**(3), 177–185 (1991)
8. Ross, T.Y., Dollár, G.: Focal loss for dense object detection. In: Proceedings of the IEEE Conference on Computer Vision and Pattern Recognition, pp. 2980–2988 (2017)
9. Varghese, R., Sambath, M.: Yolov8: a novel object detection algorithm with enhanced performance and robustness. In: 2024 International Conference on Advances in Data Engineering and Intelligent Computing Systems (ADICS), pp. 1–6. IEEE (2024)
10. Zhu, M., Zhang, S., Zhong, Y., Lu, P., Peng, H., Lenneman, J.: Monocular 3d vehicle detection using uncalibrated traffic cameras through homography. In: 2021 IEEE/RSJ International Conference on Intelligent Robots and Systems (IROS), pp. 3814–3821. IEEE (2021)
11. Zhu, X., Yin, Z., Shi, J., Li, H., Lin, D.: Generative adversarial frontal view to bird view synthesis. In: 2018 International Conference on 3D Vision (3DV), pp. 454–463. IEEE (2018)
12. Kim, Y., Kum, D.: Deep learning based vehicle position and orientation estimation via inverse perspective mapping image. In: 2019 IEEE Intelligent Vehicles Symposium (IV), pp. 317–323. IEEE (2019)
13. Yang, W., et al.: Projecting your view attentively: monocular road scene layout estimation via cross-view transformation. In: Proceedings of the IEEE/CVF conference on Computer Vision and Pattern Recognition, pp. 15536–15545 (2021)

In-Depth Evaluation and Analysis of Hyperspectral Unmixing Algorithms with Cognitive Models

Shunan Deng[1], Jinchang Ren[1,2(✉)], Rongjun Chen[1], Huimin Zhao[1], and Amir Hussain[3]

[1] School of Computer Science, Guangdong Polytechnic Normal University, Guangzhou 510665, China
jinchang.ren@ieee.org
[2] National Subsea Centre, Robert Gordon University, Aberdeen AB21 0BH, UK
[3] School of Computing, Edinburgh Napier University, Edinburgh, UK

Abstract. This paper evaluates several representative algorithms on real datasets. It analyzes Vertex Component Analysis (VCA), Total Variation Regularized Reweighted Sparse Nonnegative Matrix Factorization (RSNMF), Sparse Hyperspectral Unmixing (HU) with Mixed Norms, and Hapke Model with Convolutional Neural Network (HapkeCNN). Results indicate that VCA achieves high accuracy in endmember extraction and is widely applicable, RSNMF perform well with fewer, distinct endmembers, Sparse HU with Mixed Norms estimates abundances effectively without high-precision endmembers, addressing endmember variability, and HapkeCNN excels in nonlinear fitting and noise resistance, validating the effectiveness of cognitive models in unmixing tasks.

Keywords: Hyperspectral Unmixing · Endmember Variability · Noise Resistance · Cognitive Modelling

1 Introduction

Hyperspectral remote sensing technology is essential in various fields [1–3]. However, due to the limited spatial resolution of hyperspectral sensors, pixels often contain mixed information from multiple materials [4]. This makes research on hyperspectral unmixing (HU) techniques crucial. Early unmixing studies focused on the Linear Mixing Model (LMM), which assumes that each pixel's spectrum is a linear combination of endmember spectra [5]. The abundance coefficients, representing the proportion of each material, must meet two constraints: (1) the Abundance Non-negativity Constraint (ANC); (2) the Abundance Sum-to-One Constraint (ASC) [6].

The linear assumption of the LMM simplifies modeling and solving but overlooks spectral variability and nonlinear mixing effects. To address these issues, researchers have proposed more complex models, such as the Generalized Linear Mixing Model (GLMM) [7, 8], the Generalized Bilinear Model (GBM) [9, 10], and the Hapke model

[11]. The GLMM introduces additional flexibility to handle endmember variability, while the GBM is a simplified nonlinear model that considers second-order interactions between endmembers, making it suitable for moderately complex nonlinear mixing scenarios. The Hapke model, grounded in optical scattering theory, simulates the multiple scattering and reflection behaviors of incident light among particles, thereby characterizing the reflective spectral properties of materials. As hyperspectral unmixing techniques continue to evolve, the application of Artificial Neural Networks (ANN) [12–14], based on cognitive model principles, is becoming increasingly prevalent in this field. ANN, by mimicking the way the human brain processes information, can effectively capture complex nonlinear relationships and interdependencies in high-dimensional spectral data. This is similar to how cognitive models simulate human cognitive processes. To evaluate the effectiveness of ANN in unmixing tasks, this study selects three traditional unmixing algorithms based on the LMM—Vertex Component Analysis (VCA) [15], Total Variation Regularized Reweighted Sparse Nonnegative Matrix Factorization (RSNMF) [16], and Sparse hyperspectral unmixing (HU) with Mixed Norms—alongside one ANN-based algorithm, Hapke Model and Convolutional Neural Network (HapkeCNN) [17], for a unified experimental evaluation.

This paper undertakes the following research:

(1) Representative algorithms from various fields were selected for a unified experimental study, covering both traditional methods and modern neural network-based approaches.
(2) It verifies the effectiveness of Artificial Neural Networks (ANN) in advancing the field of hyperspectral unmixing.

The structure of this paper is as follows: Sect. 2 provides a detailed description of the algorithms employed in this study; Sect. 3 compares and analyzes the performance of these algorithms; Sect. 4 presents an in-depth discussion of the experimental results.

2 Unmixing Algorithms

The three representative categories of traditional unmixing methods are geometric models, non-negative matrix factorization, and sparse regression. VCA, RSNMF, and Sparse HU with Mixed Norms are the key algorithms within these categories, respectively. Additionally, HapkeCNN is used to evaluate the performance of ANN in nonlinear unmixing tasks.

2.1 Traditional Unmixing Methods

LMM: In the LMM, the spectral reflectance of each pixel is assumed to be a weighted average of the reflectance of different surface materials. This assumption is physically reasonable in many applications, especially when no complex interactions occur between materials, such as in remote sensing. LMM offers several advantages, including computational efficiency and clear interpretability. As a result, linear models for HU have developed a strong theoretical foundation and various solution algorithms, such as Non-negative Matrix Factorization (NMF) and sparse regression. The mathematical expression of LMM is as follows:

$$Y = EA + \varepsilon \tag{1}$$

$$y_j = \sum_{i=1}^{n} e_i a_{i,j} + \varepsilon_j \tag{2}$$

where $Y \in R^{L \times P}$: hyperspectral data, $y_j \in R^L$: spectrum of a single pixel, $E \in R^{L \times n}$: endmember spectral library containing n endmembers, L: number of bands, $P = u \times v$: number of pixels, $e_i \in R^L$: the i-th endmember, $A \in R^{n \times P}$: abundance matrix, $a_j \in R^n$: abundance of the j-th pixel, ε: overall noise, ε_j: noise of a single pixel.

VCA: VCA exploits two facts: 1) Endmembers are the vertices of a simplex, and 2) The affine transformation of a simplex is also a simplex. The algorithm iteratively projects the data in the direction orthogonal to the subspace spanned by the already found endmembers to identify new endmember locations [15].

The basic steps of the algorithm are as follows:

(1) Dimensionality reduction: Assuming the number of endmembers is n, dimensionality reduction is performed using Principal Component Analysis (PCA) or Singular Value Decomposition (SVD) to reduce the data to n or n-1 dimensions.
(2) Initialization of the projection direction: The direction vector for the first projection is set as $z = [0, \ldots 0, 1]$.
(3) Iterative projection: All pixels are projected onto the direction vector z, with extreme points representing new endmembers. The data is then projected orthogonally to the subspace of the identified endmembers, and new extreme points are found. This process repeats until all endmembers are identified.

RSNMF: RSNMF introduces a reweighted sparse regularization term into the NMF model [16], where the weights of the sparse regularization term are continuously updated during the abundance matrix iteration process. Additionally, the algorithm incorporates total variation (TV) regularization to capture the piecewise smooth structure of the abundance map, enhancing the robustness of TV-RSNMF to noise. The TV regularizer is as follows:

$$\|a\|_{TV} = \sum_{i=1}^{u-1} \sum_{j=1}^{v-1} \{|a_{i,j} + a_{i+1,j}| + |a_{i,j} - a_{i,j+1}|\}$$
$$+ \sum_{i=1}^{u-1} |a_{i,v} - a_{i+1,v}| + \sum_{j=1}^{v-1} |a_{u,j} - a_{u,j+1}| \tag{3}$$

$$\|A\|_{HTV} = \sum_{i=1}^{L} \|FA^i\|_{TV} \tag{4}$$

where A^i: the row vector form of the i-th band of the hyperspectral image, $F : R_K \to R_{u \times v}$: a function that transforms the vector of the i-th band into a two-dimensional image.

Let K represent the total number of pixels, where $u \times v = K$, indicating that the pixels are arranged in u rows and v columns.

The formula for calculating the weight matrix W is shown as follows:

$$W^{q+1} = 1/(|A^q + eps|) \tag{5}$$

where q: the number of iterations, eps: a non-negative constant.

The objective minimization function is shown as follows:

$$min_{M,A} \frac{1}{2} \|Y - MA\|_F^2 + \alpha \|W \odot A\|_1 + \beta \|A\|_{HTV} \tag{6}$$

where α, β: regularization parameter

Sparse HU with Mixed Norms: The algorithm first extracts endmember bundles from the hyperspectral image through grouped sampling (i.e., a subset of pixels is selected for the VCA experiment to identify endmembers) [21]. It then employs group sparse regularization and mixed norms to address the HU problem. The number of groups is equal to the number of endmembers, the mixed norm is defined as follows:

$$\|a\|_{G,p,q} = \left(\sum_{i=1}^{n} \|a_{G_i}\|_p^q \right)^{1/q} \tag{7}$$

where a is the abundance vector, a_{G_i} is the abundance sub-vector of the i-th group, p controls the intra-group sparsity, and q controls the inter-group sparsity.

The objective minimization function is shown as follows:

$$min_A \frac{1}{2} \|Y - MA\|_F^2 + \lambda \|A\|_{G,p,q} \tag{8}$$

2.2 ANN Unmixing Method

Hapke Model: The Hapke model is a reflection model based on the theory of optical physical scattering, capable of effectively describing the complex light reflection behavior produced by particle surfaces. The bidirectional reflectance $R(\omega)$ of the Hapke model can be expressed as:

$$R(\omega) = \frac{\omega \mu_0}{4\pi(\mu + \mu_0)} \left[(1 + B(g))p(g) + H(\omega, \mu_0)H(\omega, \mu) - 1 \right] \tag{9}$$

$$B(g) = \frac{B_0}{(1 + \tan(g/2)h)} \tag{10}$$

$$H(\omega, \mu) = \frac{(1 + 2\mu)}{(1 + 2\mu\sqrt{1 - \omega})} \tag{11}$$

where ω: average single scattering albedo, $B(g)$: backscattering function, for surfaces with low albedo, $B_0 \approx 1$, $H(\omega, \mu)$: multi-directional scattering function, μ_0: cosine of the incident angle, μ: cosine of the outgoing angle, g: phase angle, $p(g)$: phase function of the medium, h is a parameter which depends on particle spacing.

HapkeCNN: The HapkeCNN algorithm leverages the fact that the reflectance of materials in the SSA (Single Scattering Albedo) space can be regarded as a linear mixture [17]. It transforms the reflectance of hyperspectral images into SSA and then applies linear operations to address the original nonlinear problem. The corresponding mathematical expressions are as follows:

$$W = R^{-1}(E), Y = R(EA) + \varepsilon \qquad (12)$$

where $R(*)$: the function that converts SSA to reflectance using the Hapke model, $R^{-1}(*)$: the function that converts reflectance back to SSA using the Hapke model.

The HapkeCNN constructs a CNN for estimating endmembers and abundances. The network consists of multiple convolutional layers, batch normalization layers, and nonlinear activation functions (Leaky ReLU), employing skip connections to mitigate the vanishing gradient problem. The loss function of the HapkeCNN algorithm includes the following three components: 1) a nonlinear data fitting term, 2) a reconstruction error term, and 3) a geometric constraint term. The loss function is expressed as follows:

$$min_{Y,\widehat{Y},A,E} \frac{1}{2} \|Y - R\left(R^{-1}(E)A\right)\|_F^2 + \frac{\alpha}{2}\|Y - \widehat{Y}\|_F^2 + \lambda TV\left(R^{-1}(E)\right) \qquad (13)$$

$$TV(W) = \sum_{i,j=1}^{n} \frac{1}{2}\|W_i - W_j\|_2^2 \qquad (14)$$

where Y and $\widehat{Y}$ denote the input data and reconstructed data, respectively.

3 Experiments and Results

3.1 Experimental Procedure

Jasper Ridge Dataset: The image has a resolution of 100×100 pixels and contains 198 bands, covering a spectral range from 400 nm to 2500 nm with a spatial resolution of approximately 20 m. The dataset includes four materials: #1 vegetation, #2 water, #3 soil, and #4 road.

Selected the Root Mean Square Error (RMSE) and the Spectral Angle Distance (SAD) as evaluation metrics. Their mathematical expressions are as follows:

$$RMSE = \frac{1}{nP}\sqrt{\|A - \widehat{A}\|^2} \qquad (15)$$

$$SAD = \frac{1}{P}\sum_{i=1}^{P} arccos\left(\frac{e_i^T \cdot \hat{e}_i}{\|e_i\|\|\hat{e}_i\|}\right) \qquad (16)$$

Key Parameters: RSNMF: $[\alpha, \beta] = [0.01, 0.02]$ with 100 iterations. HapkeCNN: $\mu = 1$, $\mu_0 = 1$, $\alpha = 0.0001$, $\lambda = 0.1$, learning rate of Adam optimizer $= 0.001$, with 8000 iterations. Sparse HU with mixed norms: $\lambda = 0.1$, $q = 10$, with 1000 iterations. Parameter selection is based on the references.

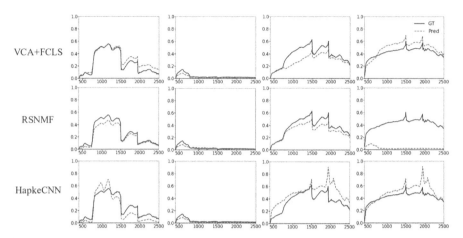

Fig. 1. The black solid line represents the ground truth, while the red dashed line represents the estimated endmembers. (Color figure online)

3.2 Analysis

VCA+FCLS: VCA + FCLS: As shown in Table 1, VCA achieves the best performance in terms of SAM values for endmember extraction, thanks to its geometric theory advantage. However, the extraction of the soil endmember is relatively poor. This discrepancy is mainly due to VCA's reliance on pure pixels for endmember extraction, which fails to account for spectral variability. Using Fully Constrained Least Squares (FCLS) for abundance estimation, as depicted in Fig. 2, the accuracy of the abundance estimation is low, indicating the presence of significant nonlinear components in the real scene. FCLS struggles to model the complex nonlinear relationship between endmember spectra and abundances.

RSNMF: As shown in Table 1, the SAM and RMSE of this algorithm are not particularly outstanding. However, as seen in Fig. 1, the spectral curves of the first three endmembers align closely with the ground truth (GT), with most peaks and troughs corresponding well with the GT. If only the SAM of the first three endmembers is considered, the accuracy reaches 0.1077, outperforming VCA. However, due to the inherent disadvantages of non-negative matrix factorization algorithms—including the non-uniqueness of the solution and the uncertainty of finding a global optimum—it is difficult to extract accurate solutions for all endmembers. From Fig. 2, it can be observed that the abundance maps of RSNMF exhibit noticeable noise. Upon analysis, it is speculated that this is caused by the impact of the weighted sparse regularization term on abundance smoothing.

Sparse HU with Mixed Norms: This algorithm focuses primarily on abundance estimation and achieved the best RMSE value in this experiment. Notably, Sparse SU with Mixed Norms does not rely on the ground truth (GT) endmember spectra; it is able to achieve high-precision abundance estimation using only the endmember bundles extracted by VCA, particularly excelling in handling endmember variability. However, one limitation of this algorithm is its sensitivity to the grouping strategy, as different

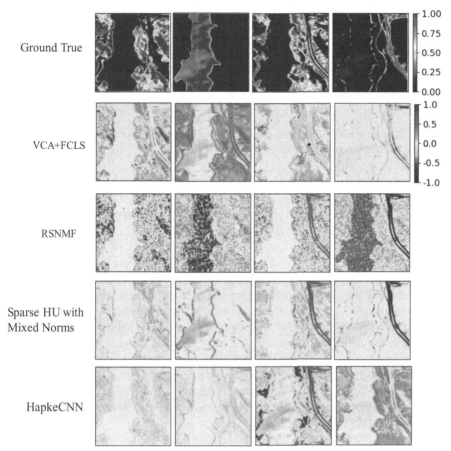

Fig. 2. The abundance maps of each algorithm illustrate the difference between the estimated abundance and the ground truth. When the estimated abundance is higher than the ground truth, it is shown in red; when the estimated abundance is lower than the ground truth, it is shown in blue. (Color figure online)

Table 1. Comparing the SAD values and RMSE values of different methods.

	VCA + FCLS	RSNMF	HapkeCNN	Sparse HU with mixed norms
SAM	**0.1095**	0.2018	0.1189	/
RSME	0.2004	0.3162	0.1830	**0.1476**

grouping methods significantly affect the results. As shown in Table 3, the accuracy fluctuated between 0.1476 and 0.3602, showing considerable variability without a clear pattern. Therefore, determining an appropriate grouping strategy remains a challenge.

Table 2. HapkeCNN results under different noise levels.

SNR	10	20	30	50
SAM	0.1551	**0.1135**	0.1184	0.1188
RSME	0.2177	0.1837	0.1833	**0.1829**

Table 3. Results under different groupings. The first four used sampling without replacement, while the last four used sampling with replacement.

Number of groups	4/25%	5/20%	8/12.5%	10/10%	10/50%	10/75%	10/90%	20/50%
RSME	**0.1476**	0.2446	0.2376	0.1833	0.1552	0.3602	0.1739	0.3239

HapkeCNN: HapkeCNN is the only algorithm in this paper that utilizes a neural network. Although its SAM and RMSE metrics are not the best, its overall performance is the highest. Thanks to the powerful fitting capability of the neural network, this algorithm demonstrates excellent noise resistance. As shown in Table 2, even after adding noise (SNR 20, SNR 30, SNR 50), the calculation accuracy is not significantly affected. In addition, the algorithm performs well in estimating the abundance of large homogeneous areas. Figure 2 shows that the abundance map corresponding to #2 water in the ground truth (GT) spectrum exhibits noticeable variations in the red regions, indicating that the same material exhibits some spectral variability in different geographic environments. HapkeCNN accurately captures this phenomenon, with almost no overestimation or underestimation of the water abundance.

4 Conclusion

This paper presents an in-depth comparative analysis of various spectral unmixing algorithms. The results indicate that VCA+FCLS and RSNMF each have their strengths in endmember extraction and spectral curve fitting, but they exhibit certain limitations when addressing endmember variability and nonlinear scenarios. Sparse HU with Mixed Norms excels in abundance estimation, although its sensitivity to grouping strategies requires further investigation and optimization. With the powerful fitting capabilities of neural networks, HapkeCNN demonstrates exceptional noise resistance and abundance estimation performance. This demonstrates that the introduction of ANN has made hyperspectral unmixing algorithms more intelligent and adaptive in extracting and analyzing spectral information, fully showcasing the potential of cognitive models in data processing. In the future, the application of cognitive models in other hyperspectral image domains, such as feature extraction [22], nondestructive quantitative measurement [23], and hyperspectral image quality assessment [24], can be further explored.

Acknowledgements. This work was supported in part by the Guangdong Province Key Construction Discipline Scientific Research Ability Promotion Project (2022ZDJS015, 2021ZDJS025),

Special Projects in Key Fields of Ordinary Universities of Guangdong Province under Grant 2021ZDZX1087, and the Guangzhou Science and Technology Plan Project under Grants (2024B03J1361, 2023B03J1327).

References

1. Chen, J., Zhao, M., Wang, X., Richard, C., Rahardja, S.: Integration of physics-based and data-driven models for hyperspectral image unmixing: a summary of current methods. IEEE Signal Process. Mag. **40**, 61–74 (2023). https://doi.org/10.1109/MSP.2022.3208987
2. Li, Y., Ren, J., Yan, Y., Ma, P., Assaad, M., Gao, Z.: ABBD: accumulated band-wise binary distancing for unsupervised parameter-free hyperspectral change detection. IEEE J. Sel. Top. Appl. Earth Observ. Remote Sens. **17**, 9880–9893 (2024). https://doi.org/10.1109/JSTARS.2024.3407212
3. Sun, G., et al.: Combinational shadow index for building shadow extraction in urban areas from Sentinel-2A MSI imagery. Int. J. Appl. Earth Obs. Geoinf. **78**, 53–65 (2019). https://doi.org/10.1016/j.jag.2019.01.012
4. Borsoi, R.A., et al.: Spectral variability in hyperspectral data unmixing: a comprehensive review. IEEE Geosci. Remote Sens. Mag. **9**, 223–270 (2021). https://doi.org/10.1109/MGRS.2021.3071158
5. Feng, X.-R., Li, H.-C., Wang, R., Du, Q., Jia, X., Plaza, A.: Hyperspectral unmixing based on nonnegative matrix factorization: a comprehensive review. IEEE J. Sel. Top. Appl. Earth Observ. Remote Sens. **15**, 4414–4436 (2022). https://doi.org/10.1109/JSTARS.2022.3175257
6. Wang, M., et al.: Tensor decompositions for hyperspectral data processing in remote sensing: a compre-hensive review. IEEE Geosci. Remote Sens. Mag. **11**, 26–72 (2023). https://doi.org/10.1109/MGRS.2022.3227063
7. Imbiriba, T., Borsoi, R.A., Moreira Bermudez, J.C.: Generalized linear mixing model accounting for endmember variability. In: 2018 IEEE International Conference on Acoustics, Speech and Signal Processing (ICASSP), pp. 1862–1866. IEEE, Calgary (2018). https://doi.org/10.1109/ICASSP.2018.8462214
8. Zhou, J., Sun, W., Meng, X., Yang, G., Ren, K., Peng, J.: Generalized linear spectral mixing model for spatial–temporal–spectral fusion. IEEE Trans. Geosci. Remote Sensing. **60**, 1–16 (2022). https://doi.org/10.1109/TGRS.2022.3188501
9. Wang, X., Zhong, Y., Zhang, L., Xu, Y.: Blind hyperspectral unmixing considering the adjacency effect. IEEE Trans. Geosci. Remote Sens. **57**, 6633–6649 (2019). https://doi.org/10.1109/TGRS.2019.2907567
10. Halimi, A., Altmann, Y., Dobigeon, N., Tourneret, J.-Y.: Nonlinear unmixing of hyp-erspectral images using a generalized bilinear model. In: 2011 IEEE Statistical Signal Processing Workshop (SSP), pp. 413–416. IEEE, Nice (2011). https://doi.org/10.1109/SSP.2011.5967718
11. Hapke, B.: Bidirectional reflectance spectroscopy: 1. Theory. J. Geophys. Res. **86**, 3039–3054 (1981). https://doi.org/10.1029/JB086iB04p03039
12. Fu, H., et al.: HyperDehazing: A hyperspectral image dehazing benchmark dataset and a deep learning model for haze removal. ISPRS J. Photogramm. Remote Sens. **218**, 663–677 (2024). https://doi.org/10.1016/j.isprsjprs.2024.09.034
13. Li, Y., et al.: CBANet: an end-to-end cross-band 2-d attention network for hyperspectral change detection in remote sensing. IEEE Trans. Geosci. Remote Sens. **61**, 1–11 (2023). https://doi.org/10.1109/TGRS.2023.3276589

14. Li, Y., Ren, J., Yan, Y., Sun, G., Ma, P.: SSA-LHCD: a singular spectrum analysis-driven lightweight network with 2-D self-attention for hyperspectral change detection. Remote Sens. **16**, 2353 (2024). https://doi.org/10.3390/rs16132353
15. Nascimento, J.M.P., Dias, J.M.B.: Vertex component analysis: a fast algorithm to unmix hyperspectral data. IEEE Trans. Geosci. Remote Sensing. **43**, 898–910 (2005). https://doi.org/10.1109/TGRS.2005.844293
16. He, W., Zhang, H., Zhang, L.: Total variation regularized reweighted sparse nonnegative matrix factorization for hyperspectral unmixing. IEEE Trans. Geosci. Remote Sens. **55**, 3909–3921 (2017). https://doi.org/10.1109/TGRS.2017.2683719
17. Rasti, B., Koirala, B., Scheunders, P.: HapkeCNN: blind nonlinear unmixing for intimate mixtures using hapke model and convolutional neural network. IEEE Trans. Geosci. Remote Sens. **60**, 1–15 (2022). https://doi.org/10.1109/TGRS.2022.3202490
18. Ekanayake, E.M.M.B., et al.: Constrained nonnegative matrix factorization for blind hyperspectral unmixing incorporating endmember independence. IEEE J. Sel. Top. Appl. Earth Observ. Remote Sens. **14**, 11853–11869 (2021). https://doi.org/10.1109/JSTARS.2021.3126664
19. Xu, X., et al.: Curvelet transform domain-based sparse nonnegative matrix factorization for hyperspectral unmixing. IEEE J. Sel. Top. Appl. Earth Observ. Remote Sens. **13**, 4908–4924 (2020). https://doi.org/10.1109/JSTARS.2020.3017023
20. Qu, K., Li, Z., Wang, C., Luo, F., Bao, W.: Hyperspectral unmixing using higher-order graph regularized NMF with adaptive feature selection. IEEE Trans. Geosci. Remote Sens. **61**, 1–15 (2023). https://doi.org/10.1109/TGRS.2023.3275740
21. Drumetz, L., Meyer, T.R., Chanussot, J., Bertozzi, A.L., Jutten, C.: Hyperspectral Image unmixing with endmember bundles and group sparsity inducing mixed norms. IEEE Trans. Image Process. **28**, 3435–3450 (2019). https://doi.org/10.1109/TIP.2019.2897254
22. Ma, P., et al.: Multiscale superpixelwise prophet model for noise-robust feature ex-traction in hyperspectral images. IEEE Trans. Geosci. Remote Sens. **61**, 1–12 (2023). https://doi.org/10.1109/TGRS.2023.3260634
23. Yan, Y., Ren, J., Sun, H., Williams, R.: Nondestructive quantitative measurement for precision quality control in additive manufacturing using hyperspectral imagery and machine learning. IEEE Trans. Industr. Inform. **20**, 9963–9975 (2024). https://doi.org/10.1109/TII.2024.3384609
24. Ma, P., Ren, J., Gao, Z., Li, Y., Chen, R.: Hyperspectral imagery quality assessment and band reconstruction using the prophet model. CAAI Trans. Intell. Tech. (2024). https://doi.org/10.1049/cit2.12373

Effective Gas Classification Using Singular Spectrum Analysis and Random Forest in Electronic Nose Applications

Yuntao Wu[1], Jinchang Ren[1,2(✉)], Rongjun Chen[1], Huimin Zhao[1], and Amir Hussain[3]

[1] School of Computer Science, Guangdong Polytechnic Normal University, Guangzhou 510665, China
jinchang.ren@ieee.org
[2] National Subsea Centre, Robert Gordon University, Aberdeen AB21 0BH, UK
[3] School of Computing, Edinburgh Napier University, Edinburgh, UK

Abstract. This paper focuses on the gas data processing and pattern recognition modules of detection and proposes a time-domain feature extraction method based on Singular Spectrum Analysis (SSATFE). This method effectively captures and reduces the dimensionality of gas signals, making the data more manageable for classification. To evaluate the extracted features, tree-based classifiers like decision trees and random forests were used. These models handle non-linear relationships well and are robust to noise, which is common in gas sensor data. Experiments on the UCL dataset show the classifiers' ability to use the key components extracted by the SSA method for accurate gas detection.

Keywords: Gas classification · Singular spectrum analysis · Random forest

1 Introduction

Gas leaks in industrial settings often lead to explosions, fires, and pollution, posing a serious safety concern [1, 2]. This is a complex issue involving the identification and monitoring of various gases. Traditional sensors are typically limited to detecting specific gases and are inadequate for addressing multi-gas leaks, while manual detection methods are unsafe. Therefore, automated gas detection technologies are essential for the timely prevention of accidents and ensuring safety [3]. The complexity of gas detection arises primarily from the diversity of gases and the distinct responses of sensors to different gas types. In 2000, researchers applied a variety of classical algorithms to address gas detection problems, including Principal Component Analysis (PCA) [4], Linear Discriminant Analysis (LDA) [5], k-Nearest Neighbors (KNN) [6], Support Vector Machines (SVM) [7, 8], and Random Forests (RF) [9]. The common feature of these classical algorithms is their relatively fixed frameworks and few parameters, which make them simple in design and widely applicable. With the rise of artificial intelligence in recent years, Artificial Neural Networks (ANNs), known for their flexible architecture and strong performance, have gained widespread use. Furthermore, deep learning models such as

Convolutional Neural Networks (CNN) [10, 11], Long Short-Term Memory Networks (LSTM) [12], Residual Networks (ResNet) [13], and Transformers [14, 15] have also been employed in the pattern recognition systems of electronic noses, yielding promising results. Although these methods have achieved some success in specific datasets and application scenarios, inconsistencies in data preprocessing, feature extraction methods, and classifier selection often lead to significant variations in the conclusions drawn from different studies.

This study focuses on preprocessing gas classification data by introducing Singular Spectrum Analysis (SSA) [16], followed by the application of tree classifiers, such as Random Forest (RF), on the processed data. SSA, as an effective time series processing method, can decompose time series signals, remove noise, and extract useful information. By employing SSA for denoising and smoothing, we extract features and apply various classifiers for gas classification tasks. This approach not only enhances classification performance but also simplifies the data processing workflow. Tree models, such as Random Forests, were chosen due to their ability to automatically select important features and their robustness in handling noise and data imbalance, coupled with low computational overhead and strong performance. Furthermore, there exists a certain connection between tree models and cognitive models: tree models partition data hierarchically, akin to the human cognitive process of progressively analyzing and deconstructing problems, thus effectively managing complex nonlinear data structures. Therefore, the combination of features extracted through SSA with the classification capability of tree models can further enhance the accuracy and efficiency of gas detection. Gas detection methods are diverse, ranging from traditional statistical approaches to modern machine learning and deep learning techniques.

Previous studies have proposed various preprocessing methods; however, they still exhibit some shortcomings. For instance, although Discrete Wavelet Transform (DWT) [17] is commonly used for data filtering and denoising with significant effectiveness, it has limitations in capturing the complex nonlinear patterns of gas signals. Such methods often focus on the low-frequency and high-frequency components of signals while neglecting the underlying complex dynamic features present in gas data, resulting in inadequate extraction of effective information to improve recognition accuracy. Furthermore, most studies fail to effectively distinguish important signals from noise, potentially leading to information loss during feature extraction, thereby impacting the accuracy of classification results. This study addresses these issues by introducing Singular Spectrum Analysis (SSA) as a preprocessing method. SSA not only smooths gas signals but also removes noise by decomposing the most significant components, thereby retaining features crucial for classification. In this way, we can more efficiently extract useful information from gas data, improving both classification accuracy and real-time performance. The main contributions of this paper are as follows:

1) Propose a time-domain feature extraction method based on Singular Spectrum Analysis (SSATFE) and combine it with a Random Forest classifier for gas detection. SSA serves as a smoothing technique, extracting the primary trend components from the gas data.

2) Apply the Random Forest classifier to the preprocessed data. Compared to traditional tree-based models such as decision trees, Random Forest demonstrates strong performance in handling gas detection tasks.

2 Background Knowledge

2.1 Singular Spectrum Analysis (SSA)

Singular Spectrum Analysis (SSA) is a non-parametric time series analysis technique that can decompose a signal into several independent components, extracting meaningful patterns and trends. The decomposition process of SSA mainly consists of the following four steps [18]:

Embedding: The time series of length N is transformed into a trajectory matrix through a sliding window, where the window size is L (with $0 < L < N/2$), forming a trajectory matrix of size $L \times K$ (where $K = N - L + 1$).

Singular Value Decomposition (SVD): Singular value decomposition is performed on the trajectory matrix, decomposing it into several rank-1 component matrices. First, the covariance matrix S is computed as formula (1):

$$S = XX^T \qquad (1)$$

Then, eigenvalue decomposition is carried out on this matrix to obtain the eigenvalues λ_i and the corresponding eigenvectors μ_i. Through these components, the main patterns and noise in the time series can be extracted.

Grouping: Based on the magnitude and nature of the singular values, the decomposed component matrices are regrouped into categories such as trend, periodic components, and noise. Each group of component matrices is then summed to obtain different reconstructed matrices corresponding to the respective categories.

Diagonal Averaging: Diagonal averaging is performed on the reconstructed trajectory matrices, converting them into new time series. The elements along each diagonal are averaged to generate the final reconstructed time series.

2.2 Random Forest (RF)

Random Forest is a supervised ensemble learning algorithm that constructs multiple decision trees and combines their voting results to produce the final classification. During training, Random Forest randomly selects samples and features to build multiple independent decision trees, and the final classification result is determined through majority voting. This algorithm has strong resistance to overfitting, can handle missing values, and provides feature importance evaluation. It is suitable for processing various types of data, though its model size can be large, and interpretability is relatively weak [19, 20].

The entire algorithm can be divided into the following three steps:

Bootstrap Sampling Method: Multiple sample sets are generated from the training dataset through bootstrap sampling with replacement. Each sample set is used to build a decision tree, referred to as a Bootstrap sample.

Building a Random Forest Model: For each Bootstrap sample, a CART decision tree is constructed. Each tree randomly selects a subset of features and uses the Gini index (Gini) as the splitting criterion to find the optimal split point and then build the decision tree. The Gini index is calculated using the following formula (2):

$$Gini(D_v) = 1 - \sum_{l=1}^{l} p_l^2 \qquad (2)$$

where p_l represents the probability that a selected sample belongs to the l-th class. The Gini coefficient reflects the probability that two randomly selected samples from Bootstrap sample belong to different classes. The smaller the Gini coefficient, the higher the probability that the selected samples belong to the same class, and the lower the impurity of Bootstrap sample.

Simple Voting: Each decision tree independently makes predictions, and the final class of the sample is determined through a simple majority vote across all the decision trees. The class with the most votes is selected as the predicted result.

3 Data and Methods

This paper presents an SSATFE detection method designed to address the challenges of gas detection. The framework utilizes SSA to reconstruct the principal component of the gas signal, followed by the extraction of four time-domain statistical features to capture the complex patterns of the gas. Finally, a Random Forest classifier is employed to perform the classification, completing the gas detection task. The method flow is illustrated in the diagram below Fig. 1.

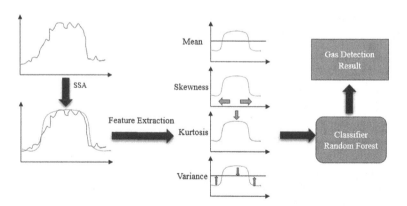

Fig. 1. SSATFE Gas Detection Method Diagram

3.1 Experimental Dataset

In this study, we used an open-source gas dataset published on the UCL Machine Learning Repository [21]. The dataset is collected from 9 sensor array systems, each consisting of 8 MOX gas sensors, totaling 72 sensors. However, only data from the 4th sensor array is used in this paper. A schematic diagram of the setup is shown in Fig. 2. The data acquisition system is set up in a custom wind tunnel with dimensions of 2.5 m × 1.2 m × 0.4 m. To simulate a real open environment, the sensors were placed in 6 different positions relative to the gas source for data collection. The raw data is divided into 8 gas types (excluding butanol and carbon monoxide) and 6 subsets based on sensor array positions (P1, P2, ..., P6). A total of 18,000 time series measurements were recorded, focusing on 10 high-priority chemicals, including ethanol, acetone, ammonia, benzene, ethylene, methane, methanol, and toluene.

The experiment was conducted under three wind speeds: 0.10 m/s, 0.21 m/s, and 0.34 m/s, as well as five different sensor heater voltages: 4.0 V, 4.5 V, 5.0 V, 5.5 V, and 6.0 V. Each combination was measured 20 times at different moments. Each measurement records a full cycle of sensor response and recovery, divided into three phases: Phase 1 (0–20 s) involves blowing clean air at a constant wind speed, Phase 2 (21–200 s) involves randomly selecting and releasing one of the 10 chemicals directly into the wind tunnel, and Phase 3 (201–260 s) stops the gas release while maintaining the same wind speed for ventilation. The sensors sampled at a rate of 100 Hz, but for easier subsequent data processing, the median value of the data collected each second is used as the sensor's reading for that second, reducing the sampling rate from 100 Hz to 1 Hz.

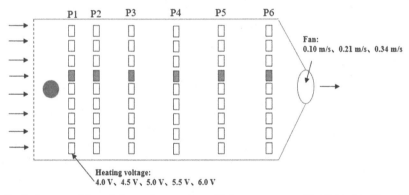

Fig. 2. A diagram of the wind tunnel apparatus with the sensor array highlighted in blue. (Color figure online)

3.2 Data Processing

The downsampled data is used as the raw input. First, the input data is normalized to eliminate the potential inconsistencies between different sensor parameters. In this study, the raw data is divided by the maximum reading of the gas sensors. The UCI public dataset uses 12-bit gas sensors, so the raw input is standardized by:

$$\hat{x}_n = \frac{x_i}{2^{12}} \tag{3}$$

where x_i is the value in the sequence x_n, $n \in [1, N]$, and $\hat{x}_n$ is the normalized input.

Next, Singular Spectrum Analysis (SSA) is applied to the standardized input data $\hat{x}_n$, with a window size of 10. Given that the downsampled gas data has a total length of 260, the window size should be greater than 0 and less than 130. However, an excessively large window size may lead to over-smoothing, which is why a window size of 10 is chosen in this study. It is assumed that each singular value corresponds to a group, and the first four singular value components are extracted to reconstruct the time series. This reconstructed time series is represented as the vector y_n, which is formed by removing irrelevant components. At this point, the sample size is 260 × 8, where 8 represents the number of sensor channels. If input directly into the random forest classifier (with each channel treated as a feature), the number of corresponding values for each feature is too large. Therefore, this study conducts time-domain feature extraction for each channel, resulting in four features: mean, skewness, kurtosis, and sample variance. The mean, skewness, kurtosis, and sample variance are four commonly used statistical features in time series analysis, each providing different perspectives to describe the distribution and characteristics of the data [22].

Mean: The mean represents the average response level of the sensor to a particular gas during the measurement. It is a statistic that describes the overall location of the data and helps to understand the overall level of the sequence, defined by

$$\bar{y}_n = \frac{1}{N} \sum_{i=1}^{N} y_i \tag{4}$$

where y_i is the reconstructed value, $i \in [1, N]$, and $\bar{y}_n$ is the mean.

Skewness: Skewness is an important measure of the asymmetry in a data distribution and quantifies the degree to which a sequence is skewed relative to the mean. If the skewness is positive, it indicates that the tail on the right side of the mean is longer; conversely, if the skewness is negative, it indicates that the tail on the left side is longer.

$$y_{skew,n} = \frac{1}{N} \sum_{i=1}^{N} \left(\frac{y_i - \bar{y}_n}{\sigma} \right)^3 \tag{5}$$

where σ is the standard deviation of the sequence y_n, and $y_{skew,n}$ is the skewness.

Kurtosis: For gas sensors, kurtosis can reveal whether the sensor shows a sharp response to sudden changes in gas levels or whether there are significant outliers as:

$$y_{Kurtosis,n} = \frac{1}{N} \sum_{i=1}^{N} \left(\frac{y_i - \bar{y}_n}{\sigma} \right)^4 \tag{6}$$

Effective Gas Classification Using Singular Spectrum Analysis 289

Sample Estimated Variance: Sample estimated variance is particularly important for analyzing gas sensor data, as it can indicate whether the sensor's response to different gases is consistent or shows significant fluctuations, defined below:

$$y_{Variance,n} = \frac{1}{N} \sum_{i=1}^{N} (y_i - \bar{y}_n)^2 \qquad (7)$$

After feature extraction, the sample size is reduced from 260 × 8 to 4 × 8, and finally, it is flattened into a 1D vector for input into the classifier.

4 Results and Discussion

4.1 Singular Spectrum Analysis Results

In the singular spectrum analysis, the window size is set to 10. At this point, the reconstructed time series obtained by extracting the largest singular component from the raw data still shows a high correlation with the original data, see in Fig. 3. After calculation, the Pearson correlation coefficient between the original data and the SSA-processed data shown in Fig. 3 is 0.9935, indicating that a large amount of feature information is retained while effectively smoothing the data. The SSA-reconstructed data is then subjected to time-domain feature extraction and input into the random forest classifier. In this study, the random forest classifier is implemented using the scikit-learn toolkit, with key parameters including the number of decision trees and the impurity measure. The impurity measure used is the Gini coefficient, and the number of decision trees is optimized by testing multiple values to find the best range. As shown in Fig. 4, setting different numbers of decision trees results in varying accuracies.

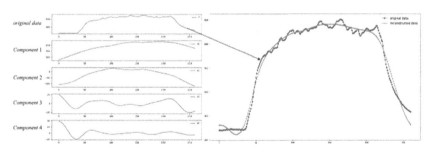

Fig. 3. Data collected from Sensor 1 on Board 4 for one instance of ammonia. The left figure shows the original data and the first four components from top to bottom, while the right figure compares the original data with the reconstructed data.

As shown in Fig. 4, the highest accuracy on the test set is achieved when the number of decision trees is set to 1000, indicating that the classification task performs best with this configuration. In the following comparison experiments, whenever the random forest classifier is used, the number of decision trees is fixed at this optimal value.

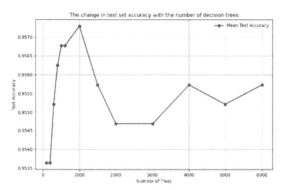

Fig. 4. The change in accuracy with the number of decision trees.

4.2 Comparison Results

First, we compared different classifiers, including Decision Tree (DT) and Random Forest (RF). The training set was used to train classification models with both classifiers, and the test set was used to evaluate the models on four metrics: Accuracy, Precision, Recall, and F1_Score. Among these, Accuracy, Precision, and Recall are the average values derived from 10 experiments, while F1_Score is calculated based on Precision and Recall. The formula for calculating the F1_Score is shown below:

$$F1\_Score = 2 \times \frac{Precision \times Recall}{Precision + Recall} \tag{8}$$

and a comparison of the results is presented in Table 1:

Table 1. Comparison of Classification Performance between Decision Tree and Random Forest

Classifier	Accuracy	Precision	Recall	F1_Score
DT	0.9271	0.9278	0.9271	0.9274
RF	0.9557	0.9569	0.9557	0.9562

The Random Forest (RF) outperforms the Decision Tree (DT) across all evaluation metrics (Accuracy, Precision, Recall, and F1_Score). This suggests that Random Forest is more effective at handling the data for this specific classification task, likely due to its ensemble learning nature, which helps reduce overfitting and improve the model's generalization ability.

The data processed by the TSSA method proposed in this paper were input into two classifiers, and the average values of the four metrics were compared across 10 experiments. The comparison results are shown in Table 2. Additionally, the SSATFE method was compared with other methods, and under consistent dataset usage, the performance of the other methods was inferior to SSATFE.

Table 2. Performance comparison of various methods.

Method	Accuracy	Precision	Recall	F1_Score
DT	0.9271	0.9278	0.9271	0.9274
RF	0.9557	0.9569	0.9557	0.9562
MCNA [23]	0.9498	\	\	\
SSATFE + DT	0.9531	0.9547	0.9531	0.9538
SSATFE + RF	**0.9760**	**0.9762**	**0.9760**	**0.9761**

In Table 2, the first two methods simply input the raw data directly into the classifiers, while the latter two combine the proposed method of applying Singular Spectrum Analysis (SSA) followed by feature extraction in the time domain. Both classifiers show an improvement in classification performance after effective data processing, with the final results indicating that the proposed method, combined with Random Forest, performs exceptionally well in completing the classification task. It is evident from the comparison that SSA can effectively extract the main information from gas signals, and the statistical features in the time domain can summarize the key information extracted from the gas data well. Ultimately, the random forest classifier successfully completes the gas detection task.

5 Conclusion

This paper proposes a method for gas detection based on Singular Spectrum Analysis (SSA) combined with time-domain feature extraction, effectively completing the intended classification task. From the comparison results in Tables 1 and 2, it can be observed that effective smoothing techniques and feature extraction methods significantly contribute to improving the accuracy of tree-based models. In this study, common time-domain feature extraction was used, but the extraction of the key information from gas features is not yet comprehensive. Future work will focus on exploring more effective feature extraction methods as well as real-time implementation of the algorithms.

In recent years, deep learning has demonstrated significant potential in time series classification tasks [24, 25]. This paper will continue to explore the application of deep learning in gas detection tasks in subsequent research, aiming to further investigate its value in long-term sequence prediction.

Acknowledgements. This work was supported in part by the Guangdong Province Key Construction Discipline Scientific Research Ability Promotion Project (2022ZDJS015, 2021ZDJS025), Special Projects in Key Fields of Ordinary Universities of Guangdong Province under Grant 2021ZDZX1087, and the Guangzhou Science and Technology Plan Project under Grants (2024B03J1361, 2023B03J1327).

References

1. Ultrafast, Highly Sensitive, and Selective Detection of p-Xylene at Room Temperature by Peptide-Hydrogel-Based Composite Material. ACS Appl. Polym. Mater. https://pubs.acs.org/doi/abs/10.1021/acsapm.9b00594. Accessed 24 Oct 2024
2. Cai, C., Mo, J., Lu, Y., et al.: Integration of a porous wood-based triboelectric nanogenerator and gas sensor for real-time wireless food-quality assessment. Nano Energy **83**, 105833 (2021). https://doi.org/10.1016/j.nanoen.2021.105833
3. Sharma, A., Khullar, V., et al.: Gas detection and classification using multimodal data based on federated learning. Sensors **24**, 5904 (2024). https://doi.org/10.3390/s24185904
4. Li, H., Luo, D., Sun, Y., GholamHosseini, H.: Classification and identification of industrial gases based on electronic nose technology (2019)
5. Freddi, S., Vergari, M., Pagliara, S., Sangaletti, L.: A chemiresistor sensor array based on graphene nanostructures: from the detection of ammonia and possible interfering VOCs to chemometric analysis. Sensors. **23**, 882 (2023). https://doi.org/10.3390/s23020882
6. Several ML Algorithms and Their Feature Vector Design for Gas Discrimination and Concentration Measurement with an Ultrasonically Catalyzed MOX Sensor | ACS Sensors, https://pubs.acs.org/doi/abs/10.1021/acssensors.2c02159. Accessed 15 Oct 2024
7. Djeziri, M.A., Djedidi, O., Morati, N., Seguin, J.-L., et al.: A temporal-based SVM approach for the detection and identification of pollutant gases in a gas mixture. Appl. Intell. **52**, 6065–6078 (2022). https://doi.org/10.1007/s10489-021-02761-0
8. Harakeh, A., et al: New approach for gas identification using supervised learning methods. In: MATEC Web Conference, vol. 261, p. 06004 (2019). https://doi.org/10.1051/matecconf/201926106004
9. Rehman, A.U., Bermak, A.: Heuristic random forests (HRF) for drift compensation in electronic nose applications. IEEE Sens. J. **19**, 1443–1453 (2019). https://doi.org/10.1109/JSEN.2018.2881745
10. Ren, X., Wang, Y., Huang, Y., Mustafa, et al: A CNN-based E-nose using time series features for food freshness classification. IEEE Sens. J. **23** (2023)
11. Xiong, L., et al.: Image presentation and effective classification of odor intensity levels using multi-channel electronic nose technology combined with GASF and CNN. Sens. Actuators, B Chem. **395**, 134492 (2023). https://doi.org/10.1016/j.snb.2023.134492
12. Nguyen, Q.D.N., Le, H.V.A., Nakano, T., Tran, T.H.: Wine quality assessment through lightweight deep learning: integrating 1D-CNN and LSTM for analyzing electronic nose VOCs signals. ACI (2024). https://doi.org/10.1108/ACI-10-2023-0098
13. Wei, G., Liu, X., He, A., Zhang, W., Jiao, S., Wang, B.: Design and implementation of a ResNet-LSTM-ghost architecture for gas concentration estimation of electronic noses. IEEE Sens. J. **24**, 26416–26428 (2024). https://doi.org/10.1109/JSEN.2024.3424312
14. Wang, Y., et al: Determination of the quality of tea from different picking periods: an adaptive pooling attention mechanism coupled with an electronic nose. Food Chem. (2024). https://doi.org/10.1016/j.foodchem.2023.137307
15. Wang, Z., et al.: PWDformer: deformable transformer for long-term series forecasting. Pattern Recogn. **147**, 110118 (2024). https://doi.org/10.1016/j.patcog.2023.110118
16. Yan, Y., et al.: PCA-domain fused singular spectral analysis for fast and noise-robust spectral–spatial feature mining in hyperspectral classification. IEEE Geosci. Remote Sens. Lett. **20**, 1–5 (2023). https://doi.org/10.1109/LGRS.2021.3121565
17. Wijaya, D.R., Sarno, R., Zulaika, E.: Noise filtering framework for electronic nose signals: an application for beef quality monitoring. Comput. Electron. Agric. **157**, 305–321 (2019). https://doi.org/10.1016/j.compag.2019.01.001

18. Chiu, J.-E., Tsai, C.-H.: On-line concurrent control chart pattern recognition using singular spectrum analysis and random forest. Comput. Ind. Eng. **159**, 107538 (2021). https://doi.org/10.1016/j.cie.2021.107538
19. Breiman, L., Friedman, J., Olshen, R.A., Stone, C.J.: Classification and Regression Trees. Chapman and Hall/CRC, New York (2017). https://doi.org/10.1201/9781315139470
20. Breiman, L.: Random forests. Mach. Learn. **45**, 5–32 (2001). https://doi.org/10.1023/A:1010933404324
21. Vergara, A., et al.: On the performance of gas sensor arrays in open sampling systems using inhibitory support vector machines. Sens. Actuat. B Chem. **185**, 462–477 (2013). https://doi.org/10.1016/j.snb.2013.05.027
22. Li, J.W., et al.: An innovative EEG-based emotion recognition using a single channel-specific feature from the brain rhythm code method. Front. Neurosci. **17** (2023). https://doi.org/10.3389/fnins.2023.1221512
23. Lightweight neural network for gas identification based on semiconductor sensor. IEEE J. Mag. https://ieeexplore.ieee.org/document/9650878. Accessed 24 Oct 2024
24. Chen, R., et al.: Rapid detection of multi-QR codes based on multistage stepwise discrimination and a compressed MobileNet. IEEE Internet Things J. **10**, 15966–15979 (2023). https://doi.org/10.1109/JIOT.2023.3268636
25. Li, X., et al: Siamese residual neural network for musical shape evaluation in piano performance assessment. In: 31st European Signal Processing Conference (2023). https://doi.org/10.23919/EUSIPCO58844.2023.10289901

Author Index

B
Bruevich, Maria II-178

C
Chen, Jiacong I-177
Chen, Jie II-1
Chen, Jinlan II-104
Chen, Rongjun I-20, II-273, II-283
Chen, Weiqi I-136
Chen, Yanxiang I-30
Chen, Yuan II-1, II-11

D
Dai, Xiaodong I-104
Dashtipour, Kia II-159, II-178
Deng, Shunan II-273
Diao, Liangjin I-177
Ding, Jonathan II-117
Ding, Zhuanlian II-41
Dong, Xingbo I-166
Dong, Yiting I-249

E
Efunwoye, Ibukunoluwa Oluwabusayo II-159

F
Fan, Chen I-1
Feng, Guanyuan I-20
Feng, Mingchen II-201
Fu, Yanping I-126

G
Gao, Fei I-1
Gao, Yuefang II-62
Gogate, Mandar II-159, II-178
Gunathilake, Nilupulee A. II-178

H
Han, Meng I-269, II-82, II-188, II-233
Hao, Li II-62
Hao, Wei I-82
Hao, Wangli I-269, II-82, II-233
He, Jian I-259
He, Yuan II-93
Hou, Wanzhen I-93
Hou, Yanqi I-177
Hu, Guyue II-93
Hu, Jingfei I-197
Hu, Wei I-10, I-136
Hu, Xiaojing I-62
Hu, Xinyuan II-233
Huang, Chengchuang II-188
Huang, Fuchuan I-259
Huang, Lili I-208
Huang, Yuesheng I-20
Hussain, Adeel II-159, II-178
Hussain, Amir I-1, II-159, II-178, II-273, II-283

J
Jia, Congcong I-166
Jia, Wei I-259
Jiang, Bo II-1, II-11
Jiang, Fengling II-159, II-178
Jiang, Hongfan II-138
Jiang, Longteng I-218
Jiao, Hui II-117
Jin, Xin I-218
Jin, Yuguang I-52
Jin, Zhe I-166
Jing, Yang I-41

L
Li, Anzhen II-201
Li, Ao II-168
Li, Chenglong I-155
Li, Chuanfu I-208

Li, Dazhi II-62
Li, Fuzhong I-269, II-233
Li, Guopping I-30
Li, Huihui I-136
Li, Jiawen I-20
Li, Ping II-222
Li, Qilang II-104
Li, Shiqi I-208
Li, Weiyi I-249
Li, Xin II-188
Li, Xinhui II-168
Li, Xinwei II-117
Li, Xuetao II-52
Li, Yi I-155
Li, Yiming I-197
Li, Yufei I-187
Li, Ziyu II-21
Ling, Chen I-20
Lingkang, Gu I-41
Liu, Fengxiang II-252
Liu, Lei I-155
Liu, Shixi I-62
Liu, Tianjin I-1
Liu, Wenhao II-41
Liu, Xiaoyong I-136
Liu, Xinyu II-72
Liu, Yijian I-218
Liu, Yudian I-62
Liu, Yuqing II-1, II-11
Liu, Zhenfei II-211
Lu, Enmeng I-229
Lu, Feixiang I-218
Lu, Jianwei II-72
Lu, Xiaofei II-82
Lu, Yongyi II-243
Luo, Bin II-159, II-178
Luo, Ye II-72
Lv, Jujian I-20
Lv, Zhao II-168

M
Ma, Leilei I-52
Ma, Qingchuan I-104, I-115
Ma, Yong II-31
Ma, Yuanmin II-1
Mai, Kaizhan II-62
Mengqi, Wu I-41
Mi, Pingping II-62
Mi, Xin II-222

Mu, Chaofan II-52

P
Pang, Mengyin II-211

Q
Qin, Jinghui II-243
Qin, Liwen II-201
Qin, Sizhe I-239
Qin, Weijie II-201
Qing, Shufan II-201

R
Rao, Ji II-72
Ren, Jie II-222
Ren, Jinchang II-159, II-178, II-201, II-273, II-283
Ren, Ximing I-20
Ren, Yongjian II-82

S
Shang, Dongxu I-72
Shang, Shoulai I-82
Shen, Guobin I-249
Shen, Longfeng I-177
Shen, Zhiya II-104
Sheng, Cunyu I-259
Shi, Chao II-222
Shi, Meilin II-201
Shu, Hao I-269, II-82, II-233
Su, Deyu I-239
Sun, Dengdi II-41, II-52
Sun, Lingma I-208
Sun, Meijun II-211
Sun, Yinqian I-229

T
Tan, Wenhao II-148
Tang, Jin I-115, II-262
Tao, Liang I-72
Tao, Wan II-127
Tu, Zijian I-239

W
Wang, Futian II-252
Wang, Hongbo II-222
Wang, Hua I-197
Wang, Huabin I-72
Wang, Jianfei II-72

Author Index

Wang, Jing I-10
Wang, Jun I-1
Wang, Leijun I-20
Wang, Lina II-211
Wang, Linbo II-148
Wang, Peixian I-20
Wang, Qianhao I-229
Wang, Shiao I-104, I-115
Wang, Wenzhong II-262
Wang, Xiao I-104, I-115, II-252
Wang, Xiaoran II-127
Wang, Xiuli II-72
Wang, Xuzhen II-262
Wang, Yifeng I-104, I-115
Wang, Zheng II-211
Wang, Zi I-239
Wei, Yubin I-187
Wu, Chenming I-218
Wu, Xia II-21
Wu, Xiaojun II-138
Wu, Xinyu I-20
Wu, Yanqiang I-10, I-136
Wu, Yuntao II-283
Wu, ZeHong II-243

X

Xi, Shanlin II-21
Xiang, Dawei II-31
Xiao, Zhanhao I-136
Xu, Guosheng I-115
Xu, Minghui I-126
Xu, Song II-211
Xu, Tianyang II-138
Xu, Wantong II-72
Xu, Youle I-177

Y

Yan, Ning I-115

Yan, Zheng II-148
Yang, Lianqiang I-166
Yang, Qingquan I-104, I-115
Yang, Tianxia I-126
Yang, Yiming II-31
Yang, Zhijing II-243
Yin, Lei I-280
Yu, Shan II-93
Yuan, Hongrui I-72

Z

Zeng, Yan II-82
Zeng, Yi I-229, I-249
Zhang, AnLong I-280
Zhang, Jiajing II-104
Zhang, Jiawei II-201
Zhang, Jicong I-197
Zhang, Jilin II-82, II-188
Zhang, Jingjing I-72
Zhang, Linglin II-72
Zhang, Mengya I-155
Zhang, Qian I-229
Zhang, Qiang II-127
Zhang, Qintao II-188
Zhang, RunMei I-280
Zhang, Wannan I-147
Zhang, Xiaojun II-117
Zhang, Yepeng I-82
Zhang, Yiwen I-187
Zhao, Dongcheng I-249
Zhao, Haifeng I-52, I-93, I-126
Zhao, Huimin II-273, II-283
Zhao, Wei I-177
Zheng, Tong I-104
Zhou, Die II-11
Zhou, Shiwei I-93
Zhouxiang, Xia I-41
Zhu, Xuefeng II-138
Zhu, Ying II-188